WATER LOSS MANAGEMENT
TOOLS AND METHODS FOR DEVELOPING COUNTRIES

Water Loss Management:
Tools and Methods for Developing Countries

DISSERTATION

Submitted in fulfilment of the requirements of
the Board for Doctorates of Delft University of Technology and of
the Academic Board of UNESCO-IHE Institute for Water Education for the
Degree of DOCTOR
to be defended in public
on Monday, June 4, 2012, at 15:00 hours
in Delft, The Netherlands

by

Harrison .E. Mutikanga
born in Kisoro, Uganda

Master of Science in Sanitary Engineering
UNESCO-IHE, the Netherlands

This dissertation has been approved by the promoter
Prof. dr. K. Vairavamoorthy

Members of the Examination Committee:

Chairman	Rector Magnificus Delft University of Technology
Vice Chairman	Rector UNESCO-IHE, Delft
Prof. dr. K. Vairavamoorthy	UNESCO-IHE/TU Delft, Promoter
Prof. dr. L. Rietveld	Delft University of Technology
Prof. dr. D.Butler	University of Exeter, Exeter, UK
Prof. dr. S. Berg	University of Florida, Gainesville, USA
Dr. S.K. Sharma	UNESCO-IHE, Delft
Dr. W.T. Muhairwe	National Water & Sewerage Corporation, Uganda
Prof. dr. M. Kennedy	UNESCO-IHE/TU Delft, (reserve)

CRC Press/Balkema is an imprint of the Taylor & Francis Group, an informa business

© 2012, H.E. Mutikanga

Published by:
CRC Press/Balkema
PO Box 447, 2300 AK Leiden, the Netherlands
e-mail: Pub.NL@taylorandfrancis.com
www.crcpress.com - www.taylorandfrancis.co.uk - www.ba.balkema.nl

ISBN 978-0-415-63648-3 (Taylor & Francis Group)

Dedication

This thesis is dedicated to the family of my late uncle, Canon Eriya K. Nkundizana.
For giving us love and education against all odds.

Acknowledgements

Many entities and individuals have contributed funding, time, energy, ideas, data, insight, encouragement and good advice to me during the study, for which I would like to take this opportunity to express my appreciation.

First and foremost, I am grateful to the Netherlands Government for the scholarship that facilitated this study under the Netherlands Fellowship Program (NFP). I wish to thank my employer, National Water and Sewerage Corporation (NWSC-Uganda) for allowing me time to work on this research and providing support in many ways during the study and in particular the former Managing Director Dr. William Muhairwe for his personal encouragement to pursue this study. Special thanks also go to the Acting Managing Director Eng. Alex Gisagara and the General Manager for Kampala Water Eng. Sekayizzi Andrew for their encouragement and support in many ways.

I would also like to extend my thanks to Prof. Kala Vairavamoorthy for accepting to be my Promoter and for sharing his expertise and valuable time with me. A special note of thanks must go to Dr. Saroj K. Sharma, a dear friend and mentor. He was the key driver responsible for shaping this research work. He made great effort to review and edit each chapter and helped with Dutch translations for the abstract and propositions. This thesis would not have been possible without his valuable guidance, expertise, suggestions and untiring encouragement.

I would like to thank two MSc research students: Silas C. Akita (UNESCO-IHE, Delft) and Olivia R. Nantongo (Makerere University, Kampala) for their contribution to this PhD research and for patiently enduring my never ending questions and comments. We all learned a lot during our joint research efforts. Thanks to Dr. Eleanor Wozei for accepting to supervise Olivia's work and for your wise guidance during the research work.

I would also like to extend thanks to Prof. Kala's core research group at Birmingham University, UK (Jotham Sempewo, Zhou Yi, Danguang Huan) and University of South Florida, USA (Krishna Khatri, Jochen Eckart, Seneshaw Tsegaye, and Kebreab Ghabremichael) for their valuable exchange of information and support. The time I spent at both Universities was interesting, memorable and fruitful.

I would like to extend thanks to all people who provided relevant research literature for this study, including Maureen Hodgins, Water Research Foundation (USA); Allan Wyte, RTI (USA); Edgar Johnson, GHD (Australia); Allan Lambert, ILMSS, (UK); Malcolm Farley (UK); Dr. Enrique Cabrera Jr. and Dr. Francisco Arregui (ITA, Universidad Politécnica de Valencia); Prof. Helena Alegre, LNEC (Lisbon); Dorothy Kobel, Cape Town University; Kenneth Muniina, NWSC; Dr. Apollo Tutensigensi, Leeds University; Dr. Dan Tindiwensi and Prof. Jackson Mwakali of Makerere University, Kampala.

Thanks to the many colleagues at NWSC who helped me in various ways especially during field data collection, questionnaire survey, workshops, and brainstorming sessions. I would like to particularly thank Emmanuel Ameda, Feriha Mugisha, Gordon Yiiki, Brian Agaba, Sam Kikomeko, Susan Nakuti, Herbert Mujuni, Mubbala Timothy, Bigabwa Moses, Tom Buyi and Dr. Frank Kizito. Special thanks also go to Eng. Dan Kyobe and Sonko Kiwanuka for their encouragement and support in many ways.

Thanks to the academic, participants and office staff at UNESCO-IHE who were always providing encouragement and assistance in many ways. I would like to particularly thank Dr. Nemanja Trifunovic, Dr. Assela Pathirana, Tanny van der Klis, Jolanda Boots, Anique Alaoui-Karsten and my PhD colleague Saeed Baghoth. Special thanks to the Ugandan and East African participants in the last four years for the good company, making my stay in Delft very comfortable and memorable.

I would also like to extend my appreciation to the members of the PhD awarding committee for their time, invaluable comments and suggestions.

I am very grateful to the family members and in particular my dear mother for your prayers and encouragement. To my cousin sister Aphia, I am indebted to you for the support during my visits to the UK and assistance in accessing various research materials. Special thanks to my sister's in-law Pam and Brenda for all the support and encouragement. To my children Mark, Malcolm and Diana for cheering me up when work was dull. To my little daughter Keza "baby sister" who was born during the PhD study, I am grateful for all the love, happiness and joy you brought. I know I should have been more available and I can't imagine how much you missed me. Lastly, I would like to thank my dear wife and friend Sheila for your inspiration, understanding, unwavering love, and encouragement throughout the study and thesis documentation. Thank you all for your patience and endurance during my absence at home but it was worthwhile.

Those who have contributed towards this dissertation are far too many to be mentioned individually. To all of you, I am truly grateful and may God bless you abundantly. Above all, I would like to thank God the Almighty for making all this possible in four years.

Table of Contents

List of Figures

List of Tables

List of Acronyms and Abbreviations

ADB   Asian Development Bank

AHP   Analytical Hierarchy Process

AMR   Automated Meter Reading

AI   Artificial Intelligence

AL   Apparent Losses

ALC   Active Leak Control

ALR   Awareness, Location and Repair

ANN   Artificial Neural Network

API   Apparent Loss Index

AWWA American Water Works Association

AZP   Average Zonal Point

BABE   Bursts and Background Estimates

BCC   Banker-Charnes-Cooper Model

CAPEX Capital Expenditure

CARL   Current Annual Real Losses

CBA   Cost-Benefit Analysis

CCR   Cooper-Charnes-Rhodes Model

CEO   Chief Executive Officer

COLS   Corrected Ordinary Least Squares

CP   Critical Point

CRS   Constant Returns to Scale

DDA   Demand Driven Analysis

DEA   Data Envelopment Analysis

DM   Decision Maker

DMA   District Mater Area

DMU   Decision Making Unit

DP   Dynamic Programming

DSS   Decision Support System

DST   Decision Support Tool

DWD   Directorate of Water Development

EAs   Evolutionary Algorithms

EC   Evaluation Criteria

ELL   Economic Level of Leakage

ELWL   Economic Level of Water Losses

EM   Evaluation Matrix

EPS   Extended Period Simulation

FAVAD Fixed and Variable Area Discharge

FIS   Fuzzy Inference System

GAs   Genetic Algorithms

GAIA   Geometrical Analysis for Interactive Assistance

GDM   Group Decision Making

GIS   Geographical Information System

GRG   Generalized Reduced Gradient

HH   Household

IBNET   International Benchmarking Network for Water and Sanitation Utilities

ICF   Infrastructure Condition Factor

IDAMC Internally Delegated Area Management Contracts

ILI   Infrastructure Leakage Index

ISO   International Standards Organization

IT   Information Technology

ITA   Inverse Transient Analysis

IWA   International Water Association

IWMM   Integrated Water Meter Management

KWDS   Kampala Water Distribution System

LCC   Life Cycle Costs

LFC   Low Flow Controllers

LHE   Leakage Handling Efficiency

LP   Linear Programming

LTM   Linear Theory Method

MCDA   Multi-criteria Decision Analysis

MD   Managing Director

MDGs   Millennium Development Goals

MENA   Middle East and North Africa

MLR   Multiple Linear Regression

MNF   Minimum Night Flow

MOEA   Multi-objective EA

MOO   Multi-Objective Optimization

MUR   Meter Under-registration

MWA   Metropolitan Water Works Authority

MWE   Ministry of Water and Environment

NDF   Night Day Factor

NHM   Network Hydraulic Modelling

NMIARP Network Management Improvement and Action Research Project

NLP	Non-Linear Programming		SFA	Stochastic Frontier Analysis
NPV	Net Present Value		SIV	System Input Volume
NRM	Newton-Raphson Method		SLP	Successive Linear Programming
NRW	Non-Revenue Water		SQP	Sequential Quadratic Programming
NSGA	Non-dominated Sorting GA		SWLMP	Strategic Water Loss Management Planning
NWSC	National Water and Sewerage Corporation (Uganda)		TE	Technical Efficiency
O & M	Operation and Maintenance		TFP	Total Factor Productivity
OFWAT	Office of Water Services (UK)		TOTEX	Total Expenditure
OPEX	Operating Expenditure		UARL	Unavoidable Annual Real Losses
ORP	Optimal Replacement Period		UBOS	Uganda Bureau of Statistics
PAS	Performance Assessment System		UNICEF	United Nations Children's Fund
PDD	Pressure-Dependent Demand		VEWIN	Association of Dutch Drinking Water Companies
PI	Performance Indicator			
PM	Pressure Management		VRS	Variable Returns to Scale
PMZ	Pressure Management Zone		WDS	Water Distribution System
POs	Private Operators		WHO	World Health Organization
PPP	Public Private Partnerships		WLA	Water Loss Assessment
PRV	Pressure Reducing Valve		WLM	Water Loss Management
RL	Real Losses		WMA	Weighted Meter Accuracy
RW	Revenue Water		WTP	Water Treatment Plant
SCADA	Supervisory Control and Data Acquisition		WRM	Water Resources Management
SE	Scale Efficiency			

Abstract

Access to adequate quantity of safe water is a fundamental human need. However, according to WHO and UNICEF Joint Monitoring Programme of UN MDGs, 884 million people in the world do not have access to improved water supply sources, almost all of them in the developing countries of Africa, Asia and Latin America. Ironically, significant amounts of safe drinking water continue to be wasted in urban water distribution systems of the developing countries. According to the World Bank, nearly 45 million m^3 of water is lost daily as leakage in water distribution systems of the developing countries – enough to serve about 200 million people. Furthermore, the World Bank estimates that close to 30 million m^3 of water is delivered everyday to customers but not invoiced due to metering inaccuracies, theft, billing errors and corruption by utility employees. This costs water utilities in the developing countries about US $6 billion every year.

Water losses not only represent economic loss and wastage of a precious scarce resource but also pose public health risks. Every leak is a potential intrusion point for contaminants in case of a drop in network pressures. Leakage also often leads to service interruption and customer complaints, is costly in terms of energy losses and increases the carbon footprint of the service provider. These problems are likely to be compounded in the future as a result of the widening gap between ageing water supply infrastructure and investment, rapid population growth, poor management practices, poor governance, and more extreme events as a consequence of climate change. These unprecedented pressures coupled with diminishing water resources and increasing costs of supplying water, have led regulatory bodies and water service providers to consider seriously urban water demand management and conservation measures. The high water losses in water distribution systems present an excellent opportunity of "un-tapped" water resources that have already been treated to drinking water standards and could be recovered cost effectively. To recover water losses requires understanding why, where and how much water is lost, and developing appropriate intervention measures. The main objective of this research have been to develop a decision support toolbox, which provide tools and methodologies required to help water utilities in the developing countries on how to assess, quantify and minimize water losses in their distribution systems.

Water utilities in developing countries are struggling to provide customers with a reliable level of service, often via aged water distribution infrastructure, with data-poor networks and restricted budgets. As a result, some of the techniques and methods used for water loss management in developed countries cannot be applied directly in developing countries. There are no appropriate tools and methodologies which are applicable or specifically suitable for water loss management in developing countries. In this context, this research aim was to develop appropriate tools and methodologies to aid water utilities in the developing countries make decisions on how to improve water distribution efficiency. Examining existing water loss management tools and methodologies and their applicability in developing countries, development of the water distribution systems performance assessment system, investigating water meter performance in the case study water distribution systems, development of methodologies for assessment of apparent losses in urban water distribution systems, pressure management strategy planning for leakage reduction, and application of the concept of multi-criteria decision making (MCDA) for evaluating and prioritizing water loss reduction strategy options were the main focus areas of this study.

The research approach included a comprehensive review of the state-of-the-art tools and methods for water loss management with the aim of identifying knowledge gaps and research needs. The main review findings included: (i) there are several tools and methodologies being applied for water loss management and they vary from simple managerial tools such as performance indicators to highly sophisticated evolutionary optimization methods for leak detection, (ii) high variation in water losses from 3% of system input volume in the developed countries to 70% in the developing countries, (iii) the existing tools and methodologies either can't be directly applied or do not fully address all aspects of water loss management in the developing countries, (iv) most of the tools and methodologies developed focus on leakage and little work has so far been done on apparent losses which are significant in the water distribution systems of the developing countries, (v) pressure management is a powerful and cost-effective strategy for leakage management, (vi) no clear methodology for prioritizing water loss reduction strategies, and (vii) no clear methodology for undertaking the analysis of economic levels of water losses. The literature review revealed that knowledge gaps do exist and there is need for developing more appropriate tools and methodologies that holistically address the unique system characteristics of the water distribution systems in the developing countries.

Understanding the condition of the water distribution system is a key factor in minimizing water losses. Although real-time in-service pipeline inspection is the direct ideal method, it is costly and out of reach for most water utilities in the developing countries. Alternative indirect assessment of water distribution systems based on the water balance and performance indicators seem to be more practical. The International Water Association (IWA) and the American Water Works Association (AWWA) have developed a standard water balance methodology and an array of performance indicators for water loss management. Whereas the IWA/AWWA water balance methodology and performance indicators provide a good foundation, they are insufficient and not directly applicable to water distribution systems in the developing countries. They require large amounts of reliable data that is costly and hardly generated by the resource constrained water utilities of the developing countries. In this study, a methodology based on the IWA/AWWA-PI concept for selecting and establishing new PIs has been developed. The methodology was applied to select 11 PIs from the IWA/AWWA menu and develop 14 new water loss management performance indicators. The performance indicators were tested in some water distribution systems in Uganda and found to be suitable for assessing water distribution system efficiency. However, the usefulness of the results depends heavily on data accuracy. In this study a procedure for estimating the underlying uncertainty in the water balance input data and how this uncertainty propagates in the non-revenue water (NRW) indicator was established as well as measures on how to minimize the uncertainty in the reported NRW figures. In the absence of performance benchmarks, Data Envelopment Analysis (DEA), a linear programming technique was applied to establish a *Pareto-efficient* frontier as a benchmark against which the performance of 25 water utilities in Uganda are evaluated and utility rankings established. The results indicate high technical inefficiencies (40-65%) in the water distribution systems with significant potential for water savings estimated at 42,600 m^3/d. The water utility rankings could serve as a catalyst for reducing the high inefficiencies observed in the Ugandan water distribution systems.

In this study, a methodology for assessing different components of apparent losses has been developed to help understand the magnitude of the problem and develop appropriate intervention measures to minimize the associated revenue losses. The methodology was then applied to the Kampala water distribution system and found suitable for estimating different components of apparent losses. The results indicate high global metering inaccuracies (-22% ± 2%) and illegal use (-10% ± 2%) expressed as a percentage of revenue water. Meter reading errors (-1.4% ± 0.1%) and data handling and billing errors (-3.5% ± 0.5%) were found to be low. Guidelines have also been established for assessment of apparent losses in water utilities of the developing countries with insufficient resources and data limitations to carryout in-depth assessment. The influence of system characteristics, operating practices, private elevated storage tanks, sub-metering, low flow rates, and water use profiles on meter accuracy was also investigated. The major findings indicate high meter failure rate (6.6%/year), average reduction in revenue water registration of 18.0% due to sub-metering, more than US $700,000 of revenue loss every year due low flow rates. The average meter under-registration due to the combined effect of the ball-valve and meter ageing of domestic water meters was found to be about of 67.2%. Based on this knowledge, a model for optimal meter replacement and guidelines for optimal meter sizing and selection based on demand profiling and economic optimization techniques have been developed to help minimize the associated utility revenue losses.

Pressure management in conjunction with district metered areas (DMAs) and network hydraulic modeling have proven to be powerful engineering tools for reducing leakage in many developed countries. Despite their apparent success, these tools have not had wide application in the developing countries mainly due to: (i) inadequate cost-benefit information to support management decision making in implementation of pressure management policies, and (ii) lack of and/or inadequate network zoning. In this study, a decision support tool for predicting the potential net benefits of implementing a pressure management scheme and help justify investment decisions was developed. In order to give users confidence in the outcomes of the planning tool, network hydraulic modeling was applied to validate the effectiveness of the decision support tool. Both methods were then applied to predict the potential net benefits of pressure management for a DMA in the Kampala water distribution system. Predictions based on the decision support tool and network hydraulic modeling indicate that reducing average zonal pressure by 7 m could result into water savings of 254 m^3/day and 302 m^3/day respectively without compromising customer service levels. The results indicate that the predicted water savings using both techniques compare fairly well. This is equivalent to more than €56,000 annual net benefits. Although conservative in its predictions, the decision support tool will be a valuable tool for engineers and decision-makers planning to implement pressure management strategies in the developing countries with inadequate resources for establishing the computationally demanding network hydraulic models.

Sustainable water loss control is a multi-dimensional problem that requires application of strategic planning techniques based on multiple criteria balanced performance. Although various water loss reduction strategies do exist, deciding on which option to choose amidst often conflicting multiple objectives and different interests of stakeholders is a challenging task for water utility managers. In this study, an integrated multi-criteria decision-aiding framework methodology for strategic planning of water loss management was developed. The PROMETHEE outranking method of the MCDA family was applied within the framework in evaluating and prioritizing water loss reduction strategy options for the Kampala water distribution system. The method was selected for its transparency in the decision-making process and capability to deal with imprecise data that is practically available for water loss management in the developing countries. The method focused on achieving the greatest financial-economic, environmental, public health, technical and socio-economic benefits based on seven criteria (revenue generation, investment cost, O & M costs, water savings, water quality, supply reliability and affordability) and preferences of the decision makers. A strategic plan that combines (i) selective mains and service lines replacement, (ii) pressure management, and (iii) improved speed and quality of repairs as priorities was the best compromise solution for the Kampala water distribution system. This study demonstrated that decision theory coupled with operational research techniques could be applied in practice to solve complex water loss management planning problems more sustainably.

In summary, this study has developed a decision support toolbox (tools and methodologies) for water loss management in developing countries. The toolbox comprises the following key components:

1. The performance assessment system (PAS) for water distribution accountability (Water balance and performance indicator computational tool; Guidelines for estimating the uncertainty in the water balance model; and DEA-as a multiple-measure performance evaluation and benchmarking tool).
2. The integrated water meter management (IWMM) framework tool to aid water utilities minimize revenue losses due to metering inaccuracies (demand profiling; optimal selection, sizing and replacement tools; guidelines for estimating water loss due to meter under-registration and failure).
3. A methodology for assessing apparent losses in urban water distribution systems; guidelines for assessing apparent losses in data-poor water distribution systems; guidelines for quantifying and recovery of apparent losses due to metering inaccuracies at ultra-low flow rates.
4. A decision support tool (PM-COBT) for pressure management planning to control leakage. The tool performs simulation analysis of leakage reduction potential under different setting scenarios for pressure reducing valves.
5. A multi-criteria decision-aiding framework methodology for strategic water loss management planning (evaluating and prioritizing alternative strategy options).

Although the toolbox has been tested and validated using water distribution systems in Uganda, the tools and methodologies therein are generic and easily adaptable to suit other WDSs in the developing countries. It is envisaged that this thesis will be an "advocacy document" that promotes good stewardship of water resources (specifically water distribution system efficiency) and sustainable delivery of water supply services in the developing countries. This thesis provides a comprehensive resource on tools and methods needed to tackle the water loss challenges and will be of interest to practitioners, policy makers, researchers, regulators, and financial institutions working to reduce losses in water distribution systems particularly in the developing countries.

Keywords: Decision support tools and methods; Developing countries; Urban water distribution systems; Water and revenue losses accountability; Water conservation.

Chapter 1 - INTRODUCTION

Parts of this chapter are based on the publication:

Mutikanga, H.E, Sharma, S.K, and Vairavamoorthy, K (2009). "Water Loss Management in Developing Countries: Challenges and Prospects". *Journal of American Water Works Association*, **101**(12), 57-68.

Summary

Globally, many countries are grappling with the dilemma of increasing water demand and diminishing water resources. The irony however, is that many water utilities particularly in the developing countries continue to operate inefficient water distribution systems (WDSs) with significant amount of water and revenue losses. There are various factors that contribute to water losses such as ageing infrastructure, high pressures, external and internal pipeline corrosion, poorly designed and constructed WDSs, metering errors, illegal use, and poor operation and maintenance practices. Since water loss is inevitable, many tools and methods for minimizing water loss in the distribution system have been developed and applied over the years. However, water losses continue to be elusive to manage and control and are considerably high in most water utilities worldwide. Tracking water distribution input is further complicated by the fact that most components of the WDS are located underground. Water losses vary from 3% of system input volume (SIV) in the developed countries to 70% in the developing countries. This high contrast suggests that probably the existing tools and methodologies are not appropriate for managing water losses in the developing countries. Therefore, this study aims to develop appropriate tools and methodologies for assessing and minimizing water losses in urban WDSs of the developing countries to ensure good stewardship of water resources and promote sustainable delivery of water services.

1.0 Introduction

This Chapter provides an overview of the scale of non-revenue water (NRW) and/or water losses and challenges of water loss management in urban WDSs and motivation for the research on the one part, and a summary description of the thesis layout and chapter linkages on the other part. The first part is presented in sections 1.1 to 1.4. The second part is presented in section 1.5 and sets the scene for Chapter 2.

1.1 Global overview of water loss management (WLM)

Urban WDSs are often "buried and forgotten" until when they manifest into leaks and bursts causing significant economic, environmental and social costs. The efficiency of WDSs is measured by the difference between SIV and water delivered to customers and billed (revenue water) commonly referred to as non-revenue water (NRW) (Lambert and Hirner 2000). NRW is made up of water losses (real and apparent losses) and authorized unbilled consumption such as water for fire fighting and flushing mains. The quantity of water lost is a measure of the operational efficiency of a WDS (Wallace 1987). High levels of water losses are indicative of poor governance and poor physical condition of the WDS (Male et al. 1985; McIntosh 2003) and the costs of system inefficiency are transferred to customers via high water tariffs (Park 2006).

Water and revenue losses are a major problem for water utilities worldwide. The amount of water lost from WDSs is astounding. According to the World Bank study, NRW from WDSs worldwide is estimated at a staggering 48 billion m^3 per year costing water utilities about US $14 billion every year (Kingdom et al. 2006). The same report indicates that about 55% of the global NRW by volume occurs in the developing countries. The provision of adequate water supply to the rapidly growing population amidst such high water losses will continue to be a major challenge facing many countries worldwide. According to WHO/UNICEF (2010), 884 million people in the world do not have access to improved water supply, almost all of them in the developing regions. This challenge is likely to be exacerbated by the rapidly increasing urban population in that region. Half of the world's population – 3 billion people – live in

3

urban areas and are projected to reach 5 billion by 2030 (Feyen et al. 2009). The stark reality is that the World's water resources are finite and limited and can no longer sustain this rate of growth unless used wisely. The high water losses in WDSs present "un-tapped" water resources that can be recovered cost effectively. These untapped and wasted resources are already treated to drinking water standards and energized to provide adequate pressure to reach the consumers.

Water losses not only have economic and environment dimensions but also public health and social dimensions as well. Leakage, often leads to service interruption, is costly in terms of energy losses and may cause water quality contamination via pathogen intrusion (Almandoz et al. 2005; Colombo and Karney 2005; Karim et al. 2003). The American Water Works Association (AWWA) estimates that 5-10 billion kWh of electricity generated annually in the USA is wasted in energizing water that is either lost as leakage or used but not paid for (AWWA 2003). Due to the water-energy nexus, a WDS which is water-inefficient is also energy-inefficient. Water loss increases the carbon footprint of the water utility operator with adverse effect on climate change (Cabrera et al. 2010).

1.1.1 Water losses in some developed countries

In the Netherlands, low leakage levels in the range 3-7% of distribution input have been reported (Beuken et al. 2006). The WDSs in the Netherlands are probably the most efficient in the world. In the USA about 22 million m^3 of water is lost per day or categorized as public use/loss (USGS 1998). The average NRW in the USA is 15% but range from 7.5% to 20% (Beecher 2002). In the USA, losses due to main breaks are on the same order of magnitude as annual flood losses, which are estimated at more than $2 billion of property damage (Grigg 2007). In the UK, often perceived to be leaders in leakage management, about 3 million m^3 per day is lost as leakage and has remained relatively stable at about 20-23% of water delivered in the past decade (OFWAT 2010). Most companies in the UK are operating at economic levels of leakage (ELL) based on current tools, techniques and technologies. In Italy, NRW levels range from 15 to 60% with an average of 42% (Fantozzi 2008). In Portugal, NRW averages 34.9% but varies from less than 20% to more than 50% (Marques and Monteiro 2003). In Greece's Larissa city, NRW has been estimated at 6 million m^3/year (or 34% of SIV) (Kanakoudis and Tsitsifli 2010). In Australia, for a data set of 10 water systems, NRW varies from 9.5 to 22%, with an average of 13.8% (Carpenter et al. 2003). In Canada's Ontario province, as much as $1 billion worth of drinking water disappears into the ground every year from leaky municipal water pipes, and leakage varies from 7% to 34% of water distribution input (Zechner 2007). The fact that water utilities and municipalities are losing such large amounts of water from WDSs undermines their efforts in promoting water conservation and efficient use of water with negative environmental, economic and social impacts.

1.1.2 Water losses in some developing countries

Whereas most water utilities in the developed world have made considerable efforts over recent years to improve WDS efficiency via reduction of NRW, progress in the developing countries is painfully slow. In Asian cities, the Asian Development Bank reports NRW in the ranges of 4.4% of total water supply (PUB, Singapore) to 63.8% (Maynilad, Manila) (ADB 2010) and 50-65% of NRW is due to apparent losses (McIntosh 2003). In Africa NRW figures ranging from 5% (Saldanha Bay, South Africa) to 70% (LWSC, Liberia) have been reported (WSP 2009). In Latin American water utilities NRW of 40-55% of water delivered have been reported (Corton and Berg 2007). In Brazil, water losses average 39.1% of water supplied, equivalent to almost 5 billion m^3 of water lost every year (Cheung and Girol, 2009).

According to the World Bank study, nearly 45 million m^3 of water is lost daily as leakage (enough to serve nearly 200 million people) and close to 30 million m^3 of water is delivered everyday to customers but not paid for due to metering inaccuracies, theft and corrupt utility employees, costing water utilities about US $6 billion every year (Kingdom et al. 2006). Clearly, this is unacceptable, that where water utilities are starving for additional revenue to expand services to the poor and where water is heavily rationed that it is also heavily wasted. This is likely to be compounded by the high rate of infrastructure deterioration which will result in greater loss of treated and energized drinking water. The impact of poorly managed urban WDSs coupled with increasing global change pressures (urbanization, climate change, population growth) is likely to result in extreme scarcity scenarios. In the Middle East and North Africa Region (MENA), countries such as Tunisia and Algeria are experiencing absolute water scarcity with less than 500 m^3/person/year of freshwater (Baroudy 2005). In East Africa, Kenya falls below the freshwater water poverty line, defined by experts as 1,000 m^3/person/year (Qdais 2003). By the year 2025, Tanzania and Uganda will be approaching the critical levels (WRM 2005). A major paradigm shift in the way water resources are managed in the developing countries is urgently required to avert the looming water scarcity.

1.1.3 Challenges and prospects for WLM in developing countries

According to WHO and UNICEF (2010), as many as 343 million people in Africa, 477 million people in Asia and 38 million people in Latin America and the Caribbean do not have adequate water supplies. The trend towards urban living is particularly acute in Africa. According to Cohen (2006), Africa's urban population is projected to more than double, from 295 million in 2000 to 1.5 billion in 2030, with over 72% of the urban population living in slums. Meeting the United Nations Millennium Goal 7 (target 10) of halving by 2015, the number of people without adequate water and sanitation amidst high water losses in WDSs of the developing countries will continue to be a major challenge for the 21st century. In the wider context of urban water demand management, public water suppliers have a major responsibility to manage water responsibly and efficiently on the supply side by reducing water losses from the distribution systems to complement customer-side demand management of using water efficiently. This will assist speed up service delivery while ensuring sustainability of urban water services in the urban areas of developing countries.

Other major challenges of WLM include ageing infrastructure, inadequate resources, poor governance issues, inadequate asset management, poorly designed WDSs, insufficient reliable data for WDS performance evaluation, intermittent supply and kleptomania for water. The prospects lie in increasing capacity building of water utility employees, research, performance based contracting, emerging new equipment and technology for leak detection, and increasing dissemination of emerging "state-of-the-art" tools and methodologies for water loss reduction and performance improvement of utility water services (Alegre et al. 2006; Arregui et al. 2006; AWWA 2009; Berg 2010; Butler and Memon 2006; Cabrera Jr et al. 2011; Fanner et al. 2007a; Fanner et al. 2007b; Farley and Trow 2003; Thornton et al. 2008; Wu et al. 2011). More information on challenges and prospects for water loss management in developing countries can be found in Mutikanga et al. (2009) and Sharma and Vairavamoorthy (2009).

1.2 Water Loss Management in Uganda

In the Ugandan Water and Sanitation Sector, an urban area is defined as a gazetted town or centre with a population of more than 5,000 people. In June 2010, there were 137 Urban Councils in Uganda, classified as Kampala city, 13 municipalities and 123 Town Councils. The 23 large towns (> 15,000 people) are under the jurisdiction of National Water and

Sewerage Corporation (NWSC), a public utility established in 1972 and 85 small towns (5,000 to 15,000 people) with operational piped water supply are under Local Governments supported by the Directorate of Water Development (DWD) in the Ministry of Water and Environment (MWE). The NWSC large towns are managed under internally delegated area management contracts (Mugisha et al. 2008) while small towns are managed by local private operators under management contracts with the Local Governments (MWE 2010). The urban population in the 137 urban councils is estimated at 4.7 million people (15% of Uganda's population) with 2.9 million people (66% of the urban population) residing in the NWSC's managed towns. The overall urban water coverage is estimated at 67% ranging from 53% in small towns to 74% in the large towns. The total NRW for the financial year 2009/2010 was estimated at an average of 20% (or 605, 161 m^3) for the small towns and 33.3% (or 23,460,729 m^3) for the NWSC large towns costing the sector over US $21 million (about 80% of the urban water sector annual budget) (MWE 2010).

1.2.1 Kampala Water Distribution System (KWDS)

Kampala is the capital city of Uganda and is the biggest branch utility managed by NWSC. Population estimates indicate that about 2.5 million inhabitants live within the service area with 1.21 million in Kampala District (UBOS 2002). The annual population growth rate since 1991 census is 3.8% making Kampala one of the fastest growing cities in the world. Like in most developing countries, water supply infrastructure development has not kept pace with population growth and has resulted in water shortages and low pressures in most parts of the distribution system. To bridge the gap, NWSC has made significant investments in augmentation projects (increasing water production) rather than investing in water loss reduction projects. The development of the KWDS commenced in the year 1928 with the construction of the present Gaba I Water Treatment Plant (WTP). The design capacity of the plant is 72,000 m^3/d. Due to hydraulic constraints, the plant is currently producing an average of 29,000 m^3/day. The Gaba II WTP was commissioned in 1992. The design capacity is 80,000 m^3/day but current production is 40,000 m^3/d on average. The Gaba III WTP with a design capacity of 80,000 m^3/d was commissioned in 2007 and is currently producing an average of 75,000 m^3/d. This makes a daily total production of 144,000 m^3/d (NWSC 2010). However, this falls short of the estimated current demand of 200,000 m^3/day and the predicted demand of 342,361m^3/d in 2025 (Poyry 2009). This situation is exacerbated by the high water losses in the KWDS with NRW estimated at 43% of SIV (or 60,274 m^3/d) (NWSC 2009).

The KWDS encompasses a service area of about 715 km^2 covering the city and its peri-urban areas. It serves about 150,000 customer service connections (83% domestic and 14% commercial) through 2,253 Km of pipelines (NWSC 2010). The old pipelines (over 50 years) are mainly made of steel and cast iron while the newer parts of the mains consist of ductile iron (DI) and plastic pipe materials. Water transmission and distribution pipe diameters range from 50 mm to 900 mm. The network is divided into two pressure zones (high and low level) with pressures varying from as low as zero (no water) to as high as 150 m in the low-lying areas. The system of Kampala includes 12 booster stations that supply higher parts of the network. There are 25 reservoirs located in various parts of the system with total storage capacity of about 60,000 m^3 which is less than half a day's water demand. Almost all customers have private water storage tanks as a safeguard against supply interruptions. Most customer complaints are related to supply interruptions, low pressure, and often poor water quality. Apparent losses are high with rampant illegal use of water (average of 939 confirmed cases every year), high metering inaccuracies and meter failure. Network maintenance is reactive other than proactive. The condition of the network has deteriorated over the years, due to poor operating practices, inadequate strategic asset management and investments

(Mutikanga et al. 2009). The high number of pipe failures reported during the year 2010 (average of 1175 breaks/100 km) provides some indication of the condition of the underlying infrastructure. This is rather very high compared to average figures reported in Sub-Saharan Africa of 800/100 km/year (Banerjee and Morella 2011). These figures are 100 times higher than the Netherlands average of 8/100 km/year (Vloerbergh and Blokker 2009), where strategic asset management has received much more attention in recent decades. Pipeline systems having an average annual pipe break ratio per 100 km of less than 40 are considered to be in an acceptable state (Pelletier et al. 2003). The NRW trends for the last six years are shown in Figure 1.1 and the water balance for 2010 calendar year is presented in Figure 1.2.

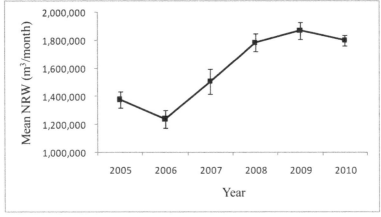

Figure 1.1NRW trends for KWDS

Parameter	Water Loss Components	Unit	Quantity	Error Margin (95% confidence level)
System Input Volume		m³	52,499,787	±7.0%
Revenue Water		m³	30,891,487	
Non-Revenue Water		m³	21,608,300	±15.3%
Water Losses		m³	21,319,631	±7.9%
	Real Losses	m³	11,863,566	±7.5%
	Apparent Losses	m³	9,456,065	±2.5%
	Customer Metering Errors	m³	8,726,065	±1.0%
	Unauthorised Consumption	m³	730,000	±2.3%

Figure 1.2 KWDS Water Balance for year 2010

From Figure 1.1, NRW has been on the increasing trend since 2006 and remains high in the KWDS. The increasing trend could be explained by: (i) commissioning of a new water treatment plant (Gaba III) in 2007 that literally meant pumping more water in a leaking system, and (ii) attributed to inadequate strategic planning to reduce NRW sustainably. Water loss management is based more on a reactive approach (reacting after failure has occurred) rather than proactive, where failure is predicted and prevented before it occurs. Reducing water losses requires a coherent action that addresses technical, operational, institutional, financial, and administrative issues (Vairavamoorthy and Mansoor 2006). The high uncertainty in the NRW figures is mainly driven by the high volumes of SIV generated from inaccurate meters that are hardly calibrated. The low uncertainty in 2010 was mainly due to replacement of some old SIV bulk meters.

The NRW was estimated based on the IWA/AWWA water balance methodology (AWWA 2003; Lambert and Hirner 2000) and the apparent loss assessment methodology (Mutikanga et al. 2011). The NRW for 2010 was estimated at 41.2% (or 21.6 million m^3), which is equivalent to 384 L/conn/d. This is relatively high compared to the best practice African benchmark of 300 L/conn/d (WSP 2009) and 112 L/conn/d for Phnom Penh Water Supply Authority, one of Asia's best utility (ADB 2007). In financial terms, a conservative estimate of US $21 million is lost every year based on the current NWSC tariff of UShs. 2088 per m^3 (or 0.91 US $/$m^3$) (MWE 2010). Using the least service level (public standpipe) with a water demand of 20 L/c/d for the urban poor in Kampala (Poyry 2009), the volume of NRW could potentially serve about 3 million poor people within the KWDS service area. The high levels of NRW in WDSs highlight the need for the research.

1.3   The Need for the Research

Reducing water losses not only conserves a scarce natural resource but also improves utility financial viability (increased revenue and reduced repair and energy costs), deferment of capital expenditure for new sources and system expansion to keep pace with increasing demand, saves energy, reduces carbon emissions, thus mitigating climate change impacts and fostering sustainability. For developing countries, reducing water losses (section §1.1.2) by half would avail over 22 million m^3/d – enough water to serve over 100 million people and water utilities would be able to recoup about US $3 billion every year that could be used to improve service coverage particularly for the urban poor. These figures highlight the importance of the research with aim of solving the water loss problem in urban WDSs of the developing countries.

In order to reduce water losses and improve efficiency of delivering water to customers, the condition of the WDS needs to be very well understood and decision-makers (DMs) need to solve the problem of how much water is being lost, where and why? Although direct real-time assessment methods such as in-line inspections are ideal, their high costs practically limit their application in most water utilities of the developing countries. In such cases, indirect performance measures such as the water balance and performance indicators (PIs) should be considered. Whereas a range of performance assessment and water loss control manuals are available (Alegre et al. 2006; AWWA 2009) and do provide a good foundation for water loss reduction, the tools and methods proposed therein do not fully address the unique characteristics of WDSs in developing countries and therefore cannot be directly applied. In addition, the most widely used indicator for water distribution efficiency is percentage NRW. This PI is misleading as it is heavily influenced by consumption which has nothing to do with the condition and operation of the WDS. Another problem is that most WDS performance measures widely used in the developed countries such as the unavoidable annual real losses (UARL) and the infrastructure index (ILI) (Lambert et al. 1999), are dubious in the context of most developing countries with financial constraints to effectively undertake active leakage control and reduce leakage to the least technically possible levels. The assumptions used in deriving the UARL empirical formula do practically breakdown in situations of developing countries. The accuracy of the water balance input data and the uncertainty propagation into the final NRW indicator is another research area that has not been fully addressed yet critical for meaningful interpretation of NRW figures. Furthermore, there are no standard benchmarks for performance improvement of WDS efficiency that take into account multiple inputs and outputs. In such cases, application of frontier-based benchmarking techniques would be required.

Whereas significant research has been carried out to address leakage in WDSs, little work has been done so far on apparent losses. Currently, assessment of apparent losses is based on rules-of-thumb. Although attempts have been made to assess components of apparent losses, they have only focused on metering inaccuracies in WDSs of the developed countries (Lund 1988; Noss et al. 1987; Richards et al. 2010). Water meter performance in WDSs of the developing countries that are not so well managed and provide water irregularly is still not very well understood. Generally, apparent loss control in urban water supply systems is in its infancy, and much work remains to bring it to par with available real loss interventions (AWWA 2003).

It is now widely acknowledged that pressure management is the most cost-effective and efficient tool for leakage management. However, pressure management is hardly applied in the developing countries and where attempts have been made (Babel et al. 2009; McKenzie et al. 2004), optimal solutions have not been provided. There is need to identify implementation barriers and to develop appropriate intervention measures that promote adoption of pressure management policies in the developing countries and optimal intervention measures to maximize leakage reduction.

Sustainable water loss control is a complex problem with economic, environmental, social and public health dimensions. Although various water loss reduction strategies do exist, deciding on which option to choose amidst often conflicting multiple objectives and different interests of stakeholders is a challenging task for water utility managers. This is further complicated in developing countries with either imperfect data or lack of it. The development of a well-structured decision-aid framework that includes stakeholder preferences would be a valuable tool to assist water utility managers in evaluating and prioritizing water loss reduction strategy options.

Clearly, knowledge gaps still exist with respect to water loss control in water distribution systems particularly in the developing countries. This research seeks to bridge the knowledge gaps by developing appropriate tools and methodologies for minimizing water losses in not so well managed water distribution systems of the developing countries.

1.4   Objectives of the Study

As highlighted in the aforementioned need for research, this research aims to develop a decision support toolbox (tools and methodologies) for assessing and minimizing water losses in distribution systems of the developing countries. The specific objectives of the research are:

1. To develop an appropriate performance assessment system for evaluation and efficiency improvement of urban water distribution systems in the developing countries, specifically focusing on water loss reduction and to validate its effectiveness by application to a real-developing world case study.

2. To investigate water meter performance in the Kampala water distribution system and develop generic intervention tools for minimizing the associated revenue losses in the developing countries.

3. To develop a methodology for assessing apparent losses in urban water distribution systems based on field data and investigations in the Kampala water distribution

system; and investigate the apparent water losses caused by metering inaccuracies at ultralow flow rates.

4. To develop a decision support tool for pressure management planning to control leakage in urban water distribution systems of the developing countries by application of economic analysis and network hydraulic modeling techniques.

5. To develop a multi-criteria decision-aiding framework methodology for strategic water loss management planning in developing countries and evaluate its effectiveness in prioritizing water loss reduction strategy options by application to the Kampala water distribution system.

1.5 Outline of the Thesis

The thesis is organized in eight Chapters, each dealing with a particular aspect of water loss management, thus contributing to the achievement of the study objectives. A brief overview of the structure is given below.

Chapter 1 introduces the subject and provides insight into the problem of water distribution losses and highlights the importance of water loss management. It gives an overview of the magnitude of water losses in both the developing and developed countries and specifically for the Ugandan water sector with emphasis on the Kampala water distribution system. The need for the research is also highlighted followed by the aims and objectives of the study. Lastly, the summary outline of the thesis is presented.

Chapter 2 presents a comprehensive review of the state-of-the-art of methods and tools applied for WLM in WDSs. The methods and tools are critically discussed; knowledge gaps identified and opportunities for future research and applications are considered.

Chapter 3 deals with water accountability and efficiency measurement of WDSs. A performance assessment system is developed consisting of a methodology for selecting and developing water loss management PIs, a water loss assessment computational tool (water balance and array of PIs), and guidelines for estimating uncertainty in the water balance model. The performance assessment system is applied and validated using data from Ugandan WDSs. The challenges of establishing a PI culture in developing countries are discussed. The DEA-linear programming (LP) benchmarking methodology is presented and its application to improve WDS efficiency is illustrated using 25 branch water distribution utilities in Uganda. A comparison of PI and DEA-based benchmarking approaches is presented and policy implications of the study are discussed.

Chapter 4 investigates customer water meter performance in the KWDS with focus on small meters of size 15 mm. The influence of different factors affecting in-service water meter performance such as elevated storage service tanks, sub-metering, meter model, usage and/or age, water use profiles, network pressures and particulates in water are examined. Guidelines for estimating water loss due to metering inaccuracies and failure are established. Demand profiling techniques for optimal meter sizing and selection are discussed. The meter sampling and testing procedures are also discussed. Finally, a model for optimal water meter replacement period based on operational research and economic optimization techniques is developed.

Chapter 5 deals with apparent water losses. A new methodology for assessing apparent losses in urban WDSs is developed. The method is validated using KWDS operational data, field measurements and investigations. The method is then applied to assess the different apparent loss components in the KWDS. A framework for estimating apparent losses in WDSs for developing countries with inadequate resources for detailed field investigations and data collection is established. The apparent losses caused by ultralow flow rates for 15 mm meter sizes are investigated and the procedure for quantifying the associated revenue losses is established. The factors influencing the level of apparent losses are discussed and appropriate intervention strategies are proposed.

Chapter 6 presents a new decision support tool (DST) for predicting benefits of pressure management for leakage control required for planning and implementing pressure management strategies in WDSs of the developing countries. An algorithm based on non-linear programming (NLP) techniques for evaluating the different water savings for different pressure reducing valve (PRV) settings is presented. Network hydraulic modeling is applied to validate the results of the DST. The economic model used for evaluating the benefits of pressure management in conjunction with the DST and network hydraulic modeling is also presented. The approaches are validated using data for a real DMA in the KWDS. The limitations of the DST are discussed as well.

Chapter 7 presents an integrated multi-criteria decision-aiding framework methodology for strategic WLM planning. The PROMETHEE outranking method of the MCDA family applied to prioritize water loss reduction strategy options for the KWDS including stakeholder preferences is discussed. This chapter applies some of the performance measures and water loss reduction options identified in Chapter 2. The challenges of applying the framework methodology in the KWDS are also discussed.

Finally, Chapter 8 presents the main conclusions of this study and makes recommendations for future research.

The general structure and the way the chapters link to each other is shown in Figure 1.3.

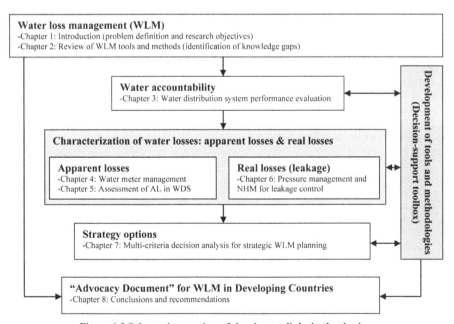

Figure 1.3 Schematic overview of the chapter links in the thesis

Additionally, the thesis includes three appendices at the end: (i) appendix A provides the pressure management DST computer code based on visual basic, (ii) appendix B provides the discrete MCDA questionnaire survey forms used during the stakeholder preference elicitation process and (iii) appendix C provides the results of the questionnaire survey highlighting preferences on the relative importance (weights) of performance measures (objectives and criteria) and preference thresholds needed for the evaluation model (matrix).

The next Chapter presents a review of tools and methodologies for WLM and identifies research gaps that justify the work carried out in subsequent Chapters of the thesis.

1.6   References

ADB. (2007). "Phnom Penh Water Supply Authority, An Exemplarly Water Supply Utility in Asia." Asian Development Bank (ADB) (http://www.adb.org/water/actions/cam/PPWSA.asp) (retrieved on 30th June 2011)

ADB. (2010). "Every Drop Counts: Learning from Good Practices in Eight Asian Cities." Asian Development Bank, Manila.

Alegre, H., Baptista, J. M., Cabrera, E. J., Cubillo, F., Hirner, W., Merkel, W., and Parena, R. (2006). *Performance Indicators for Water Supply Services, IWA Manual of Best Practice*, IWA Publishing.

Almandoz, J., Cabrera, E., Arregui, F., Cabrera Jr, E., and Cobacho, R. (2005). "Leakage Assessment through Water Distribution Network Simulation." *Journal of Water Resources Planning and Management*, 131, 458-466.

Arregui, F., Jr., C. E., and Cobacho, R. (2006). *Integrated Water Meter Management* IWA Publishing, London.

AWWA. (2003). "Committee report: Applying worldwide BMPs in water loss control." *Journal AWWA*, 95(8), 65-79.

AWWA. (2009). "Water Audits and Loss Control Programs: AWWA Manual M36." American Water Works Association, Denver, USA.

Babel, M. S., Islam, M. S., and Gupta, A. D. (2009). "Leakage Management in a low-pressure water distribution network of Bangkok." *Water Science and Technology:Water Supply*, 9(2), 141-147.

Banerjee, S. G., and Morella, E. (2011). "Africa's Water and Sanitation Infrastructure." The World Bank, Washington, DC, USA.

Baroudy, E., Lahlou, A.A., and Attia, B. (2005). *Managing Water Demand: Policies, Practices, and Lessons from Middle East and North Africa*, IWA Publishing/IDRC.

Beecher, J. A. (2002). "Survey of State Agency Water Loss Reporting Practices." AWWA, Colorado, USA.

Berg, S. (2010). *Water Utility Benchmarking: measurements, methodologies and performance incentives* IWA Publishing, London.

Beuken, R. H. S., Lavooij, C. S. W., Bosch, A., and Schaap, P. G. (2006). "Low leakage in the Netherlands Confirmed." *Proceedings of the 8th Annual Water Distribution Systems Analysis Symposium (ASCE)*, Cincinnati, USA, 1-8.

Butler, D., and Memon, F. A. (2006). *Water Demand Management*, IWA Publishing, London.

Cabrera, E., Pardo, M. A., Cobacho, R., and Cabrera Jr, E. (2010). "Energy Audit of Water Networks." *Journal of Water Resources Planning and Management*, 136(6), 669-677.

Cabrera Jr, E., Dane, P., Haskins, S., and Theuretzbacher-Fritz. (2011). *Benchmarking Water Services: Guiding water utilities to excellence*, IWA Publishing, London.

Carpenter, T., Lambert, A., and McKenzie, R. (2003). "Applying the IWA approach to water loss performance indicators in Australia." *Water Science and Technology: Water Supply*, 3(1/2), 153-161.

Cheung, P. B., and Girol, G. V. (2009). "Night flow analysis and modeling for leakage estimation in a water distribution system." Integrating Water Systems, Boxall and Maksimovic, eds., Taylor and Francis Group, London.

Cohen, B. (2006). "Urbanization in developing countries: Current trends, future projections, and key challenges for sustainability." *Technology in Society*, 28, 63-80.

Colombo, A. F., and Karney, B. W. (2005). "Impacts of Leaks on Energy Consumption in Pumped Systems with Storage." *Journal of Water Resources Planning and Management*, 131(2), 146-155.

Corton, M. L., and Berg, S. V. (2007). "Benchmarking Central American Water Utilities." Public Utility Research Centre, University of Florida, Gainesville, Florida.

Fanner, P., Sturm, R., Thornton, J., and Liemberger, R. (2007a). *Leakage Management Technologies*, Water Research Foundation Denver, Colorado.

Fanner, P., Thornton, J., Liemberger, R., and Sturm, R. (2007b). *Evaluating Water Loss and Planning Loss Reduction Strategies*, Awwa Research Foundation, AWWA;, Denver, USA; IWA, London, UK

Fantozzi, M. (2008). "Italian case study in applying IWA WLTF approach: results obtained." Water Loss Control, J. Thornton, R. Sturm, and G. Kunkel, eds., McGraw-Hill, New York, 421-432.

Farley, M., and Trow, S. (2003). *Losses in Water Distribution Networks: A Practitioner's Guide to Assessment, Monitoring and Control*, IWA Publishing, London.

Feyen, J., Shannon, K., and Neville, M. (2009). *Water and Urban Development Paradigms: Towards an integration of engineering, design and management approaches*, CRC Press, Taylor and Francis Group, Leiden.

Grigg, N. S. (2007). "Main Break Prediction, Prevention, and Control." Awwa Research Foundation, Denver, Colorado.

Kanakoudis, V., and Tsitsifli. (2010). "Results of an urban water distribution network performance evaluation attempt in Greece." *Urban Water Journal*, 7(5), 267-285.

Karim, M. R., Abbaszadegan, M., and LeChevallier, M. (2003). "Potential for Pathogen Intrusion During Pressure Transients." *Journal American Water Works Association*, 95(5), 134-146.

Kingdom, B., Liemberger, R., and Marin, P. (2006). "The Challenge of Reducing Non-Revenue Water (NRW) in Developing Countries ", The World Bank, Washington, DC, USA.

Lambert, A., and Hirner, W. (2000). "Losses from Water Supply Systems: Standard Terminology and Recommended Performance Measures." The IWA's Blue Pages:IWA's Information Source on Drinking Water Issues, London.

Lambert, A. O., Brown, T. G., Takizawa, M., and Weimer, D. (1999). "A review of performance indicators for real losses from water supply systems." *Aqua- Journal of Water Services Research and Technology*, 48(6), 227-237.

Lund, J. R. (1988). "Metering Utility Services: Evaluation and Maintenance." *Water Resources Research*, 24(6), 802-816.

Male, J. W., Noss, R. R., and Moore, I. C. (1985). *Identifying and Reducing Losses in Water Distribution Systems*, Noyes Publications, New Jersey.

Marques, R. C., and Monteiro, A. J. (2003). "Application of performance indicators to control losses-results from the Portuguese water sector." *Water Science and Technology: Water Supply*, 3(1/2), 127-133.

McIntosh, A. C. (2003). *Asian Water Supplies: Reaching the Urban Poor*, Asian Development Bank and IWA Publishing.

McKenzie, R. S., Mostert, H., and de Jager, T. (2004). "Leakage reduction through pressure management in Khayelitsha: two years down the line." *Water SA*, 30(5), 13-17.

Mugisha, S., Berg, S. V., and Muhairwe, W. T. (2008). "Using Internal Incentive Contracts to Improve Water Utility Performance: The Case of Uganda's NWSC." *Water Policy*, 9(3), 271-84.

Mutikanga, H. E., Sharma, S., and Vairavamoorthy, K. (2009). "Water Loss Management in Developing Countries: Challenges and Prospects." *Journal AWWA*, 101(12), 57-68.

Mutikanga, H. E., Sharma, S. K., and Vairavamoorthy, K. (2011). "Assessment of Apparent Losses in Urban Water Systems." *Water and Environment Journal*, 25(3), 327-335.

MWE. (2010). "Water and Environment Sector Performance Report." Ministry of Water and Environment (MWE), Kampala, Uganda.

Noss, R. R., Newman, G. J., and Male, J. W. (1987). "Optimal Testing Frequency for Domestic Water Meters." *Journal of Water Resources Planning and Management*, 113(1), 1-14.

NWSC. (2009). "Annual Performance Report (2008-2009)." National Water and Sewerage Corporation, Kampala, Uganda.

NWSC. (2010). "NWSC Annual Performance Report for Financial Year 2009-2010." National Water and Sewerage Corporation, Kampala, Uganda.

OFWAT. (2010). "Service and delivery-performance of the water companies in England and Wales 2009-10 report ", OFWAT, UK.

Park, H. J. (2006). "A Study to develop strategies for Proactive Water Loss Management," Ph.D Thesis, Georgia State University, USA.

Pelletier, G., Mailhot, A., and Villeneuve, J. P. (2003). "Modeling water pipe breaks-three case studies." *Journal of Water Resources Planning and Management*, 129(2), 115-123.

Poyry. (2009). "Update of the 2003 Feasibility Study for Kampala Water Supply. Draft Final Feasibility Report (Volume II)." NWSC, Kampala, Uganda.

Qdais, H. A. A. (2003). "Water Demand Management—Security for the MENA Region." *Seventh International Water Technology Conference, Cairo, Egypt, 1–3 April*, 5–23.

Richards, G. L., Johnson, M. C., and Barfuss, S. L. (2010). "Apparent losses caused by water meter inaccuracies at ultralow flows." *Journal of American Water Works Association*, 105(5), 123-132.

Sharma, S. K., and Vairavamoorthy, K. (2009). "Urban water demand management: prospects and challenges for the developing countries." *Water and Environment Journal*, 23, 210-218.

Thornton, J., Sturm, R., and Kunkel, G. (2008). *Water Loss Control*, McGraw-Hill, New York.

UBOS. (2002). "Population Census Report." Uganda Bureau of Statistics (UBOS), Kampala, Uganda.

USGS. (1998). "Estimated use of water in the United States in 1995, Cirlcular 1200." US Geological Survey, Denver.

Vairavamoorthy, K., and Mansoor, M. A. (2006). "Demand Management in Developing Countries." Water Demand Management, D. Butler and F. A. Memon, eds., IWA Publishing, London.

Vloerbergh, I. N., and Blokker, E. J. M. (2009). "Failure Data Analysis - a Dutch case study." Strategic Asset Management of Water Supply and Wastewater Infrastructures, H. Alegre and M. d. Almeida, eds., IWA Publishing, London.

Wallace, L. P. (1987). *Water and Revenue Losses: Unaccounted for Water*, AWWA, Denver, Colorado, USA.

WHO, and UNICEF. (2010). "Progress on Sanitation and Drinking-Water:2010 Update ", World Health Organization and UNICEF, Geneva, Switzerland.

WRM. (2005). *Water Resources Management Reform* Directorate of Water Development, Kampala, Uganda.

WSP. (2009). "Water Operators Partnerships: African Utility Performance Assessment." Water and Sanitation Program (WSP) - Africa, The World Bank, Nairobi, Kenya.

Wu, Z. Y., Farley, M., Turtle, D., Kapelan, Z., Boxall, J., Mounce, S., Dahasahasra, S., Mulay, M., and Kleiner, Y. (2011). *Water Loss Reduction*, Bentley Institute Press, Exton, Pennsylvania.

Zechner. (2007). "Time for a change." The Undergrounder. Ontario Sewer and Water Main Construction Association, Page 9.

Chapter 2 - Review of Methods and Tools for Water Loss Management

Parts of this chapter are based on:

Mutikanga, H.E, Sharma, S.K., and Vairavamoorthy, K.. (2012). "Review of methods and tools for managing losses in water distribution systems". *Journal of Water Resources Planning and Management (ASCE)* (**Accepted**).

Summary

The increasing costs of urban water supply coupled with dwindling water resources have compelled water service providers to improve efficiency of their water distribution systems by reducing water losses. Tools and methods are required to reduce these water losses. This chapter provides a comprehensive review of state-of-the-art tools and methods and their applications in water loss management. The purpose of the review is to assess knowledge gaps and identify future research needs. The review findings indicate that a sizeable number of tools and methods have been developed and applied for water loss management. They vary from simple managerial tools such as performance indicators to highly sophisticated optimisation methods such as evolutionary algorithms. Although a number of methodologies and tools have been developed, their application to WDSs in most developing countries is generally still limited due to the unique conditions that exist such as intermittent water supply, limited resources and high levels of apparent losses. The study also reveals the gap between developed methods and their applications in practice. Future research needs identified include developing methods and tools for uncertainty in flow measurements and their propagation into NRW, apparent loss intervention measures, appropriate performance indicators and benchmarking techniques for water loss management in developing countries, planning and refining economic models for pressure management, online monitoring and optimal sensor placement for leaks/bursts detection, and exploring further discrete multi-criteria analysis involving stakeholders in planning and management of water losses. Where possible, these studies should be carried out in an action-based research framework with close collaboration between water companies and research institutions.

2.0   Introduction

The previous Chapter introduced the subject, highlighted the challenges of water loss management (WLM) and presented the objectives of the study. This Chapter makes a critique of the existing methods and tools for WLM and identifies research knowledge gaps. The subsequent Chapters will bridge some of the knowledge gaps by developing appropriate tools and methods for WLM in developing countries. This chapter begins by introducing the research conceptual framework (Figure 2.1).

The rest of the Chapter is organized as follows. Section 2.1 presents the basic definitions and terminologies used in WLM. The research methodology is outlined in Section 2.2. Section 2.3 reviews tools and methods for leakage management. Tools and methods for apparent losses assessment and reduction are reviewed in section 2.4. Section 2.5 reviews optimization methods applied in WLM. Multi-criteria decision analysis, online monitoring and event detection methods are reviewed in sections 2.6 and 2.7 respectively. Section 2.8 reviews performance benchmarking methods and tools applied to WLM. Future research needs are presented in section 2.9 and conclusions based on the review are drawn in section 2.10.

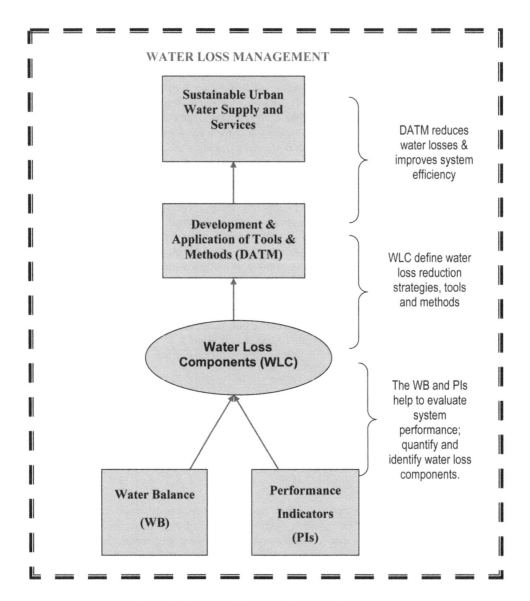

Figure 2.1 Conceptual framework for water loss management

2.1 Definitions and Terminologies

The definitions and terminologies used in this study for WLM are those that have been developed by the International Water Association (IWA) task forces on PIs and water losses (Alegre et al. 2006; Lambert and Hirner 2000) and adopted by the AWWA (AWWA 2003) and are now used widely in practice (Thornton et al. 2008). The terminologies used are shown in Figure 2.2. Of much interest to this study is the NRW which is defined as the difference between system input volume and revenue water. It is comprised of water losses and unbilled authorised consumption (water for flushing mains, fire fighting, sewer jetting etc.). This study mainly focuses on the water loss component of NRW. Water loss is made up of two components: real losses (RL) and apparent losses (AL). RL refers to the annual volumes lost through all types of leaks and breaks on mains, service reservoirs (including overflows) and service connections, up to the point of customer metering. AL are the non-physical losses that include customer meter under-registration, unauthorised use, meter reading and data handling errors.

System Input Volume (corrected for known errors)	Authorised Consumption	Billed Authorised Consumption	Billed Metered Consumption	Revenue Water
			Billed Unmetered consumption	
		Unbilled Authorised Consumption	Unbilled metered consumption	Non-revenue water (NRW)
			Unbilled unmetered consumption	
	Water losses	Apparent Losses	Unauthorised consumption Customer metering inaccuracies and data handling errors	
		Real losses	Leakage on transmission and distribution mains	
			Losses at utility's storage tanks	
			Leakage on service connections upto customer metering point	

Figure 2.2 The IWA/AWWA water balance methodology

2.2   Research Methodology

The research methodology applied in this study was an extensive literature review based on articles published in academic journals, conference proceedings and text books on methods and tools applied in WLM. The library databases were used to search high ranking journal articles with valuable information on techniques used for WLM.

2.3   Leakage Management

Leakage management involves assessment, detection and control. A typical integrated leakage management model is shown in Figure 2.3. Bursts and leaks will increase as major parts of the WDS continue to deteriorate unless all the four leakage management activities (shown by the arrows) in Figure 2.3 are carried out to an appropriate extent. In order to effectively manage leakage it must be accurately assessed.

2.3.1   Leakage assessment methods

Leakage assessment refers to tools and methods used to quantify the volume of leakage. There are basically three techniques that have been widely used for leakage assessment and are categorised as follows:

1. Mass (or volume) balance methods (Water balance/audit).
2. Network Hydraulic Modelling (NHM) simulations.
3. Flow statistical analysis.

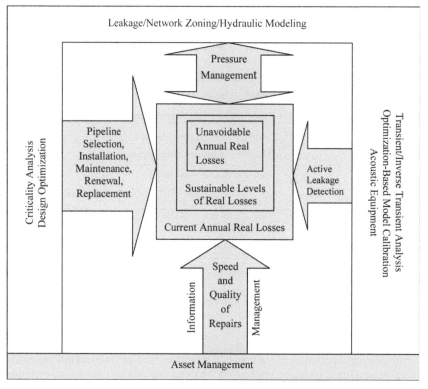

Figure 2.3 Integrated leakage management techniques
(Source: Adapted from Wu et al. 2011)

2.3.1.1 Mass (or volume) balance methods

The mass (or volume) water loss methodology is based on the principle that the SIV must at any one time in point be equal to the sum of the water consumed and the change in volume inventory (storage and pipelines) if there is no leakage in the system.

$$V_{SIV} = V_C + \Delta V + V_L \tag{2.1}$$

where V_{SIV} = metered system input volume, V_C = metered water use, ΔV = system storage volume and V_L = system leakage. This water balance methodology simplifies the rather complex task of keeping track of water supplied in WDSs. Currently, there are two main water balance methodologies used for assessing water losses: (i) the IWA/AWWA standardized water balance methodology (Figure 2.2) and (ii) the UK water balance methodology presented in Figure 2.4 (Farley and Trow 2003; Lambert 1994). The water balance/audit process is an effective tool for systematic accounting of water supply and consumption.

Distribution Input (DI)				
Water Taken (WT)				**Distribution Losses (DL)**
Water Taken (WT)			Distribution Operational Use (DOU)	Distribution Losses (DL)
Water Delivered through Supply Pipes (WDS)			Miscellaneous Water Taken (WTM)	Distribution Losses (DL)
Measured (WDSM)	Unmeasured Use (WDSU)	Unmeasured Supply Pipe Losses (WDSL)	Miscellaneous Water Taken (WTM) [Legally and illegal, Meter under-registration]	Distribution Losses (DL)

Figure 2.4 The UK water balance methodology
(Source: Lambert, 1994)

The UK water balance defers from the IWA/AWWA methodology in the following aspects:

- terminologies used are different e.g. the term "apparent losses" is not used in the UK methodology;
- in the UK, metered consumption is corrected for errors by including meter under-registration (MUR), thus under-estimating NRW;
- In the UK, illegal water use is categorized under unbilled authorized consumption. The focus is mainly on assessing leakage.
- the UK water balance methodology provides details on how to reliably measure and quantify un-metered domestic consumption and leakage using background and burst estimates (BABE) (Lambert and Morrison 1996) and MNF techniques (Farley and Trow 2003).

However, the major drawback of both methodologies is that most components are estimated using not well-defined standard techniques. This ambiguity leaves room for water utility managers to manipulate and mask NRW figures and its components (Brothers 2001). In explaining why a decade ago (2000-01), leakage levels in the UK were less than recent levels (2009-10), the Office of Water Services-OFWAT (now the Water Services Regulatory Authority) attributed it to a change in how Severn Trent and Thames Water used to assess their water balance data. Both companies were previously under-reporting leakage levels (OFWAT 2010). In addition, the methods give a quick "top-down" snapshot of the magnitude of water losses in the system but do not exactly pinpoint where the losses are occurring. Some countries have also reported difficulties in using the IWA/AWWA water balance methodology directly, for example in Greece where there is a minimum charge of water used (20 m^3), using billed metered consumption in the water balance may be misleading (Kanakoudis and Tsitsifli 2010).

The water balance input data from flow instruments is subject to measurement errors and uncertainties. These uncertainties and their propagation into output results must be quantified and reported to give credibility into the water balance estimates. The goal is to improve the quality of measurements and the reported NRW figures. Guidelines on how to quantify the uncertainties have recently been established (ISO 2008) and many researchers have attempted to quantify the water balance uncertainties (Herrero et al. 2003; Mutikanga et al. 2010a; Sattary et al. 2002; Stent and Harwood 2000). Although guidelines have been established and attempts made to quantify the uncertainty in the water balance, the practice of uncertainty analysis in decision making is still very limited. It is always not clear to most practitioners how uncertainty analysis will improve decision making. In addition, uncertainty analysis is expensive and time consuming and is perceived to be useful in the academia (Shrestha 2009).

Despite the drawbacks, the IWA/AWWA water balance methodology and its array of PIs is a valuable tool that provides standardized terminologies and definitions for performance benchmarking, both locally and internationally. During the past decade, several water balance software tools have been developed to promote use of the IWA/AWWA standardised water balance methodology. However, these software are too costly (typically priced from \$ 50,000 to \$100,000) and out of reach of most cash-strapped water utilities in the developing countries. They include Aquafast software developed for the water research foundation in the USA (Fanner et al. 2007b), Aqualibre developed by Bristol Water Utility (UK) (Liemberger and McKenzie 2003) and SIGMA developed by Instituto Tecnologico del Agua (ITA) of the Universidad Politecnica de Valencia, Spain (Alegre et al. 2006).

2.3.1.2 Minimum night flow (MNF) method

In order to refine the top-down water balance, the bottom-up approach that includes field investigations such as MNF is required. MNF measurements for leakage assessment are usually carried out in District Metered Areas (DMAs). DMAs are discrete supply zones within a network with each DMA consisting of about 500-3,000 connections (Farley and Trow 2003). They are mainly used for monitoring and leak detection in WDSs. The MNF is the lowest flow supplied to a hydraulically isolated supply zone, usually measured between mid-night and 5 am (Wu et al. 2011). During night time water use is at its lowest and pressures in the network are relatively high and a significant amount of flow measured during the hour of MNF is likely to be leakage. To calculate the leakage at MNF time [Q_L (t_{MNF})], customer legitimate night time use must be accurately assessed and deducted from measured flow into the DMA [Q_{DMA} (t_{MNF})] at time of MNF (Equation 2). The hourly leakage rate (Q_L,t) throughout the day is then calculated by multiplying the Night-Day-Factor (NDF) with the leakage rate at MNF based on fixed and variable area discharge (FAVAD) principles that explain the pressure-leakage relationships (Lambert 2002; May 1994).

$$Q_L (t_{MNF}) = Q_{DMA} (t_{MNF}) - Legitimate \ Night\text{-}Time \ Uses \qquad (2.2)$$

$$Q_L (t) = Q_L, (t_{MNF}) \ x \ [P(t)/P(t_{MNF})]^{NI} \qquad (2.3)$$

Where, $Q_L (t)$ is the leakage rate at the hour t (t ≠ t_{MNF}), t_{MNF} is the MNF hour, $Q_L, (t_{MNF})$ is the leakage rate at the MNF hour, $P(t)$ is the average hourly nodal pressure at the hour t (t ≠ t_{MNF}), $P(t_{MNF})$ is the average hourly nodal pressure at the MNF hour, NI is the pressure exponent. Studies have indicated that the NI values range from 0.5 to 2.3 depending on type of leak and pipe material (Greyvenstein and van Zyl 2007). In a recent study, Van Zyl and Cassa (2011), have shown that the leakage exponent NI does not provide a good characterization of the pressure response of a leak, and different leakage exponents result for the same leak when measured at different pressures. Although the MNF method is the most widely applied, it has the following limitations:

- It does not exactly reveal how this leakage is distributed in the network;
- It is not very effective for systems with irregular supply and un-zoned networks;
- It relies heavily on accurate estimation of the expected night flows;

In Amman city (Jordan) with severe water rationing conditions, MNF analysis was reported to be dubious and not informative (Decker 2006). In order to be able to apply MNF techniques for leakage assessment, Decker (2006) suggested that the best approach would be to progressively have limited DMAs which receive water regularly. For more effective leakage assessment, hybrid methods that combine network hydraulic modeling, MNF and top-down water balance approaches have been reported (Cheung and Girol 2009). There are many systems that now automate the monitoring of night flows such as Netbase (Burrows et al. 2000) and TakaDu (Armon et al. 2011), thus saving time and errors arising out of manual interpretation.

Annual real losses can also be derived from first principles using component analysis (Lambert 1994; Wu et al. 2011). This approach uses basic infrastructure data (mains length, number of service connects, etc); infrastructure condition factor (ICF) for background leakage; average flow rates and runtimes of different types of leaks (background, reported and unreported) on different components of the WDS (mains, service lines, reservoirs, etc).

2.3.1.3 Network hydraulic modelling (NHM) as a leakage assessment tool

WDSs are often very large and complex consisting of several kilometres of pipes of varying sizes and materials, storage reservoirs, pumps and various appurtenances. These systems are very difficult to understand and require large amounts of data for their analysis. NHM is one tool that has evolved over time to help engineers understand and manage WDSs. NHM involves using computer and mathematical models to predict the behaviour of the WDS and are routinely used for operational investigations, planning tasks and network design purposes (AWWA 2005). Like all mathematical models, WDS model parameters require calibration before useful results can be obtained from simulation. Calibration is a process of fine tuning a model until it simulates field conditions within acceptable limits. Guidelines for WDS model calibration have recently been established (Speight et al. 2010) but challenges still remain between theory and real world applications (Savic et al. 2010).

The most used network hydraulic modelling software that is freely available is EPANET 2 (Rossman 2000). Its hydraulic solver uses the gradient method with an open source code that allows extended modifications. Network simulation software provides the capability to mathematically replicate the non-linear dynamics of a WDS by solving the governing set of quasi-steady state hydraulic equations that include conservation of mass and energy within a loop. For leakage management and control, the NHM can be applied for many purposes, including network zoning and re-zoning, modelling leakage as pressure-dependent demand (PDD), pressure management planning, evaluating pipe renewal and replacement alternatives (Wu et al. 2011).

The conventional network hydraulic solvers analyze WDSs based on the assumption that nodal demands are fixed and independent of network pressures commonly referred to as demand-driven analysis (DDA). These assumptions are increasingly being challenged and new modelling techniques and algorithms are emerging (Giustolisi et al. 2008; Wu et al. 2010). DDA is only appropriate when WDSs are simulated under normal conditions with adequate pressures which in practice is not always the case e.g. during mains failure or in irregular water supply systems. Leakage is often implicitly included in nodal demands during design of WDSs which is also not realistic. Leakage is a type of PDD and must be explicitly considered in order to simulate hydraulic characteristics. This realisation motivated researchers to develop techniques for realistically modelling leakage in WDSs.

Germanopoulous (1985) was the first to report the inclusion of leakage terms in WDS models. He applied empirical functions to relate users demand and leakage to network pressures and included the functions in the mathematical formulation of the network analysis problem. The pressure-consumption relationship for a given node is expressed as:

$$C_i = C_i^k a_i e^{-b_i P_i / P_i^k} \qquad (2.4)$$

where, P_i = pressure at node i; C_i=the consumer outflow at node i; C_i^k= the nominal consumer demand; and a_i, b_i, P_i^k = constants for the particular node. C_i^k is the outflow normally provided to consumers assuming that the pressures in the system are adequate. , P_i^k corresponds to the nodal pressure at which a given proportion of C_i^k is known to be provided. The network model includes leakage using Equation 2.5.

$$V_{ij} = c_l \left(L_{ij} P_{ij}^{av} \right)^{N1} \qquad (2.5)$$

where, V_{ij} = leakage flow rate from the pipe connecting nodes i and j; c_l = a constant depending on the network; L_{ij} = pipe length, P_{ij}^{av}= average pressure along the pipe and $N1$ is the pressure exponent and in this case 1.18 was used and it derives from experimental data.

Vela et al. (1991) extended the method by incorporating pipe size and condition parameters as shown in Equation 2.6.

$$V_{ij} = c_l \left(L_{ij} D_{ij}^d e^{a\tau} P_{ij}^{av} \right)^{N1} \qquad (2.6)$$

where D and τ are pipe diameter and age respectively; d is 1 for (D < 125 mm) and is -1 for (D >125 mm); and a is a leakage shape parameter which is difficult to determine. The only draw-back about this methodology is the required data of field measurement required to determine the values of a_i, b_i, P_i^k for each node. This may be too costly and out of reach for most water utilities especially in the developing countries. In addition, the method assumes leakage flow to be uniformly distributed along a pipe. This assumption of uniform distribution of background leakage does not seem to be valid from a practical point of view, as pipe joints and fittings are not continuously located along a pipe.

As an alternative, leakage can now be modeled as orifice flow based on emitter hydraulics in EPANET 2 (Rossman 2007). Leakage along a pipe is allocated to the connected nodes in a hydraulic model. The emitter nodes allow leakage to be modeled using appropriate pressure-dependent outflow relationships as shown in Equation 2.7.

$$Q_{i,l}(t) = K_i [P_i(t)]^{N1} \qquad (2.7)$$

where $Q_{i,l}$ (t) is the leakage aggregated at node i at time t; P_i (t) is the pressure at node i at time t and K_i is the emitter coefficient for the node i, and a positive K_i is an indicator of leakage demand at node i. In addition, the method assumes that the leakage to be included in the model has been assessed accurately which is often not the case in practice. Uncertainties in leakage assessment could be as high as ±46% (Lambert 1994). It is also difficult to ascertain the correct emitter coefficients for pressure-deficient nodes. Negative demand is possible with negative nodal pressure and the emitter flow can increase without an upper bound as the pressure increases. Modelling leakage is still a challenging task and is influenced by other factors. Laboratory studies on soil versus orifice head have indicated that soil surrounding a leak influences leakage flow rates (Walski et al. 2006).

However, the last two decades have seen tremendous improvements in network hydraulic modeling techniques. Several researchers have assessed leakage using network hydraulic simulations that fully incorporate leakage as PDD (Almandoz et al. 2005; Burrows et al.

2003; Giustolisi et al. 2008; Tabesh et al. 2009; Tucciarelli et al. 1999). Other than Tabesh et al. (2009) and Burrows et al. (2003) who evaluated their methodologies on real case studies in Iran and UK, the other methodologies have not been demonstrated on real WDSs to evaluate their effectiveness. Although the methodology proposed by Tabesh et al. (2009) to evaluate water losses concurs with the IWA/AWWA water balance methodology, it had the following shortcomings:

- In assessing legitimate night use during the hour of MNF they used default values (6% active population at night with 10 L/head/h) suggested by McKenzie (1999) for South African Cities, which are not likely to be valid for Iranian towns;
- They introduced new terminologies of "operational error" and "management error" in assessment of apparent losses that are not part of the IWA/AWWA water balance methodology and are likely to cause confusion;
- In assessing meter inaccuracies they did not consider user demand profiles yet meter accuracy is a function of water used at different flow rates (Arregui et al. 2006b). In addition, they did not indicate which meter testing standards were used as different standards are likely to produce different meter accuracy results;
- The network model was calibrated off-line with a short-term sample of hydraulic data, thus reflect the network characteristics that were prevalent at the time.

Leakage analysis using network hydraulic modelling could be improved by automated data transfer using continuous online hydraulic measurements (Machell et al. 2010). In developing countries use of network hydraulic modeling is still limited by the high cost of collecting up to date data to drive simulations and inadequate human resources capacity for sophisticated modeling (Trifunovic et al. 2009). Another technique currently being applied by most researchers to assess leakage is flow statistical analysis.

2.3.1.4 Leakage assessment using statistical techniques

Leakage estimation using statistical techniques has been attempted by various researchers (Arreguin-Cortes and Ochoa-Alejo 1997; Buchberger and Nadimpalli 2004; Jankovic-Nisic et al. 2004; Palau et al. 2012). Buchberger and Nadimpalli (2004) proposed a leak screening method which is an intermediary between the simple water audit and the complex hydraulic model. The method infers a range of minimum and maximum leakage flow rates based on the behaviour of several sequential sample statistics computed from continuous measurements of the main flow into a DMA. The limitation of the method is that it has neither been tested in the field nor does it pinpoint the location of the individual leaks in the network as rightly acknowledged by the authors. Arreguin-Cortes and Ochoa-Alejo (1997) applied stratified random sampling techniques and leak flow gauging to assess leakage in 15 Mexican cities indicating that majority of leaks were on customer service lines. The method is prone to considerable errors based on bias of the formation of strata, although the random sampling procedure minimizes the bias. In their study of water consumption data and statistical analysis in southeast of England, Jankovic-Nisc et al. (2004) proposed a methodology for optimal positioning of flow meters of a monitoring time step and recommended a smaller size of DMA with 250 properties for effective leak and burst detection. They urge that pipes supplying large DMAs are less sensitive to changes of demand and therefore any sudden burst or background leakage that is the same order of magnitude as domestic consumption would be difficult to detect. Palau et al. (2012) applied a multivariate statistical technique, called principle component analysis, for burst detection in urban WDSs. The advantage of the method is that it allows for a sensitive and quick analysis without use of computationally demanding mathematical algorithms. The technique can also be used to detect other abnormal flow conditions in the network such as illegal use of water.

Whereas the aforementioned methods for leakage assessment (MNF, NHM and statistical analysis) are valuable tools for prioritizing zones with high leakage rates, they do not provide information about the location of leaks. Leak detection must be carried out to pin-point the exact location of leaks to facilitate repairs.

2.3.2 Leak detection methods

Leak detection is the "narrowing down" of a leak to a section of a pipe network while leak location refers to "pin-pointing" the exact position of the leak (Pilcher et al. 2007). Leakage can be minimised by increasing efficiency of leak detection via shortening the awareness, location and repair (ALR) times as illustrated conceptually in Figure 2.5. There are various tools and methods for leak detection and control ranging from simple visual inspections to sophisticated real-time and/or on-line monitoring systems (Smith et al. 2000).

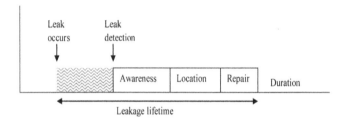

Figure 2.5 Life cycle of a leak

The leak detection methods can be categorised under three main groups:

- Passive (offline) observation and surveillance;
- Acoustic methods;
- Hydraulic methods.

2.3.2.1 Passive observation and surveillance

Traditionally, leak detection is carried out using patrol men who walk the entire length of pipelines searching for visible leaks that could be eclipsed by other objects such as swamps, vegetation etc. Utilities also often rely on the public information to report visible leaks and bursts. This technique is generally simple to apply but is slow and not effective. To be effective it must be complimented by other methods.

2.3.2.2 Acoustic leak detection

Acoustic devices such as listening rods, stethoscopes, ground microphones, noise loggers and more recently leak noise correlators (computer-based) are widely used in practice for leak detection and location (Clark 2012; Hartley 2009). These devices rely on sound and vibration signals induced by leaks impacting the soil from pipelines under pressure. Leak noise frequency is usually between 20 to 250 Hz and is localized near the origin of the leak. As human listening is limited to about 50 Hz, new acoustic devices are equipped with sound amplifiers such as geophones that look like a stethoscope (Smith et al. 2000). The effectiveness of acoustic leak detection methods has been successfully demonstrated for both metallic and plastic pipes (Fuchs and Riehle 1991; Hunaidi and Chu 1999). Leaks with low flow rates between 0.05 and 3.5 m^3/h could be accurately localized (Fuchs and Riehle 1991). However, plastic pipes were reported to strongly attenuate sound and leak frequency signal is mostly below 50 Hz making it hard to detect (Hunaidi et al. 2000). The disadvantages of acoustic correlators are that their effectiveness depends on the operator's experience, size of the leak, knowledge of exact pipe location and are limited to smaller size diameters and depth of less than 1.8 m. For larger diameter pipelines (> 300 mm), in-service pipeline methods (tethered and swimming smart balls) using acoustic sensors have been developed (e.g. Sahara) and reported to be very effective in detecting and pinpointing leaks in pressurized pipelines (Mergelas and Henrich 2005; Ong and Rodil 2012). Their additional advantage is that they do provide information on condition of the pipeline necessary for strategic asset management. The disadvantages of the Sahara in-service pipeline method is that it is limited to larger diameter pipes, requires depressurizing mains to enable removing the sensor from inside the pipeline and may not be suitable for mains with bends and in locations that are not easily accessible by vehicles. Other smaller equipment such as JD7 have started to emerge (Wu et al. 2011). Additional problems of leak detection in large diameter mains could be found in Hamilton and Krywyj (2012).

Sensors equipped with multi-parameter measurements (flow, pressure and noise) are now available for network monitoring and leak localisation (Koelbl et al. 2009a). Multi-parameter measurements is a promising methodology that is likely to be an alternative to DMAs for leakage monitoring (Farley 2012).

Other non-acoustic methods such as ground-penetrating radar, infrared thermography, and tracer gas have been found promising (Fanner et al. 2007a; Hunaidi et al. 2000). However, none of these methods are in wide scale use and their effectiveness is not very well understood. In addition, their costs are likely to be inhibitive for wide scale application. In general, manual site surveying and acoustic techniques are time consuming and costly, labour intensive and often imprecise. Automatic detection through the analysis of hydraulic measurements from permanently installed sensors is a more cost-effective method which can provide a rapid response to the on-set of a burst or leak event (Ye and Fenner 2011).

2.3.2.3 Hydraulic leak detection methods

Hydraulic leak detection methods use hydraulic characteristics (flows and pressures) to detect, locate and quantify leaks in pressurised pipelines. These methods can be categorised into three main groups depending on flow conditions in the pipelines:

- Hydrostatic methods (steady-state)
- Transient Methods (un-steady state)
- Inverse Analysis Methods (steady state or unsteady-state)

The earlier discussed mass-balance methods could also fall under hydraulic leak detection methods.

2.3.2.3.1 Hydrostatic Leak Detection Methods

Hydrostatic leak detection methods are based on the premise of a decline in pipeline pressure due to the presence of leakage. These methods are commonly used in practice during commissioning of new pipelines to detect presence of leaks by sealing pressurised mains at both ends and observing for any pressure drops. They are also used in locating leaks in DMAs in a process commonly referred to as "step-testing". According to Wu et al. (2011), a step test is the process of successively closing valves within a zone to isolate sections of pipes in turn and then recording the corresponding reduction in flow on the meter. A large reduction in flow is indicative of a leak within the isolated section. Their main disadvantage is shutting-off supply to isolated parts of the network and may be ineffective in networks with few isolation valves. It is also manually intensive and relatively expensive and provides information to a relatively small part of the network.

2.3.2.3.2 Transient leak detection

Leak detection based on transients is a promising cost-effective and non-intrusive technique (Ferrante and Brunone 2003). A sudden burst or leak in a pipeline causes pressure reduction inducing a negative pressure wave (transient) which travels in opposite directions from the break point and is reflected at the pipeline boundaries. The analysis of this leak-induced wave is the premise for leak detection and location using transient-based methods. The timing of the initial and reflected pressure waves combined with the knowledge of the system wave speed enables the location of the leak. The observed pressure signal can be analysed in either the time or the frequency domain. The magnitude of the pressure wave provides an estimate of the leak size. In practice, controlled transients could be generated by closure of network valves or opening and closing fire hydrants.

There are several methods for modelling transients in literature with the most common being pressure discrepancy and dynamic volume balance. Both approaches use the analysis of discrepancies between simulated and measured data as leak indicators. According to Martins and Seleghim Jr. (2010), leak detection and location is best realised using transient pressure wave based methods while the mass balance method is more capable for quantifying the leak flow rate more accurately and of detecting gradually developing (progressive) leaks. However, some studies have inspected existing pipeline abnormalities using pressure waves associated with hydraulic transients (Sattar and Chaudhry 2008). The two methods do actually complement each other.

The leak reflection method (or time domain reflectometry-TDR) has been reported as the simplest transient technique for determining the size and location of a leak (Brunone 1999). The detection of change in the measured pressure due to leak reflection is not trivial. The cumulative sum (CUSUM) change algorithm has been applied to automatically evaluate leak reflections and minimize ambiguity associated with simple visual inspection of the transient trace to locate bursts and leaks in single pipelines (Lee et al. 2007). The main advantage of the methods lies in their simplicity. However, the methods have mainly been limited to laboratory experiments using single pipelines and rely heavily on the accuracy of the transient model to detect small pressure signals. However, some promising results from controlled laboratory environments (Misiunas et al. 2005) to actual transmission mains have started to emerge (Misiunas et al. 2006).

In order to accurately detect leak reflections without necessarily having very precise transient signals, several researchers have devised various approaches such as frequency domain techniques (Covas et al. 2005; Lee et al. 2006; Lee et al. 2005; Mpesha et al. 2001); impulse response (Kim 2005); transient damping (Nixon et al. 2006; Wang et al. 2002); wavelet analysis (Allen et al. 2011; Ferrante and Brunone 2003; Ferrante et al. 2007); cross-correlation analysis (Beck et al. 2005) and virtual distortion method (Holnicki-Szulc et al. 2005). The coupling of these transient analysis methods with inverse mathematics could make it possible to detect, locate and quantify leakage using inverse transient analysis. This approach could be a valuable tool for practical applications in real WDSs especially where accurate system data is insufficient.

2.3.2.3.3 Inverse-transients for leak detection

Inverse transient analysis (ITA) for leak detection in WDSs requires generation of a transient of acceptable magnitude, measurement of transient responses usually a pressure transient (and flow if possible) at appropriate locations in the network and a network transient simulation model in which identical transient are introduced and simulated until a best fit between measured and simulated pressure responses is obtained.

Pudar and Liggett (1992) paved way for leak detection research based on measurements of state variables (pressure and/or flows) and solving an inverse problem of the steady state network hydraulic model. However, they concluded that leak detection by calibration of the steady-state WDS is unlikely to bring satisfactory results due to limited observed information and the requisite to accurately know the pipe roughness coefficients.

Liggett and Chen (1994) developed a new leak detection method based on ITA. This novel method calibrates while determining leaks or unauthorised use simultaneously. The calibration process is facilitated by optimisation techniques with the objective function of minimizing the sum of the squared differences between measured and computed pressure responses (Equation 2.8).

$$ E = \sum_{i=1}^{M} (h_i^m - h_i)^2 \qquad (2.8) $$

where E = objective function; h_i^m = measured head; h_i = numerically modelled head; and M = total number of measurements. The model is based on the assumption that it is a good representation of the system behaviour. In contrast to steady state network hydraulic modeling, ITA can provide the large amount of data required for successful calibration. Ligget and Chen (1994) used the Levenberg-Marquardt (LM) standard optimisation technique to fit the measured data to the numerical model results. The problem with using the LM method (or any derivative-based technique) is that the search space can be quite large and it does not guarantee convergence to a global minimum. The accuracy of the inverse method is very dependent on the quality and quantity of measurements (sampling design) and optimal location of measurement sites is essential for ITA in pipe networks (Kapelan et al. 2005; Vitkovsky et al. 2003). However, sampling design methodologies are still perceived as research tools and their application is still limited in practice probably due to the sophisticated modelling and optimisation tools required for solving sampling design problems. There are principal sources of errors that can exist in the ITA method that users must be aware of including data errors (noise in measurement variables and calibration accuracy), model input errors and model structural errors (Vitkovsky et al. 2007). Recent experimental studies based on ITA have shown that it is possible to detect and locate leaks in PVC pipes with an accuracy of between 4-15% of the total pipeline length (Soares et al. 2011).

The advantages of ITA methods include low cost, non-intrusive in nature, leak detection at far distances compared to acoustic devices and are less sensitive to pipe roughness coefficients. Theoretically, the ITA method is applicable to all network configurations. Although ITA has various advantages, its application to real WDSs is still limited due to various practical challenges such as: (i) highly looped networks with various appurtenances that are likely to quickly dampen any induced transients, (ii) difficulty in differentiating transient wave reflections due to leaks and other operational events such as demand changes, pump switching and valve closure and openings, (iii) risk of induced transients causing bursts and water quality contamination due to intrusion, (iv) the problems of pressure wave speed calibration and wave reflections at nodes (i.e. noise) in complex WDSs, and (v) a good transient model must be based on initial conditions which require a calibrated steady state model, calibration must be undertaken by considering leakage as a node demand. Without good estimation of leakage as a node demand, how can the initial conditions be produced for transient analysis? It appears most of the research of using transient models for leak detection is running into a dead-end. More information on leak detection methods can be found in reviews on leak detection in pipelines (Wang et al. 2001), transient-based leak detection methods (Colombo et al. 2009) and calibration of transient models (Savic et al. 2010).

2.3.3   Leakage control techniques

Leakage control refers to the application of different tools and methods to reduce leakage volumes in WDSs. There are various reactive and active leakage control (ALC) techniques used for reducing leakage such as pressure management, mains rehabilitation and speed in executing repairs of known leaks and bursts. Although mains rehabilitation is very effective in reducing leakage, it is costly and can hardly be justified based on a single criterion of leakage reduction. Speed in repairing mains failure is an inevitable reactive approach once failure has occurred. Pressure management is the only proactive and cost-effective tool that can reduce background leakage once pipes have been laid (Savic and Walters 1995; Ulanicki et al. 2000).

Effective management of pressures in WDSs is the essential foundation of effective leakage control (Fanner et al. 2007a; Thornton et al. 2008). Understanding the pressure-leakage relationship is therefore very important. The effect of operating at different pressures is modelled by FAVAD principles (May 1994) and FAVAD modified leakage equations (Cassa et al. 2010). However, it is difficult to apply the comprehensive representation of FAVAD equations to system-wide leakage analysis using a hydraulic model. In practice, the basic FAVAD equation for analysing and predicting changes in leak flow rate (L_0 to L_1) as average pressure changes from P_0 to P_1 is (Lambert and Fantozzi 2010):

$$L_1/L_0 = (P_1/P_0)^{N1} \qquad\qquad (2.9)$$

The leakage ratio increases proportionately with the increase in the average service pressure to the power $N1$. It is important to note that it is the ratio of average pressures and assumed $N1$ exponent that influence the reliability of the predictions. This is useful for evaluating pressure management strategies for leakage reduction. For example, reducing pressure by half reduces leakage by 29%, 50%, 65% and 82% of the original rate for $N1$ values of 0.5, 1, 1.5 and 2.5 respectively. It is critical to ensure accurate pressure measurements and $N1$ values. Experimental studies have shown N1 values close to 0.5 for leaks with small round holes (both plastic and metallic pipes) and N1 values close to 1.5 for small leaks (undetectable background leakage) from joints and fittings that are quite sensitive to pressures (Fanner et al. 2007a).

Pressure reduction is usually achieved by either pressure reducing valves (PRVs) or reduced pumping heads (for variable-speed pumps). There are three types of PRVs commonly applied in practice: fixed-outlet, time-modulated and flow-modulated. The fixed-outlet PRV is the traditional control method and uses a basic hydraulically operated control valve. Advanced pressure control makes use of both the time and flow modulated PRVs. The time-modulated is simple and easy to use. It is basically a timing device that can be attached to the controlling pilot on any PRV to reduce the outlet pressure at certain times of the day (McKenzie 2001). In both the fixed-outlet and time-modulated PRVs, the controller is aimed at maintaining the

PRV outlet pressure at a fixed specified set point. This limits their effectiveness to reduce background leakage especially when demand is low. Pressure management is more efficient if the DMA pressure is controlled in response to changes in demand or flow – so called flow-modulation (Fanner et al. 2007a). The advances in computer software control and telemetry systems have made flow-modulation practical.

In the implementation of a pressure control scheme, the PRV set-point can be adjusted electronically or hydraulically. Electronic control uses flow sensors, microcontrollers and solenoid valves as actuators. Although electronic control is efficient, it may not be suitable for WDSs with harsh field conditions and irregular power supply. The hydraulic flow modulator (e.g. Aquai-Mod Controller) is a more flexible and preferred solution in practice (Li et al. 2009). During pressure control, both steady state and dynamic aspects are encountered. The steady state behaviour of the system ensures optimal background leakage reduction without violating system minimum pressure requirements. The dynamic behaviour of the system is concerned with pressure changes (oscillations) caused by interactions between modulating valves and transients in the WDS. Whereas the dynamic performance of fixed and time-modulated PRVs for pressure regulation have been studied and are well understood (Prescott and Ulanicki 2003; Prescott and Ulanicki 2008), the dynamic performance of flow-modulated pressure control is still a research area (Li et al. 2009).

Most water systems are designed to provide a minimum working pressure at all points in the system throughout the day. This means that minimum pressure occurs at some critical point in the system, which is often the highest point in the system or the point furthest from the point of supply. As maximum pressure limitations are not considered in most design criteria, so many systems have areas with excessive pressure especially during off-peak periods. There is, therefore, scope for leakage reduction by managing system pressures to the optimum levels of service in most WDSs worldwide. There are various benefits of PM that include the following (Lambert and Fantozzi 2010):

- **Water losses** - reduced surges, new leak frequencies and natural rate of rise of leakage;
- **Customer service** – better service reliability due to less water supply interruptions;
- **System deterioration** – extended useful life of infrastructure;
- **Operating and maintenance costs** – reduced pumping energy and repairs and ALC;
- **Social costs** – reduced frequency of main breaks and disruptions of road users;
- **Capital costs** – deferment of infrastructure renewal and expansion;
- **Demand management** – less consumption from pressure related uses of water

Due to the various benefits, leakage control by pressure management has attracted the attention of researchers, consultants and water utilities. Real-world applications have reported promising results with varying levels of leakage reduction by pressure management. In one of the largest PM projects in the world so far, McKenzie et al. (2004) reported leakage reduction of about 40% of water supplied (or water savings of 9 million m^3/year) in Khayelitsha, South Africa. On the Gold Coast in Australia, leakage reductions of about 50% have been reported with additional benefits of 70-90% in reduced frequency of mains failure (Girard and Stewart 2007; Waldron 2008). In Waitakere city, New Zealand, NRW reduction has been reported and 74% reduction in frequency of mains failure (Pilipovic and Taylor 2003). In Cyprus, the Water Board of Lemesos reported leakage savings of 38% and 41% reduction in frequency of mains failure (Charalambous 2008). In Bangkok city, Thailand with low network pressures, leakage reductions of up to 12.5% have been reported by reducing average network pressure by 2.4 m (Babel et al. 2009). Although the case studies indicate significant achievements, the main drawback is that they did not provide optimal solutions. In a world of scarce resources, optimal solutions are needed and further benefits could be realized by optimal pressure reducing valve (PRV) settings and location using various optimization techniques such as GAs (Awad et al. 2009).

Pressure management is hardly applied as a leakage control tool in most developing countries despite the various benefits mainly due to two reasons. The first reason is lack of decision support tools to accurately predict benefits associated with pressure management to justify the investments. It is important that such planning studies are carried out before implementing pressure management projects (Ulanicki et al. 2000). Tools and methods to predict the associated economic benefits have been recently developed (Awad et al. 2008; Gomes et al. 2011). These tools may not be directly applied in developing countries due to data and other resource constraints. The second reason is that WDSs are not well configured for effective pressure management. The recently developed network zoning tools (Gomes et al. 2012; Sempewo et al. 2008) look attractive even though they have not yet been tried out on real networks to evaluate their effectiveness. In Chapter 6, the potential of pressure management for leakage control in the KWDS is examined. More information on leakage management in WDSs can be found in a recent review by Puust et al. (2010).

2.4   Apparent Losses Management

Apparent Losses (AL) often referred to as commercial losses consist of metering inaccuracies, unauthorized use of water, meter reading errors, data handling and billing errors. They occur as a result of inefficiencies in the measurement, recording, archiving, and operations used to track water volumes in a water utility (AWWA 2009). AL have a negative impact on water utility revenues and consumption data accuracy. Whereas various tools and methodologies for assessing real loss components of the water balance have been developed, little work has been done to assess AL. According to the AWWA (2003), AL control in water supply systems is in its infancy, and much work remains to be done to bring it to par with available real loss interventions. The IWA's Water Loss Task Force developed a four component approach for effective AL management (AWWA 2009), which has been modified to include implementation tools as shown in Figure 2.6.

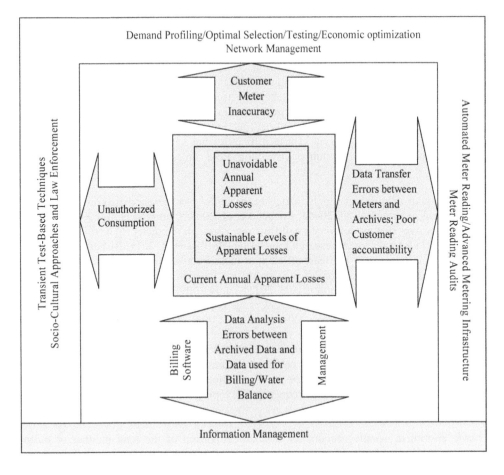

Figure 2.6 Four potential tools for active apparent loss control
[Source: Adapted from (AWWA 2009)]

2.4.1 Tools and methods for water meter management

The customer meter error is often thought to be the main cause of AL. Like all mechanical devices, water meters typically decline in accuracy with usage over time causing substantial revenue losses to the utility and gives rise to unequal billing policy (Pasanisi and Parent 2004). Most AL tools and methodologies developed are for minimizing revenue losses due to customer metering inaccuracies. Many studies have focused on water meter replacement based on meter testing, economic optimization, derivative optimization methods, and operational research techniques (Allander 1996; Arregui et al. 2003; Arregui et al. 2011; Arregui et al. 2009; Egbars and Tennakoon 2005; Lund 1988; Mutikanga et al. 2011a; Noss et al. 1987; Wallace and Wheadon 1986; Yee 1999). These studies applied regression analysis and assumed a linear relationship between meter accuracy degradation rate and age or usage. Pasanisi and Parent (2004) studied meters' degradation using a Markovian dynamic model, based on four discrete states, each one of which characterizes a more and more inaccurate metrology. Inference calculations are made in a Bayesian framework by MCMC (Markov Chain Monte Carlo) techniques. However, in practice, the regression analysis is the most widely applied due to its simplicity.

Although the tools and methodologies developed are valuable, their application in practice is rather difficult due to various reasons such as: (i) simplified assumptions made of uniform meter age and annual usage for all meters; (ii) the uncertainties related to estimation of water lost due to failed meters and the performance of meters after repair; (iii) uncertainties in predicting the in-situ meter degradation rates; (iv) uncertainties in measuring customer water use rates; (v) some methods do not take into account the time value of money; (vi) the assumption that the rate of decline of meter accuracy versus age and/or usage is linear; (vii) meter accuracy degradation rate is only a function of age and/or usage and (viii) lack of standards for testing old meters. These difficulties have been encountered in practice and during meter management studies involving field investigations (Arregui et al. 2009; Mutikanga et al. 2011b).

In WDSs with intermittent supply, metering inaccuracies are exacerbated by private elevated storage tanks (Arregui et al. 2006a; Cobacho et al. 2008; Criminisi et al. 2009). These tanks have ball-valves that induce very small flow rates through the meter. In a recent study carried out in the USA by the water research foundation (Barfuss et al. 2011; Richards et al. 2010), it was reported that meters are least efficient at measuring ultralow flows. Un-measured flow reducers (UFRs) have been reported as promising tools for reducing apparent losses due to metering (Fantozzi 2009; Fantozzi et al. 2011; Yaniv 2009). Selection of the right meter type and size for the required consumption flow range is critical in reducing apparent losses due to metering inaccuracies (Johnson 2001; Johnson 2003). Intermittent supply coupled with ageing pipeline infrastructure, poor repair practices and inappropriate metering technology have been reported to cause high meter failure rates in Kampala city, Uganda (Mutikanga et al. 2011c). Metering inaccuracies could be minimized by integrated meter management policies and strategies (meter type and selection, quality control, proper sizing and installation, optimal meter testing frequency and replacement) (Arregui et al. 2006b; Arregui et al. 2012a; Van Zyl 2011).

Large customers usually account for a small proportion of the total number of customer connections but account for 50% of total utility revenue generated including sewerage charges. Often these meters are oversized as utilities try to maximize new connection charges and sizing service lines is based on rules of thumb other than on demand profiling and hydraulic computations. van der Linden (1998) proposed five steps for resizing large meters in order to maximize revenues. The Boston Water and Sewer Commission recovered over 593,924 m^3 per year by downsizing over 400 meters of size 40 mm and larger. In addition to reducing NRW, the meter resizing effort could generate over 700,000 US $ annually (Sullivan and Speranza 1992). Commercial losses in large meters can be minimized by an integrated approach for large customers' water meter management (Arregui et al. 2012b). In Chapter 4, the water meter problems in developing countries are examined and appropriate intervention tools and methodologies for minimizing revenue losses developed.

2.4.2 Tools and methods for managing unauthorized use of water

Unauthorized water use occurs through deliberate actions of customers or other users who draw water from the system without paying for it. It occurs in many ways including illegal connections, illegal-reconnections, meter by-pass, meter tampering and abuse of fire hydrants. Unauthorized water use is a socio-technical problem that requires not only engineering solutions but also socio-cultural approaches that require changes in community behavior and attitudes toward water use as well as a strong stance against fraudulent practices of utility staff and water users. The socio-cultural approaches including working with local communities at the lowest administrative and street levels (territory management concept) have been reported to have been major drivers in reducing NRW in some Asian cities such as the east zone of Metro Manila where NRW has been reduced from 63% to 11% in the past 14 years saving over 0.6 million m^3 of water per day (Luczon and Ramos 2012). Pressure management data from WDSs can be used in algorithms for inverse calculations to detect unauthorized use (Liggett and Chen 1994). In a recent laboratory study, it has been shown that location and characteristics of illegal branches can be detected by means of fast transient tests (Meniconi et al. 2011). The effectiveness of these methods in real-world WDSs is doubtful as discussed under section (§2.3.2.3.3).

2.4.3 Tools and methods for minimising meter reading and data handling errors

Meter reading and data handling errors arise during the process of meter readings (gathered manually or automatically), data transfer to the billing system and archiving of customer consumption data. These errors can be caused by picking wrong meter readings (intentionally or accidentally), failure of Automated Meter Reading (AMR) equipment, wrongly captured data by billing assistants, erroneous system volume estimations and other policy and billing adjustment shortcomings. Leveraging the well-developed IT, metering technologies, and billing procedures and policies will be imperative for minimising these errors. Many water utilities are increasingly migrating from traditional manual meter reading to Advanced Metering Infrastructure (AMI) as a way of minimizing apparent losses due to meter reading and data handling errors (AWWA 2009). AMI systems with smart meters provide additional advantages of post-meter leak detection and management.

2.4.4 Assessing apparent water losses

In the absence of adequate data and proper methodology, most developed countries use default values or rules of thumb (e.g. AL is computed as 1% to 3% of total system input volume in Australia) which tend to be lowest values for well managed water systems, for component computation of apparent losses (Lambert 2002). However, these default values may not be appropriate for developing countries where illegal use of water is rampant and meter management policies are ineffective. In a model (BENCHLEAK) developed for leakage management by the South African Water Research Commission, McKenzie et al. (2002) applied a default value of 20% of total water losses. As rightly acknowledged by the authors, this approach was too simplistic, unrealistic and not scientific. For example in Johannesburg city, an upscale area such as Sandton is likely not to have the same level of illegal use as Soweto township and to assume a default value of 20% for AL is grossly erroneous. Seago et al. (2004) proposed a simplified approach for assessing apparent losses for South African water utilities based on age of water meters; water quality and qualitative information (very low to very high and poor to good) provided by utilities. Although this approach provides insight into the breakdown of AL components it has the following shortcomings:

i.   Metering errors were estimated based on the assumption that meters are replaced every five years in Europe. It is not age but usage that influence meter accuracy degradation rate. In addition, water quality is not the only factor that influences meter accuracy as rightly acknowledged by the authors. As a minimum, meter accuracy should be based on testing a few in-service meter samples of different age groups coupled with customer demand profiling to estimate weighted meter accuracy bands based on age.

ii. The qualitative data derived from water utility survey is likely to be biased. For example, what is very high or very low illegal connections? This depends on the illegal use control policy in place, size of city, socio-economic and cultural aspects that were not considered in this case.

iii.   AL have a direct impact on revenue water and are easily understood when expressed in terms of water sales volume rather than total losses volume.

In order to develop appropriate strategies for AL, accurate assessment of its components is required. Accurate assessment of AL for urban WDSs is still a research area. This study attempts to close the knowledge gaps in Chapter 5 by developing a methodology for assessment of AL in urban WDSs. In addition, guidelines for assessment of AL in water utilities of the developing countries with insufficient resources and data limitations to carryout in-depth assessment have been established in this study (Mutikanga et al. 2011b) and are presented in Chapter 5.

2.5   Real Loss Management using Optimization Methods

Considerable research effort has been made in developing optimization methods for optimal leak detection and control to minimize leakage in water distribution systems. The objective function of the optimization problem is either to minimize excessive pressures, and, inter alia, leakage or to directly minimize leakage. The constraints are usually the network analysis governing equations (energy conservation and mass balance) and minimum pressure requirements. The inclusion of pressure-dependent terms and terms that model the effect of valve actions into the governing equations allow the formal application of optimisation techniques. The water loss management methods developed using optimization techniques can be categorized as follows: (i) leak detection based on optimization methods, (ii) optimization of system pressure to minimize leakage, and (iii) optimization of pipe renewal and pump scheduling.

2.5.1   Leak detection based on optimization methods

Wu et al. (2010) developed a model-based optimization method for detection of leakage hotspots in water distribution systems. Leakage is represented as pressure-dependent demand simulated as emitter flows at selected nodes. The leakage detection method is formulated to optimize the leakage node locations and their associated emitter coefficients such that the differences between the model predicted and field observed values for pressure and flows are minimized. The optimisation problem is solved using genetic algorithms (GA). The optimized emitter coefficient of zero indicates no leakage, otherwise nodal leakage is identified. The greater the optimized emitter coefficient, the greater the expected leakage. This methodology has been successful tested in the UK (Wu et al. 2010) and Thailand (Sethaputra et al. 2009) to detect leakage hotspots. The major limitation of the method is that it requires a very good calibrated model and high quality data that is often not available in most water utilities particularly in the developing countries.

Several researchers have used mathematical programming techniques to minimize leakage using optimal location and/or optimal setting of flow control valves (Alonso et al. 2000; Jowitt and Xu 1990; Vairavamoorthy and Lumbers 1998). The pros and cons of the mathematical programming methods for leakage control have been documented by Vairavamoorthy and Lumbers (1998). Other leak detection methods and leak detection based on inverse transient analysis (ITA) and solving the ITA optimization problem have been presented (Kapelan et al. 2003; Vitkovsky et al. 2000).

2.5.2   Optimization of system pressure to minimize leakage

The use of search heuristic methods to which evolutionary algorithms (EAs) and GAs belong, have been adopted as alternative powerful stochastic optimization techniques to mathematical programming techniques and have proven to be very robust in solving highly non-linear and non-differentiable engineering problems including network pressure optimisation to minimize leakage (Nicklow et al. 2010). The general GA optimization framework is illustrated in Figure 2.7. The availability of open source GA libraries and freely available hydraulic solvers (e.g. EPANET) provide opportunities for developing cost-free optimization tools for variety of water distribution applications such as pressure control and optimal network designs.

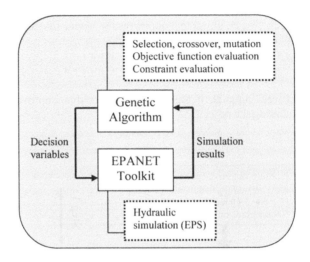

Figure 2.7 GA general Optimization framework

Savic and Walters (1995) were the first to apply GA for optimal pressure regulation to minimize leakage in water distribution systems. The optimization problem of minimizing the pressure heads is formulated with the settings of isolation valves as decision variables and minimum allowable pressures as constraints. The major drawback was that this method was not validated on a real case study network. Since then improvements and applications of GAs to solve leakage optimisation problems in WDSs have been reported by various researchers for optimal valve location (Reis et al. 1997); optimal valve setting (Araujo et al. 2006); optimal PRV flow modulation characteristic curves (AbdelMeguid and Ulanicki 2010); optimal setting of time modulated PRVs (Awad et al. 2009) and optimal storage tank levels (Nazif et al. 2010).

2.5.3   Optimization of pipeline renewal and pump scheduling

In WDSs, optimization methods have mainly been applied in scheduling rehabilitation of pipelines (Kleiner et al. 1998) and pump scheduling to minimize generation of excessive heads as alternative options for leakage minimization (Lansey and Awumah 1994). The major drawback of classical optimization methods is that they can at best find one solution in one simulation run, thus they are not convenient for solving multi-objective optimization problems.

2.5.4   Multi-objective optimization methods

Multi-objective optimization (MOO) is the process of simultaneously optimizing two or more conflicting objectives subject to certain constraints. Usually, one objective is achieved by compromising the second objective and vice-verse. A single solution which can optimize all objectives simultaneously does not exist. Instead, a set of Pareto optimal solutions is obtained as the best trade-off solution for decision-making. As most real-world optimization problems are often with more than one objective especially in the water resources domain, MOO is one of the fastest growing areas based on EAs (Deb 2001); group decision-making (Lu et al. 2007) and particularly in the water resources domain (Simonovic 2009). The general MOO framework is illustrated in Figure 2.8 and can be posed mathematically as (Price and Vojinovic 2011):

Find the vector $\mathbf{x}^* = [x_1, x_2, x_3, ..., x_n]$ that optimizes the vector function:

$$f(x) = |f(x_1), f(x_2), f(x_3), ..., f(x_n)| \text{ for } \quad \begin{aligned} g_i\,(x) &> 0 \text{ for } i = 1,2,3,..., m \\ h_i\,(x) &> 0 \text{ for } i = 1,2,3,..., p \end{aligned} \quad (2.10)$$

A solution $\mathbf{x}^*$ is Pareto optimal, if there is no other $\mathbf{x}$ that improves the solution of one objective without altering the other.

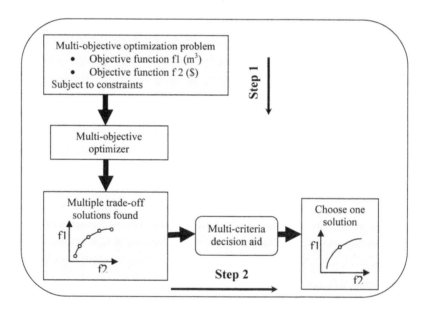

Figure 2.8 Preference-based multi-objective optimization framework framework
[Source: Adapted from Deb (2001)]

The major advantage of multi-objective EAs (MOEAs) is that they can find optimal solutions out of several trade-off alternatives with often conflicting objectives due to their population-approach (Deb 2001; Nicolini and Zovatto 2009).

MOO based on GAs has been recently applied to solve leakage problems in two very interesting and promising real-world water distribution case studies (Alvisi and Franchini 2009; Nicolini et al. 2011). Nicolini et al. (2011) demonstrated the application of sophisticated optimisation techniques for leakage management based on a real WDS in Italy. They applied a single objective GA calibration based methodology for optimal values of pipe friction factors and a leakage-dependent coefficient. The objective functions were: (i) to minimize the number of installed PRVs as a surrogate indicator for investment costs, and (ii) minimize the volume of leakage. The constraints were the governing equations, minimum pressure limits and maximum number of allowable control valves. The robustness of the procedure was evaluated using two multi-objective GA algorithms, the non-dominated sorting GA (NSGA-II) and Epsilon-MOEA. One of the optimal solutions involving the installation of four PRVs was adopted and implemented by the water utility. The water savings estimated after about three months of implementation were 281 m^3/day or 14% of system input volume. Alvisi and Franchini (2009) developed a methodology for optimal rehabilitation and leakage detection scheduling in WDSs. The objective functions were: (i) minimize the volume of leakage, and (ii) the break repair costs. The constraints were the governing equations and maximum allowable budget for proactive leakage and rehabilitation interventions. The optimizer used was the non-dominated sorting GA (NSGA-II). The procedure was found to be a very valuable utility decision support tool for apportioning the available budget between leak detection and pipe replacements, when and where to carry out leak detection and which pipes to replace and when. More detailed information on EAs could be found in a comprehensive state-of-the-art review for GAs in water resources planning and management (Nicklow et al. 2010).

In practice, a DM needs only one solution, which calls for additional higher level information. Often, such higher level information is non-technical, qualitative and experience-driven (Deb 2001). The procedure of handling MOO problems is referred to as preference-based MOO. In the parlance of management, such search and optimization problems are known as multi-criteria decision analysis (MCDA). A finite number of discrete solutions selected from the Pareto optimal set could be further ranked using MCDA as shown in Figure 2.8. These hybrid methods have been recently applied to select the optimal WDS design option from a set of optimal design solutions (Tanyimboh and Kalungi 2008).

2.6   Multi-criteria Decision Analysis (MCDA)

MCDA is a decision making technique used in solving decision problems with the following characteristics (Figueira et al. 2005; Lu et al. 2007; Simonovic 2009):

- multiple and conflicting criteria
- incommensurable criteria (different units)
- overall goal of ranking a finite number of decision options based on a family of evaluation criteria.

Guitouni and Martel (1998) described the MCDA methodology as a non-linear recursive process made up of four steps: (i) structuring the decision problem, (ii) articulating and modeling the preferences, (iii) aggregating the alternative evaluations (preferences) and (iv) making recommendations. Discrete MCDA methods include multi-attribute utility theory (MAUT), the analytical hierarchy process (AHP), compromise programming, fuzzy set analysis, TOPSIS, ELECTRE, PROMETHEE, ORESTE etc. Although MCDA techniques have been applied widely in the water resources domain (Hajkowicz and Collins 2007), their application in water loss planning and management has been limited. Clearly, WLM is a multi-criteria decision problem as it impacts on service quality, water quality, energy costs, environmental and social aspects. WLM is thus a multi-criteria problem with a high level of complexity that requires the use of MCDA techniques to select alternative strategies based on preferences of different stakeholders and often conflicting criteria. Morais and Almeida (2007) developed a strategy for leakage management based on group decision making for a hypothetical case study in the North-East of Brazil. However, no water balance was carried out in the decision process to ascertain whether the problem was leakage or apparent losses. In a recent study, a methodology framework for strategic planning and selection of based on the PROMETHEE outranking method of the MCDA family was developed and applied to select and prioritise water loss reduction options for Kampala city in Uganda (Mutikanga et al. 2011d).

Critics of MCDA say that the method is prone to manipulation, is very technocratic and provides a false sense of accuracy while proponents claim that MCDA provides a systematic, transparent approach that increases objectivity and generates results that can be reproduced (Janssen 2001; Macharis et al. 2004). There is need to explore further MCDA for controlling water losses in distribution systems. Chapter 7 evaluates and prioritizes water loss reduction options using MCDA as a decision-support tool.

2.7   Online Monitoring and Event Detection

Online monitoring and/or real-time control is gaining increasing use in water utilities as a fast response leak and burst detection protocol. The advances in technology (computerised sensors, microprocessors, telemetry, communication and software application packages) have enabled continuous gathering of flow and pressure data from water distribution systems in (near) real-time. This has led to development of systems capable of detecting and diagnosing abnormalities in water distribution systems and prompt near real-time intervention measures. One such system has been recently established in the UK under the Neptune project research consortium (Savic et al. 2008). The Neptune decision support system (DSS) is based on the analysis of real-time information derived from pressure loggers, flow meters, customer complaints and analysis of short-term water consumption forecasts (Morley et al. 2009). The DSS uses a methodology based on the Dempster-Shafer theory and combines evidence from several independent sources as indicated in Figure 2.9 (Bicik et al. 2011).

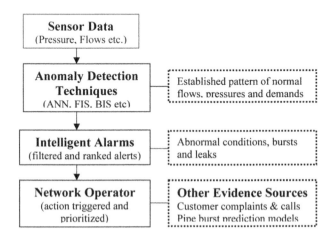

Figure 2.9 Real-time control system for pipe burst and leak detection

Water distribution sensor data (flow and pressure) usually in the form of time-series can be utilised in data-driven models for leak detection. Mounce et al. (2010) developed a method of using Artificial Neural Network (ANN) on flow and pressure data. A neural network with a mixture density network was used to predict a probability density function (PDF) of the hydraulic parameters. The PDF was then coupled to a Fuzzy Inference System (FIS) to detect leaks/bursts and other abnormal flows. The method was verified online on a real case study consisting of 144 DMAs in the UK and found to be very effective in delivering intelligent "smart alarms" for detected bursts. Alternative data-driven models coupled with Bayesian Inference Systems (BIS) have been reported in the UK with promising on-line burst detection results (Romano et al. 2009). Like all data-driven models, the disadvantage of the system is that it requires at least 2-3 months normal data for training and prediction accuracy of the artificial intelligence system.

Ye and Fenner (2011) developed a novel burst detection method based on auto-regression and adaptive Kalman filtering of hydraulic measurements. The results suggest that flow measurement data are more sensitive to burst or leak than pressure measurement data. They claim that the Kalman filtering method has advantages of computational efficiency, rapid detection rates, and does not require large quantities of training data compared to the artificial intelligence system. According to Savic et al. (2008), the artificial intelligence system is superior in detecting medium to large abnormal events as they occur (suitable for online application) while the Kalman filter based technique has potential to identify small abnormal events and provides potential for an early warning of system failure, thus the two techniques are complementary. The main disadvantage of the methods is their inability to detect existing stable leaks. To improve efficiency and cost effectiveness of leak detection and localisation, methodologies for optimal sensor placement have been developed and tested in real case study networks in the UK (Farley et al. 2010), Spain (Perez et al. 2009) and in Cyprus (Christodoulou et al. 2010).

Nazif et al. (2010) developed a hybrid model using ANNs and GAs for finding optimal storage reservoir levels to minimize network excessive pressures as a tool for leakage reduction. The model was applied to a real case study in the northwest part of Tehran metropolitan area. The results indicate that network leakage could be reduced by 30% annually when the tank levels are optimized using the proposed hybrid model. They report that the advantages of the model are reduction in runtime and ease of implementation and the disadvantage is loss of accuracy. Critics of ANNs view them as "black box" models that do not provide sufficient insight into the way they capture complex functional relationships. Support Vector Machines have been recently presented as an alternative to ANNs in detection of anomalies in water distribution systems (Mounce et al. 2011). Support Vector Machines are statistical pattern recognizers which perform similar functions to ANNs. However, they have better generalization ability and require smaller training sets than ANNs. Other methodologies based on the Self Organizing Map (SOM) (Aksela et al. 2009) and context classification (Branisavljevic et al. 2011) been presented for improved real-time data and leak detection in water distribution systems. Recent case study applications in Jerusalem (Armon et al. 2011) and Singapore (Allen et al. 2011) have demonstrated great potential to improve water distribution efficiency using online monitoring and event detection methods.

Whereas online detection and monitoring is a promising proactive methodology for network abnormal events management, it is still somehow impractical for large WDSs because of the computational burden such optimization imposes, "noise" in the sensor data and transient network effects. In addition, real-time control techniques may not be appropriate for developing countries due to the high costs of hydraulic measurement equipment, irregular supply regimes and inadequate communication infrastructure systems.

2.8 Performance Benchmarking for Water Loss Management

Performance assessment systems (PAS) and benchmarking are powerful management tools for evaluating and improving performance as has been demonstrated through their systematic use in many industries for decades (Alegre 2004) and most recently in the Ugandan construction industry (Tindiwensi 2006). However, their application to the water industry for WLM particularly in developing countries is still limited.

2.8.1 Performance Assessment Systems

PAS are used to assess the extent that management targets are met and even to evaluate the general impact of management strategies. They are used by different institutions to measure performance such as regulators (e.g. OFWAT in the UK), financial institutions (e.g. the World Bank), policy makers and utility management. The task of measuring and evaluating performance is accomplished by well defined performance indicators (PIs) (Brueck 2005; Crotty 2004). The assessment of the undertaking's performance with the use of PIs can measure the quality of service and the utility's effectiveness and efficiency; make transparent the comparison between the objectives, provide benchmarking between similar undertakings and encourage them to provide an improved service (Alegre et al. 2006). The most widely used indicators for WLM are those developed by IWA (Alegre et al. 2006; Lambert et al. 1999) and adopted by the AWWA (AWWA 2003). These indicators are presented in Table 2.1.

The most used PI for assessing water losses and target setting is % NRW. Although percentage NRW is recommended as a basic financial indicator, its main disadvantage is that it is affected by changes in consumption, which has nothing to do with the utility's WLM.

Table 2.1 IWA/AWWA performance indicators for water losses

Level	Water Resources	Operational	Financial
Basic	Inefficiency of use of water resources: real losses as a percentage of system input volume	Water losses: volume/service conn./year)	NRW: NRW as a percentage of system input volume
Intermediary		Real losses: volume/service.conn./day (when system is pressurised) Apparent losses: volume/service conn./year	
Detailed		Infrastructure leakage index (ILI)	NRW: value of NRW as a percentage of of the annual cost of running the water system

(Source: AWWA, 2003)

The other most used technical PI is the infrastructure leakage index (ILI) defined as the ratio of current annual real losses (CARL) to the unavoidable annual real losses (UARL). UARL represents the lowest technically achievable annual RL for a well maintained and managed system. The ILI is a measure of how well a distribution network is managed (maintained, repaired, and rehabilitated) for the control of real losses, at the current operating pressure. For well managed WDSs, ILI is equal to or very close to 1 and tends to increase as the system grows older. The default values used in calculating UARL are presented in Table 2.2

(Lambert et al. 1999). In a study carried out on 197 water undertakings around the world by the IWA water loss task force team, ILI figures varied from 0.3 (Austria and Netherlands) to 598 in South East Asia (McKenzie et al. 2007). They concluded that ILI is not suitable for comparing systems in developed countries and developing countries with often intermittent supply and low pressure WDSs. They suggested ILI target values of 2 for well-managed systems in the developed countries and 5 for developing countries. The analysis of the data set for 30 water utilities in South Africa (Seago et al. 2004), indicates poor correlation between NRW and ILI probably due to the fact that ILI is an indicator of leakage in the system as opposed to NRW that includes apparent losses and unbilled authorized consumption on top of leakage.

The ILI has the following major drawbacks which stem from the empirical data used to assess UARL (Lambert et al. 1999):

- It is an index and does not fulfil some of the basic requirements for PIs;
- It is a purely technical indicator that does not take into account economic factors;
- It is generally used to benchmark performance but has been found not appropriate for comparing developed countries with well managed systems and developing countries with intermittent supply and low pressure systems (McKenzie et al. 2007);
- The threshold used (0.5 m^3/h) for technically undetectable leaks is now dated as significant advances in acoustic leak detection equipment have evolved over the years which have dramatically lowered the threshold (Hunaidi and Brothers 2007) e.g. 0.25 m^3/h used in South Africa (McKenzie 2001) and accurately detected leaks of 0.05 m^3/h in Germany (Fuchs and Riehle 1991).
- It only measures performance on active leakage control – "find and fix" leakage activities and excludes efforts made using other leakage reduction options;
- The ILI of 0.3 reported in Netherlands and Austria indicates that UARL is greater than CARL reaffirming the ILI deficiencies even for water systems in the developed countries;
- It does not recognize benefits of PM as default values are calibrated at a reference pressure of 50m. ILI will not change by lowering or increasing system pressure;
- For the UARL calculation, the coefficient 0.8 L/service connection/day/ m pressure used in the equation was based on one service connection to one customer. In real practice, this may need to be revised as one service connection could serve more than one customer e.g. a block of apartments. Generally, the assumptions used for estimating components of ILI (CARL and UARL) have not been universally accepted.
- The assumptions made for the empirical formula of UARL do break down for WDSs with financial constraints and where leakage control is reactive other than proactive. In addition, data required for computation of UARL is costly and hardly gathered by water utilities in developing countries. These performance measures may not be appropriate or directly applied in the developing countries.

Table 2.2 Standard unit values used for calculating UARL

Infrastructure component	Background (undetectable) losses	Reported bursts	Unreported bursts	Total UARL
Mains	20 L/km/h*	0.124 bursts/km/year at 12 m³/h* for 3 days duration	0.006 bursts/km/year at 6 m³/h* for 50 days duration	18(L/km mains/day/m of pressure)
Service connections to edge of street	1.25 L/conn./h*	2.25/1000 conns./year at 1.6 m³/h* for 8 days duration	0.75/1000 conns./year at 1.6 m³/h* for 100 days duration	0.8 (L/conn/day/ m of pressure)
Service connections from curb-stop to meter**	0.5 L/conn./h*	1.5/1000 conns./year at 1.6 m³/h* for 9 days duration	0.5/1000 conns./year at 1.6 m³/h* for 101 days duration	25 (L/km of service conn./day/m of pressure)

*All flow rates are specified at a reference pressure of 50 m. ** Assuming average length of service connection from curb-stop to customer meter. (Source: Lambert et al. 1999)

However, ILI is still a valuable tool as an indicator of real losses in WDSs and where appropriate it should be applied in its current form or with appropriate modifications mainly focusing on improving assumptions made in estimating ILI components particularly the UARL (black box) to suit local conditions and advances made in technology over the years.

Using the ILI analogy, the apparent loss index (ALI) has been proposed (Rizzo et al. 2007; Thornton et al. 2008). The ALI is defined as the ratio of the current annual apparent losses (CAAL) to unavoidable annual apparent losses (UAAL). In absence of a reliable UAAL, a base value of 5% of water sales is recommended as a reference value (Rizzo et al. 2007). The benchmark reference value of 5% of water sales is rather high for most water utilities in developing countries. High figures for AL as percentage of water sales (revenue water) have been reported in various studies including 37% in Kampala city, Uganda (Mutikanga et al. 2011b); 33% in Lusaka, Zambia (Sharma and Chinokoro 2010); 16% in Medellin, Colombia (Garzon-Contreras and Palacio-Sierra 2007); 26.5% in Manila, Phillipines (Dimaano and Jamora 2010), and 36% in Jarkata, Indonesia (Schouten and Halim 2010). In the developed countries, the benchmark seems low for cities with universal customer metering and too high for partially metered systems. In Philadelphia city, USA with universal metering, apparent losses are estimated at 9.6% of revenue water (AWWA 2009) while in England and Wales with 37% metered households, apparent losses are estimated at 2.8% of revenue water (OFWAT 2010). AL for systems with customer storage tanks should not be compared directly with systems on direct mains pressure supply due to the ball-valve effect that amplifies AL in systems with storage tanks (Lambert 2002). There is clear need for more appropriate PIs and indices particularly to address the unique features of WDSs in the developing countries and for performance comparisons across utilities. In chapter 3 of this study, appropriate PIs and the ALI for WLM in the developing countries are proposed.

2.8.2 Performance target-setting

Water loss in urban WDSs is inevitable. However, excessive water loss is unacceptable. Between these two extremes, an optimum level of water loss exists. This optimum level is known as the economic level of water loss (ELWL). The ELWL has been defined as "that level of water losses which results from a policy under which the marginal cost of each individual activity for managing water losses can be shown to be equal to the marginal value of water in the supply zone" (Pearson and Trow 2008).

In the UK, the traditional target setting indicator for leakage is the economic level of leakage (ELL) defined as "the level at which it would cost more to reduce leakage further than to produce water from an alternative source" (OFWAT 2006). The UK economic regulator (OFWAT) sets and evaluates annual targets for water companies for various reasons but mainly to improve performance by monitoring progress towards meeting the set targets. OFWAT has powers to invoke heavy penalties for consecutive non-performance within a three-year period unless valid reasons beyond the company's control are provided. Penalties may include withdrawal of operating licences in extreme cases. ELL includes both Short-run ELL (SRELL) and Long-run ELL (LRELL). SRELL includes active leakage control and speed and quality of repair activities whereas LRELL includes PM and mains replacement. One of the limitations of SRELL is that it does not account for reduction in the number of leaks and bursts and the subsequent repair costs due to PM interventions which is likely to be an economic dominant factor. New generation approaches on how to include PM in SRELL calculations have been proposed (Fantozzi and Lambert 2007).

The calculation of ELL is complex and there is no single standardized methodology for undertaking the analysis of ELL (Fanner et al. 2007b). Critics of ELL urge that it should not only reflect the economic value of water lost but should include environmental and social aspects (Bouchart et al. 2001; Howarth 1998). In support of the ELL critics, The UK House of Lords has recently urged OFWAT to review the term "ELL" to "SLL" (sustainable levels of leakage) that incorporates environmental and social costs of leakage (HoL 2006). In response, OFWAT carried out a review to establish which environmental and social costs to include and to assess whether ELL was still suitable for target setting among other tasks (OFWAT 2006). Targets in the UK are now set based on sustainable economic level of leakage (SELL) that includes environmental and social costs while ensuring that companies are operating efficiently and providing the best value for consumers and the environment (OFWAT 2010). Alternative approaches for setting leakage targets have been proposed (Trow 2007). In other countries, the benefits of target setting have been reported as well. In Selangor, Malaysia, NRW reduction by 18,540 m^3/day has been reported in a performance-target based service contract framework using the IWA PIs (Liemberger 2002). The problem with performance-target based service contracts arises from establishing baseline data that is acceptable to both parties (the utility and the contractor). The methodologies for establishing not only ELL but economic levels of apparent losses and NRW in general are still lacking. ELL is hardly computed in water utilities of the developing countries due to data limitations. A model for estimating the optimum NRW levels for developing countries has been recently proposed (Wyatt and Romeo 2010). Clearly, further research is needed in developing simpler and more widely accepted methodologies for estimating sustainable economic levels of water losses (real and apparent).

2.8.3 Benchmarking methods

Benchmarking is a powerful management tool used for comparing one's business processes and performance metrics with the industry's best and/or best practices. Although benchmarking has been used widely in other sectors, it has recently become very popular in the water industry as indicated by numerous publications (Berg 2010; Cabrera Jr et al. 2011; van den Berg and Danilenko 2011). The International Benchmarking Network for Water and Sanitation (IBNET) program has grown into the largest publicly available water sector performance mechanism that collects, analyzes, and provides access to the information of more than 2,500 water and wastewater service providers from more than 110 countries around the world (van den Berg and Danilenko 2011). In the Netherlands, the Dutch water companies are self-regulated through voluntary benchmarking under the Association of the Dutch Drinking Water Companies (VEWIN) (De Witte and Marques 2010). Benchmarking is a valuable tool used by water utility managers, policy makers, regulators and financial institutions for different purposes with the target of improving water services and optimizing operations. There are various benchmarking methods and the most widely used in the water industry have been summarized in Figure 2.10. The methods are usually categorized as metric or process benchmarking (Berg 2006).

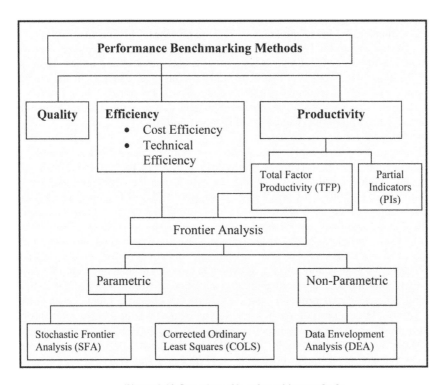

Figure 2.10 Overview of benchmarking methods

Process benchmarking is a normative tool for comparing the effectiveness of one's processes and procedures for executing different functions to those of selected peer groups. Comparisons often reveal performance gaps and help underperforming undertakings to adapt and internalize those more efficient and effective processes and procedures as appropriate. The methods used for process benchmarking are usually partial methods that deal with parts of the business such as PIs and can either be quantitative or qualitative. Its effectiveness depends on the level of information provided by different PIs. It is the most widely used in the water sector (e.g. IWA/AWWA PIs, IBNET, OFWAT, VEWIN etc.) due to its simplicity. Its disadvantage is that it does not provide any overall efficiency measure. The future paradigm of performance measurement is a multi-factor, informative and relative one (Tindiwensi 2006).

In metric benchmarking, well established empirical procedures are used by analysts to measure performance and identify performance gaps. It is more meaningful when carried out over time, tracking year-to-year changes in performance. The quantitative total methods that cover the whole business such are the most preferred for metric benchmarking. There are categorized as either parametric or non-parametric methods. The parametric methods such as stochastic frontier analysis (SFA) use econometric approaches (statistics and regression) in assessing company efficiency. The non-parametric approaches such as DEA use linear programming techniques to determine a company's efficiency frontier, which is assumed to be deterministic. The DEA and SFA methods are two main approaches used to construct production frontiers (Coelli et al. 2003). For both methods, data on the input and output quantities used by a sample of firms is required. A frontier is then fitted over the top of these data points and technical inefficiency is measured as the distance between each data point and the estimated frontier. DEA is the most widely used non-parametric method in practice (Berg and Marques 2011; Emrouznejad et al. 2008) and is increasingly being applied for measuring efficiency of water utility companies (Picazo-Tadeo et al. 2008; Picazo-Tadeo et al. 2011; Romano and Guerrini 2011). The method is simple to use, requires relatively small data sets and does not require specification of a functional form for production frontier. On the other hand SFA methods require several choices, mainly on the functional form and distribution assumptions, which both parties may find difficult to understand and communicate. Further details on these methodologies and their applications can be found in several textbooks (Cooper et al. 2000; Thanassoulis 2001; Zhu 2009) and recent studies (Abbott and Cohen 2009; Berg and Marques 2011; Walter et al. 2009).

The most remarkable example for WLM that combines PIs, target setting and benchmarking techniques is perhaps the one of England and Wales where the water industry is highly regulated. In the last two decades, leakage has been reduced from 5,112 ML/d in 1994/95 to now 3,281 ML/d (2009/10) or 9.7 m^3/km/d or 133.1 L/property/d (OFWAT 2010). This is a reduction of more than 35%. Assuming an average consumption of 150 L/c/d, the water saved (1,831 ML/d) is enough to serve more than 12 million people or the whole area served by Severn Trent, the second biggest water company in England and Wales. Bridgeman (2011), attributes this success to industry reforms in 1989, comparative competition, incentive regulation and development of more robust asset management tools and methodologies. Other water loss benchmarking studies on WLM using partial methods have been reported in various countries that include Canada (McCormack 2005), South Africa (Seago et al. 2004), New Zealand (McKenzie and Lambert 2008), Australia (Carpenter et al. 2003), in Austria (Koelbl et al. 2009b), in Asia (ADB 2010), in Africa (WSP 2009), in Latin America (Corton and Berg 2007) and internationally (Lambert 2002; McKenzie et al. 2007).

In a more recent benchmarking study on 18 water utilities in India, the analysis based on the DEA methodology reveals inefficiency in the WDSs and considerable potential for NRW reduction by 12.6% among other parameters (Singh et al. 2010). In Palestine, the efficiency of the WDSs was evaluated by applying DEA to 33 municipalities and the findings indicated that water losses were the main cause of inefficiency and network rehabilitation was required starting with the most DEA inefficient municipalities in order to minimize water losses (Alsharif et al. 2008). In a benchmarking study carried out in the USA, over 100 water utilities were analyzed using linear regression models and findings confirm that water utilities that use proactive strategies for WLM had better system efficiency (Park 2006). DEA and COLS have been used in comparative efficiency evaluations and regulation of water distribution companies in England and Wales (Cubbin and Tzanidakis 1998; Thanassoulis 2000). The major drawback in performance benchmarking is that the whole process losses credibility unless data used to define the PIs is reliable and accurate, generated in a transparent and auditable process. In addition, context information related different factors that could influence performance needs to be recognized for meaningful comparisons of performance across water companies, regionally and internationally (Skipworth et al. 1999).

Whereas, performance assessment systems and benchmarking are useful tools for evaluating and improving WDS efficiency, their application in the water industry of developing countries is still limited. The systems and tools developed may not be directly applicable to WDSs of the developing countries. In Chapter 3, appropriate tools and methodologies for

WDS performance evaluation and improvement are developed and applied to validate their effectiveness.

2.9   Future Research Needs

From the identified knowledge gaps, future research should focus on the following important areas in order to foster sustainable reduction of water distribution losses:

1. *Improving the quality of the water balance input data:* The PIs computed from the water balance such as NRW are not very useful for decision-making if data used to generate them is not reliable. The issue of data quality, uncertainty in flow measurements and uncertainty propagation into the final PIs is critical and still a research area.

2. *Assessment of apparent losses:* Although a lot of research has been undertaken for real losses, little progress has been made in the area of apparent losses. There is a need to develop more appropriate tools and methodologies to bring apparent loss interventions at par with available real loss interventions (AWWA 2003). Benchmarking indices analogous to the infrastructure leakage index (ILI) (Lambert et al. 1999) are still research areas.

3. *Solving problems in developing countries:* Water distribution systems in developing countries have peculiar technical characteristics such as poorly zoned networks, irregular supply etc. (Mutikanga et al. 2009; Sharma and Vairavamoorthy 2009) and other non-technical issues (Schouten and Halim 2010). These unique conditions demand unique tools and methods for water loss control that require further research. There are good lessons to learn from Asia particularly the unrivalled case of Phnom Penh city in Cambodia with NRW of 6.6% of total water supply (ADB 2010).

4. *Improving performance indicators:* Whereas the IWA/AWWA PIs provide a good foundation, they are insufficient for international water loss benchmarking (McKenzie et al. 2007) and not directly applicable to most water distribution systems in the developing countries. They require large amounts of reliable data that is costly and hardly generated by the resource constrained water utilities of the developing countries. There is need to develop generic methodologies for selecting, modifying and establishing new appropriate PIs based on local conditions particularly for developing countries.

5. *Pressure management:* The dynamic behaviour of water distribution systems under PRV control is still a research area particularly for multi-inlet DMAs (Li et al. 2009). Further work is also needed to test and refine the prediction models (Awad et al. 2008) for quantifying the economic benefits in order to understand fully the real impacts of pressure management on parameters such as burst reduction frequency and deferment of capital expenditure.

6. *Strategic planning:* Although various water loss reduction strategies do exist, deciding on which option to choose amidst often conflicting multiple objectives and different interests of stakeholders is a challenging task for water utility managers. Further research with aim of developing integrated multi-criteria decision-aiding framework methodologies for strategic planning of water loss management is required. Such tools are envisaged to help water utilities in

evaluating and prioritizing water loss reduction strategies particularly in the developing countries where water utilities often lack the necessary capabilities to carryout strategic planning. Further work is also needed in developing criteria for evaluating multi-objective optimizers.

7. *Online monitoring and detection:* Whereas there has been advances in online monitoring and detection equipment and technologies, real time control is still not yet fully developed and optimised for dynamic water loss reduction and further work to reduce on number of spurious alerts and detection of slowly progressive leaks and bursts (Savic et al. 2008) is still required. Guidelines on which burst detection methods (ANN systems, Support Vector Machines, Kalman Filtering etc.) to apply and when are still needed. In addition, the benefits of traditional DMAs are increasingly being challenged and they may no longer be relevant in future. Further research to investigate more open network scenarios, development and optimal placement of multi-parameter sensors (flow, pressure, water quality) for efficient leakage management and other water utility objectives is needed.

8. *Applied research:* Generally, there is a gap between theory and applications. For example, leak detection using inverse transient analysis methods has been one of the active research areas with very limited applications to water distribution systems in practice due to various reasons outlined in Wu et al. (2010). Future efforts should be focused on action-based research with close collaboration between water service providers and research institutions. Recent studies under the Neptune project in the UK have indicated good practical results based on this approach framework (Mounce et al. 2010; Savic et al. 2008; Ye and Fenner 2011) and in Italy (Alvisi and Franchini 2009; Nicolini et al. 2011). To facilitate this process, Abbott and Vojinovic (2009) have proposed the fifth generation of web-based numerical modelling as the most practical way of closing the knowledge gap between academic professionals and societies of knowledge consumers.

2.10 Conclusion

This Chapter presents a literature review of tools and methods for water loss management and their applications. The knowledge gaps and research needs have been identified and discussed. The major findings of the review include:

- Water loss management research and applications are widespread and growing; although progress in developing countries was found to be painfully slow. Water losses will continue to be one of the major challenges particularly in the developing countries for the 21st century;
- The majority of the research was in transient-based leak detection although with hardly any applications. It appears most of the research of using transient models for leak detection is running into a dead-end;
- The least researched area was the apparent loss component of water losses. Most research focused on leakage control and management, with the UK water industry being the pace-setters. More research efforts are still required to match apparent losses interventions with real losses especially in the developing countries where they are more prominent;
- There is still shortage of appropriate water loss management performance indicators for meaningful benchmarking across utilities at country level, regionally and internationally. More research effort is still required.

- Generally, there is a gap between theory and applications and future efforts should be focused on action-based research with close collaboration between water service providers and research institutions. Recent studies in the UK and Italy have demonstrated that this approach is possible and is very effective in minimizing water losses in real-world water distribution systems;

- There are various methods and tools available for WLM that include the water balance, PIs, acoustic leak detection and localization equipment, MNF analysis, flow statistical analysis, network hydraulic and optimization modelling, pressure management, integrated water meter management, advanced metering infrastructure, online monitoring and event detection, MCDA, and performance benchmarking. However, decision support guidelines on which appropriate tool or method to choose for given local conditions are still lacking and need further research.

- Whereas good progress has been made in developing tools and methods for WLM, accessibility is still limited in practice. Software remote access through web-based techniques is likely to improve access and improve communication between research institutions and service providers.

- Although more basic and novel research is popular in the academia, what is now needed is less blue sky research and more applied research.

It can be concluded that, although not exhaustive, this review could be a valuable reference resource for practitioners, policy makers, regulators and researchers dealing with WLM in distribution systems and provides a road map for future research. The subsequent chapters of this study address some of the identified knowledge and research gaps.

2.11 References

Abbott, M., and Cohen, B. (2009). "Productivity and efficiency in the water industry." *Utilities Policy*, 17, 233-244.

Abbott, M. B., and Vojinovic, Z. (2009). "Applications of numerical modelling in hydroinformatics." *Journal of Hydroinformatics*, 11(3-4), 308-319.

AbdelMeguid, H., and Ulanicki, B. (2010). "Pressure and Leakage Management in Water Distribution Systems via Flow Modulation PRVs." *Proceedings of the 11th Water Distribution System Analysis Conference (WDSA 2010)*, Tuscon, Arizona.

ADB. (2010). "Every Drop Counts: Learning from Good Practices in Eight Asian Cities." Asian Development Bank, Manila.

Aksela, K., Aksela, M., and Vahala, R. (2009). "Leakage detection in a real distribution network using a SOM." *Urban Water Journal*, 6(4), 279-289.

AL-Ghamdi, A. S., and Gutub, S. A. (2002). "Estimation of leakage in the water distribution network of the holy city of Makkah." *Journal of Water Supply: Research and Technology-AQUA*, 51(6), 343-349.

Alegre, H. (2004). "Performance Indicators as a Management Support Tool." Urban Water Supply Management Tools, L. W. Mays, ed., McGraw-Hill, New York.

Alegre, H., Baptista, J. M., Cabrera, E. J., Cubillo, F., Hirner, W., Merkel, W., and Parena, R. (2006). *Performance Indicators for Water Supply Services, IWA Manual of Best Practice*, IWA Publishing.

Allander, H. D. (1996). "Determining the economical optimum life of residential water meters." *Water Engineering and Management*, 143(9), 20-24.

Allen, M., Preis, A., Iqbal, M., Srirangarajan, S., Lim, H. B., Girod, L., and Whittle, A. J. (2011). "Real-time in-network distribution system monitoring to improve operational efficiency " *Journal AWWA*, 103(7), 63-75.

Almandoz, J., Cabrera, E., Arregui, F., Cabrera Jr, E., and Cobacho, R. (2005). "Leakage Assessment through Water Distribution Network Simulation." *Journal of Water Resources Planning and Management*, 131, 458-466.

Alonso, J. M., Fernando, A., Guerrero, D., Harnandez, V., Ruiz, P. A., Vidal, A. M., Martinez, F., Vercher, J., and Ulanicki, B. (2000). "Parallel Computing in Water Network Analysis and Leakage Minimization." *Journal of Water Resources Planning and Management*, 126(4), 251-260.

Alsharif, K., Feroz, E. H., Klemer, A., and Raab, R. (2008). "Governance of water supply systems in the Palestinian Territories: A data envelopment analysis approach to the management of water resources." *Journal of Environmental Management*, 87, 80-94.

Alvisi, S., and Franchini, M. (2009). "Multiobjective optimization of rehabilitation and leakage detection scheduling in water distribution systems." *Journal of Water Resources Planning and Management*, 135(6), 426-439.

Araujo, L. S., Ramos, H., and Coelho, S. T. (2006). "Pressure Control for Leakage Minimisation in Water Distribution Systems Management." *Water Resources Management*, 20(1), 133-149.

Armon, A., Gutner, S., Rosenberg, A., and Scolnicov, H. (2011). "Algorithmic monitoring for a modern water utility: a case study in Jerusalem." *Water Science and Technology*, 63(2), 233-239.

Arregui, F., Cabrera, E., Cobacho, R., and Garcia-Serra, J. (2006a). "Reducing Apparent Losses Caused by Meters Inacuracies." *Water Practice and Technology*, 1(4), doi:10.2166/WPT.2006093.

Arregui, F., Cabrera Jr, E., Cobacho, R., and Palau, V. (2003). "Management strategies for optimum meter selection and replacement." *Water Science and Technology: Water Supply*, 3(1/2), 143-152.

Arregui, F., Jr., C. E., and Cobacho, R. (2006b). *Integrated Water Meter Management* IWA Publishing, London.

Arregui, F. J., Cobacho, R., Cabrera Jr, E., and Espert, V. (2011). "Graphical Method to Calculate the Optimum Replacement Period for Water Meters." *Journal of Water Resources Planning and Management*, 137(1), 143-146.

Arregui, F. J., Martinez, B., Soriano, J., and Parra, J. C. (2009). "Tools for Improving Decision Making in Water Meter Management." *Proceedings of the 5th IWA Water Loss Reduction Specialist Conference*, Cape Town, SA, 225-232.

Arregui, F. J., Soriano, J., Cabrera, E., and Cobacho, R. (2012a). "Nine steps towards a better water meter management." *Water Science and Technology*, 65(7), 1273-1280.

Arregui, F. J., Soriano, J., and Gavara, F. J. (2012b). "An integrated approach for large customers water meter management." *Proceedings of the 7th IWA Water Loss Reduction Specialist Conference (CD-ROM)*, February 26-29, Manila, Philippines.

Arreguin-Cortes, F. I., and Ochoa-Alejo, H. L. (1997). "Evaluation of Water Losses in Distribution Networks." *Journal of Water Resources Planning and Management*, 123(5), 284-291.

Awad, H., Kapelan, Z., and Savic, D. (2008). "Analysis of Pressure Management Economics in Water Distribution Systems." *Proceedings of the 10th Annual Water Distribution System Analysis Conference WDSA2008, August 17-20*, Kruger National Park, South Africa, 520-531.

Awad, H., Kapelan, Z., and Savic, D. A. (2009). "Optimal setting of time-modulated pressure reducing valves in water distribution networks using genetic algorithms." Integrating Water Systems, Boxall and Maksimovic, eds., Taylor and Francis Group, London, UK, 31-37.

AWWA. (2003). "Committee report: Applying worldwide BMPs in water loss control." *Journal AWWA*, 95(8), 65-79.

AWWA. (2005). "Computer Modeling of Water Distribution Systems - M32, Second Edition." American Water Works Association, Denver, Colorado, USA.

AWWA. (2009). "Water Audits and Loss Control Programs: AWWA Manual M36." American Water Works Association, Denver, USA.

Babel, M. S., Islam, M. S., and Gupta, A. D. (2009). "Leakage Management in a low-pressure water distribution network of Bangkok." *Water Science and Technology:Water Supply*, 9(2), 141-147.

Barfuss, S. L., Johnson, M. C., and Neilsen. (2011). "Accuracy of In-Service Water Meters at Low and High Flow Rates." Water Research Foundation, Denver, CO, USA.

Beck, S. B. M., Curren, M. D., Sims, N. D., and Stanway, R. (2005). "Pipeline network features and leak detection by cross-correlation analysis of reflected waves." *Journal of Hydraulic Engineering*, 131(8), 715-723.

Berg, S. (2010). *Water Utility Benchmarking: measurements, methodologies and performance incentives* IWA Publishing, London.

Berg, S., and Marques, R. (2011). "Quantitative studies of water and sanitation utilities: a benchmarking literature survey." *Water Policy*, 13, 591-606.

Bhave, P. R., and Gupta, R. (2006). *Analysis of Water Distribution Networks*, Alpha Science International Ltd., Oxford, UK.

Bicik, J., Kapelan, Z., Mackropoulous, C., and Savic, D. A. (2011). "Pipe burst diagnostics using evidence theory." *Journal of Hydroinformatics*, 13(4), 596-608.

Bouchart, F. J. C., Salleh, H. M., Sawkins, J. W., and Jowitt, P. W. (2001). "Leakage targets and socio-economic efficiency." *Water and Environment Journal*, 15(1), 21-26.

Branisavljevic, N., Kapelan, Z., and Prodanovic, D. (2011). "Improved real-time data anomaly detecting using context classification." *Journal of Hydroinformatics*, 13(3), 307-323.

Bridgeman, J. (2011). "Water Industry Asset Management in England and Wales: success and challenges." *Water and Environment Journal*, 25(3), 318-326.

Brothers, K. J. (2001). "Water Leakage and Sustainable Supply-Truth or Consequences?" *Journal American Water Works Association*, 93(4), 150-152.

Brueck, T. M. (2005). *Developing and Implementing a Performance Measurement System*, Water Environment Research Foundation (WERF).

Brunone, B. (1999). "Transient test based technique for leak detection in outfall pipes." *Journal of Water Resources Planning and Management*, 125(5), 302-306.

Buchberger, S. G., and Nadimpalli, G. (2004). "Leak Estimation in Water Distribution Systems by Statistical Analysis of Flow Readings." *Journal of Water Resources Planning and Management*, 130, 321-329.

Burrows, R., Crowder, G. S., and Zhang, J. (2000). "Utilization of network modelling in the operational management of water distribution systems." *Urban Water*, 2(2), 83-95.

Burrows, R., Mulreid, G., and Hayuti, M. (2003). "Introduction of a fully dynamic representation of leakage into network modelling studies using EPANET." Proceedings of the International Conference on Advances in Water Supply Management, C. Maksimovic, D. Butler, and F. A. Memon, eds., Swets & Zeitlinger, Lisses, 109-118.

Cabrera Jr, E., Dane, P., Haskins, S., and Theuretzbacher-Fritz. (2011). *Benchmarking Water Services: Guiding water utilities to excellence*, IWA Publishing, London.

Carpenter, T., Lambert, A., and McKenzie, R. (2003). "Applying the IWA approach to water loss performance indicators in Australia." *Water Science and Technology: Water Supply*, 3(1/2), 153-161.

Cassa, A. M., Van Zyl, J. E., and Laubscher, R. F. (2010). "A numerical investigation into the effects of pressure on holes and cracks in water supply pipes." *Urban Water Journal*, 7(2), 109-120.

Charalambous, B. (2008). "Use of district metered areas coupled with pressure optimization to reduce leakage." *Water Science and Technology: Water Supply*, 8(1), 57-62.

Cheung, P. B., and Girol, G. V. (2009). "Night flow analysis and modeling for leakage estimation in a water distribution system." Integrating Water Systems, Boxall and Maksimovic, eds., Taylor and Francis Group, London.

Christodoulou, S., Agathokleous, A., Kounoudes, A., and Mills, M. (2010). "Wireless Sensor Networks for Water Loss Detection." *European Water*, 30, 41-48.

Clark, A. (2012). "Increasing efficiency with permanent leakage monitoring." *Proceedings of the 7th IWA Water Loss Reduction Specialist Conference*, Manila, Philippines, Feb 26-29, 2012.

Cobacho, R., Arregui, F., Cabrera, E., and Cabrera Jr, E. (2008). "Private Water Storage Tanks: Evaluating their Inefficiencies." *Water Practice and Technology*, 3(1), doi:10.2166/WPT.200825.

Coelli, T., Estache, A., Perelaman, S., and Trujillo, L. (2003). *A Primer on Efficient Measurement for Utilities and Transport Regulators*, World Bank Institute, Washington, D.C, USA.

Colombo, A. F., Lee, P., and Karney, B. W. (2009). "A selective literature review of transient-based leak detection methods " *Journal of Hydro-environment Research*, 2, 212-227.

Cooper, W. W., Seiford, L. M., and Tone, K. (2000). *Data Envelopment Analysis: A Comprehensive Text with Models, Applications, References and DEA-Solver Software.*, Kluwer Academic Publishers, Boston.

Corton, M. L., and Berg, S. V. (2007). "Benchmarking Central American Water Utilities." Public Utility Research Centre, University of Florida, Gainesville, Florida.

Covas, D., Ramos, H., and Almeida, B. A. (2005). "Standing wave difference method for leak detection in pipeline systems." *J. Hydraul. Eng.*, 131(12), 1106-1116.

Criminisi, A., Fontanazza, C. M., Freni, G., and La Loggia, G. (2009). "Evaluation of the apparent losses caused by water meter under-registration in intermittent water supply." *Water Science and Technology:WST*, 60(9), 2373-2382.

Crotty, P. (2004). *Selection and Definition of Performance Indicators for Water and Wastewater Utilities*, Awwa Research Foundation, A merican Water Works Association (AWWA), Denver, Colarado, USA.

Cubbin, J., and Tzanidakis, G. (1998). "Regression versus data envelopment analysis for efficiency measurement: an application to the England and Wales regulated water industry." *Utilities Policy*, 7, 75-85.

De Witte, K., and Marques, R. C. (2010). "Designing performance incentives, an international benchmark study in the water sector." *Central European Journal of Operational Research*, 18, 189-220.

Deb, K. (2001). *Multi-Objective Optimization using Evolutionary Algorithms*, John Wiley & Sons, Ltd Chichester, England.

Decker, C. W. (2006). "Managing water losses in Amman's renovated network: a case study." *Management of Environmental Quality: An International Journal*, 17(1), 94-108.

Dimaano, I., and Jamora, R. (2010). "Embarking on the World's Largest NRW Management Project." *Proceedings of the 6th IWA Water Loss Reduction Specialist Conference*, Sao Paulo, Brazil, CD-ROM.

Egbars, C., and Tennakoon, J. (2005). "Ipswich Water's Meter Replacement Strategy." Water Asset Management International, 19-21.

Emrouznejad, A., Parker, B., and Tavarez, G. (2008). "Evaluation of research in efficiency and productivity: a survey and analysis of the first 30 years of scholarly literature in DEA." *Journal of Socio-Economics Planning Science*, 42(3), 151-157.

Fanner, P., Sturm, R., Thornton, J., and Liemberger, R. (2007a). *Leakage Management Technologies*, Awwa Research Foundation Denver, Colorado, USA.

Fanner, P., Thornton, J., Liemberger, R., and Sturm, R. (2007b). *Evaluating Water Loss and Planning Loss Reduction Strategies*, Awwa Research Foundation, AWWA, Denver, USA

Fantozzi, M. (2009). "Reduction of customer meters under-registration by optimal economical replacement based on meter accuracy testing programme and unmeasured flow reducers." *Proceedings of the 5th IWA Water Loss Reduction Specialist Conference*, Cape Town, South Africa, 233-239.

Fantozzi, M., Criminisi, A., Fontanazza, C. M., and Freni, G. (2011). "Investigations into under-registration of customer meters in Palermo (Italy) and effect of introducing low flow controllers." *Proceedings of the 6th IWA Specialist Conference on Efficient Water Use and Management (CD-ROM)*, Dead Sea, Jordan.

Fantozzi, M., and Lambert, A. (2007). "Including the effects of pressure management in calculations of short-run economic leakage levels " *Proceedings of the 4th IWA Water Loss Reduction Specialist Conference*, Bucharest, Romania, 256-267.

Farley, B., Mounce, S. R., and Boxall, J. B. (2010). "Field Testing of an Optimal Sensor Placement Methodology for Event Detection in an Urban Water Distribution Network." *Urban Water Journal*, 7(6), 345-356.

Farley, M. (2012). "Are there Alternatives to the DMA?" *Proceedings of the 7th IWA Water Loss Reduction Specialist Conference*, Manila, Philippines, Feb 26-29, 2012.

Farley, M., and Trow, S. (2003). *Losses in Water Distribution Networks: A Practitioner's Guide to Assessment, Monitoring and Control*, IWA Publishing, London.

Ferrante, M., and Brunone, B. (2003). "Pipe system diagnosis and leak detection by unsteady-state tests: wavelet analysis." *Advances in Water Resources*, 26, 107-116.

Ferrante, M., Brunone, B., and Meniconi, S. (2007). "Wavelets for the analysis of transient pressure signals for leak detection." *Journal of Hydraulic Engineering*, 133(11), 1274-1282.

Figueira, J., Greco, S., and Ehrgott, M. (2005). *Multicriteria Decision Analysis: State of the Art Surveys*, Springer, New York.

Fuchs, H., and Riehle, R. (1991). "Ten years of experience with leak detection by acoustic signal analysis." *Applied Acoustics*, 33(1), 1-19.

Garzon-Contreras, F., and Palacio-Sierra, C. (2007). "A Case Study of Leakage Management in Medellin City, Colombia." *Proceedings of the 4th IWA Water Loss Reduction Specialist Conference*, Bucharest, Romania, 434-443.

Germanopoulos, G. (1985). "A Technical Note on the inclusion of Pressure Dependent Demand and Leakage terms in Water Supply Network Models." *Civil Engineering and Environmental Systems*, 2(3), 171-179.

Girard, M., and Stewart, R. A. (2007). "Implementation of Pressure and Leakage Management Strategies on the Gold Coast, Australia: Case Study." *Journal of Water Resources Planning and Management*, 133, 210.

Giustolisi, O., Savic, D., and Kapelan, Z. (2008). "Pressure-Driven Demand and Leakage Simulation for Water Distribution Networks." *Journal of Hydraulic Engineering*, 134(5), 626-635.

Gomes, R., Marques, A. S., and Sousa, J. (2011). "Estimation of the benefits yielded by pressure management in water distribution systems." *Urban Water Journal*, 8(2), 65-77.

Gomes, R., Marques, A. S., and Sousa, J. (2012). "Decision support system to divide a large network into suitable District Metered Areas." *Water Science and Technology*, 65(9), 1667-1675.

Greyvenstein, B., and van Zyl, J. E. (2007). "An experimental investigation into the pressure-leakage relationship of some failed water pipes." *Journal of water supply: Research and Technology-AQUA*, 56(2), 117-124.

Guitouni, A., and Martel, J. M. (1998). "Tentative guidelines to help choosing an appropriate MCDA method." *European Journal of Operational Research*, 109, 501-521.

Hajkowicz, S., and Collins, K. (2007). "A Review of Multiple Criteria Analysis for Water Resource Planning and Management." *Water Resources Management*, 21, 1553-1566.

Hamilton, S., and Krywyj, D. (2012). "The Problem of Leakage Detection on Large Diameter Mains." *Proceedings of the 7th IWA Water Loss Reduction Specialist Conference*, Manila, Philippines, Feb 26-29, 2012.

Hartley, D. (2009). "Acoustics Paper." *Proceedings of the 5th IWA Water Loss Reduction Specialist Conference*, Cape Town, South Africa 115-123.

Herrero, M., Cabrera Jr, E., and Valero, F. J. (2003). "A New Approach to Assess Performance Indicators' Data Quality." Pumps, Electromechanical Devices and Systems Applied to Urban Water Management Systems, E. Cabrera and E. Cabrera Jr., eds., Swets & Zeitlinger, Lisse, The Netherlands, 69-78.

HoL. (2006). "Water Management." House of Lords Science and Technology Committe, London. HL Report No. 191-I, Vol.1.

Holnicki-Szulc, J., Kolakowski, P., and Nasher, N. (2005). "Leakage detection in water networks." *Journal of Intelligent Material Systems and Structures*, 16(3), 207-219.

Howarth, D. A. (1998). "Arriving at the economic level of leakage: environmental aspects." *Water and Environment Journal*, 12(3), 197-201.

Hunaidi, O., and Brothers, K. (2007). "Night Flow Analysis of Pilot DMAs in Ottawa." *IWA Specialised Conference: Water Loss 2007* Bucharest, Romania, 32 - 46.

Hunaidi, O., Chu, W., Wang, A., and Guan, W. (2000). "Detecting leaks in plastic pipes." *Journal AWWA*, 92(2), 82-94.

Hunaidi, O., and Chu, W. T. (1999). "Acoustical characteristics of leak signals in plastic water distribution pipes." *Applied Acoustics*, 58, 235-254.

ISO. (2008). "Uncertainty of Measurement - Part 3: Guide to expression of uncertainty in measurement (GUM:1995)." International Organization for Standardization(ISO)/International Electrotechnical Commission (IEC), Geneva, Switzerland.

Jankovic-Nisic, B., Makismovic, C., Butler, D., and Graham, N. J. D. (2004). "Use of flow meters for managing water supply networks." *Journal of Water Resources Planning and Management*, 130(2), 171-179.

Janssen, R. (2001). "On the use of Multi-criteria Analysis in Environmental Impact Assessment in The Netherlands." *Journal of Multicriteria Decision Analysis*, 10(2), 101-109.

Johnson, E. H. (2001). "Optimal water meter selection system." *Water S. A.*, 27(4), 481-488.

Johnson, E. H. (2003). "Optimal water meter sizing and maintenance system." *Water Science and Technology: Water Supply*, 3(1-2), 79-85.

Jowitt, P. W., and Xu, C. (1990). "Optimal Valve Control in Water-Distribution Networks." *Journal of Water Resources Planning and Management*, 116(4), 455-472.

Kanakoudis, V., and Tsitsifli. (2010). "Results of an urban water distribution network performance evaluation attempt in Greece." *Urban Water Journal*, 7(5), 267-285.

Kapelan, Z. S., Savic, D. A., and Walters, G. A. (2003). "A hybrid inverse transient model for leakage detection and roughness calibration in pipe networks." *Journal of Hydraulic Research*, 41(5), 481-492.

Kapelan, Z. S., Savic, D. A., and Walters, G. A. (2005). "Optimal Sampling Design Methodologies for Water Distribution Model Calibration." *Journal of Hydraulic Engineering, ASCE*, 131(3), 190-200.

Kim, S. H. (2005). "Extensive development of leak detection algorithm by impulse response method." *Journal of Hydraulic Engineering*, 131(3), 201-208.

Kleiner, Y., Adams, B. J., and Rogers, J. S. (1998). "Selection and scheduling of rehabilitation alternatives for water distribution systems." *Water Resources Research*, 34(8), 2053-2061.

Koelbl, J., Martinek, D., Martinek, P., Wallinger, C., Fuchs-Hanusch, D., Zangal, H., and Fuchs, A. (2009). "Multiparameter measurements for network monitoring and leak localising." *Proceedings of the 5th IWA Water Loss Reduction Specialist Conference*, Cape Town, South Africa 620-627.

Koelbl, J., Mayr, E., Theuretzbacher-Fritz, H., Neunteufel, R., and Perfler, R. (2009). "Benchmarking the process of physical water loss management." *Proceedings of the 5th IWA Water Loss Reduction Specialist Conference*, Cape Town, South Africa 176-183.

Lambert, A. (1994). "Accounting for losses: the bursts and background concept." *Water and Environment Journal*, 8(2), 205-214.

Lambert, A., and Hirner, W. (2000). "Losses from Water Supply Systems: Standard Terminology and Recommended Performance Measures (IWA's Blue Pages)." The International Water Association (IWA) London.

Lambert, A., and Morrison, J. A. E. (1996). "Recent developments in application of "Bursts and Background Estimates" concepts for leakage management." *Water and Environment Journal*, 10(2), 100-104.

Lambert, A. O. (2002). "International Report: Water losses management and techniques." *Water Science and Technology: Water Supply*, 2(4), 1-20.

Lambert, A. O., Brown, T. G., Takizawa, M., and Weimer, D. (1999). "A review of performance indicators for real losses from water supply systems." *Aqua- Journal of Water Services Research and Technology*, 48(6), 227-237.

Lambert, A. O., and Fantozzi, M. (2010). "Recent Developments in Pressure Management." *Proc. of the 6th IWA Water Loss reduction Specialist Conference, June 6-9*, Sao Paulo, Brazil.

Lansey, K. E., and Awumah, K. (1994). "Optimal pump operations considering pump switches." *Journal of Water Resources Planning and Management*, 120(1), 17-35.

Lee, P., Lambert, M., Simpson, A., and Vitkovsky, J. P. (2006). "Experimental verification of the frequency response method of leak detection." *Journal of Hydraulic Research*, 44(5), 451-468.

Lee, P. J., Lambert, M. F., Simpson, A. R., Vitkovsky, J. P., and Misiunas, D. (2007). "Leak location in single pipelines using transient reflections." *Australian Journal of Water Resources*, 11(1), 53-65.

Lee, P. J., Vitkovsky, J. P., Lambert, M. F., Simpson, A. R., and Liggett, J. A. (2005). "Frequence Domain Analysis for Detecting Pipeline Leaks." *Journal of Hydraulic Engineering*, 131(7), 596-604.

Li, P., Postlethwaite, I., Prempain, E., and Ulanicki, B. (2009). "Flow modulated dynamic pressure control with Aquai-Mod controller." Integrated Water Systems, Boxall and Maksimovic, eds., Taylor and Francis Group, London, 63-69.

Liemberger, R. (2002). "Performance target based non-revenue water reduction contracts: a new concept successfully implemented in southeast Asia." *Water Supply*, 2(4), 21-28.

Liemberger, R., and McKenzie, R. (2003). "Aqualibre: A New Innovative Water Balance Software." IWA Efficient 2003 Conference, IWA, Tenerife.

Liggett, J. A., and Chen, L. C. (1994). "Inverse transient analysis in pipe networks." *Journal of Hydraulic Engineering*, 120(8), 934-955.

Lu, J., Zhang, G., Ruan, D., and Wu, F. (2007). *Multi-Objective Group Decision Making: Methods, Software and Applications with Fuzzy Set Techniques*, Imperial College Press, London.

Luczon, L. C., and Ramos, G. (2012). "Sustaining the NRW reduction strategy: The Manila Water Company Territory Management Concept and Monitoring Tools." *Proceedings of the 7th IWA Water Loss Reduction Specialist Conference*, Manila, Philippines, Feb 26-29, 2012.

Lund, J. R. (1988). "Metering Utility Services: Evaluation and Maintenance." *Water Resources Research*, 24(6), 802-816.

Macharis, C., Springael, J., De Brucker, K., and Verbeke, A. (2004). "PROMETHEE and AHP: The design of operational synergies in multicriteria analysis. Strengthening PROMETHEE with ideas of AHP." *European Journal of Operational Research*, 153, 307-317.

Machell, J., Mounce, S. R., and Boxall, J. B. (2010). "Online modelling of water distribution systems: a UK case study." *Drinking Water Engineering and Science*, 3, 21-27.

Marques, R. C., and Monteiro, A. J. (2003). "Application of performance indicators to control losses-results from the Portuguese water sector." *Water Science and Technology: Water Supply*, 3(1/2), 127-133.

Martins, J. C., and Seleghim Jr, P. (2010). "Assessment of the performance of accoustic and mass balance methods for leak detection in pipelines transporting liquids." *Journal of Fluids Engineering*, 132(1), 1-8.

May, J. (1994). "Pressure Dependent Leakage." *World Water and Environmental Engineering, October 1994*, 13.

McCormack, C. A. "Canadian Utilities Learn to Fly through Benchmarking of Water Losses." *Leakage 2005*, Halifax, Canada.

McKenzie, R. (2001). *PRESMAC: Pressure Management Program*, WRC, Report TT 152/01, South Africa.

McKenzie, R., and Lambert, A. (2008). "Benchmarking of Water Losses in New Zealand Manual." New Zealand Water and Wastes Association, Wellington.

McKenzie, R., Seago, C., and Liemberger, R. (2007). "Benchmarking of Losses from Potable Water Reticulation Systems - Results from IWA Task Team." *Proc. of the 4th IWA Specialised Water Loss Reduction Conference, September 23-26*, Bucharest, Romania, 161-175.

McKenzie, R. S. (1999). "SANFLOW user guide. South African Water Research Commission, WRC. Report TT 109/99."

McKenzie, R. S., Lambert, A. O., Kock, J. E., and Mtshweni, W. (2002). "Benchmarking of Leakage for Water Suppliers in South Africa: Users Guide for the BENCHLEAK Model." WRC, Report TT 159/01, South Africa.

McKenzie, R. S., Mostert, H., and de Jager, T. (2004). "Leakage reduction through pressure management in Khayelitsha: two years down the line." *Water SA*, 30(5), 13-17.

Meniconi, S., Brunone, B., Ferrante, M., and Massari, C. (2011). "Transient tests for locating and sizing illegal branches in pipe systems." *Journal of Hydroinformatics*, 13(3), 334-345.

Mergelas, B., and Henrich, G. (2005). "Leak locating method for pre-commissioned transmission pipelines: North American Case Studies." *Leakage 2005*, Halifax, Canada.

Misiunas, D., Lambert, M., Simpson, A., and Olsson, G. (2006). "Burst detection and location in water distribution networks." *Water Science and Technology: Water Supply*, 5(3-4), 71-80.

Misiunas, D., Vitkovsky, J. P., Simpson, A. R., and Lambert, M. F. (2005). "Pipeline break detection using pressure transient monitoring." *Journal of Water Resources Planning and Management*, 131(4), 316-325.

Morais, D. C., and Almeida, A. T. (2007). "Group Decision Making for Leakage Management Strategy of Water Network." *Resources Conservation and Recycling*, 52, 441-458.

Morley, M. S., Bicik, J., Vamvakeridou-Lyroudia, L. S., Kapelan, Z., and Savic, D. A. (2009). "Neptune DSS: A decision support system for near-real time operations management of water distribution systems." Integrating Water Systems, Boxall and Maksimovic, eds., Taylor and Francis Group, London, 249-255.

Mounce, S. R., Boxall, J. B., and Machell, J. (2010). "Development and verification of an online artificial intelligence system for detection of bursts and other abnormal flows." *Journal of Water Resources Planning and Management*, 136(3), 309-318.

Mounce, S. R., Mounce, R. B., and Boxall, J. B. (2011). "Novelty detection for time series data analysis in water distribution systems using support vector machines." *Journal of Hydroinformatics*, 13(4), 672-686.

Mpesha, W., Gassman, S. L., and Chaudhry, M. H. (2001). "Leak detection in pipes by frequence response method." *Journal of Hydraulic Engineering*, 127(2), 134-147.

Mutikanga, H., Sharma, S. K., and Vairavamoorthy, K. (2010). "A Comprehensive Approach for Estimating Non-Revenue Water in Urban Water Supply Systems " *Proceedings of the IWA World Water Congress and Exhibition* Montreal, Canada, September 19-24, CD-ROM.

Mutikanga, H., Vairavamoorthy, K., Kizito, F., and Sharma, S. K. (2011a). "Decision Support Tool for Optimal Water Meter Replacement." *Proceedings of the 2nd International Conference on Advances in Engineering Technology*, Entebbe, Uganda, February 2011, 649-655, ISBN 978-9970-214-00-7.

Mutikanga, H. E., Sharma, S., and Vairavamoorthy, K. (2009). "Water Loss Management in Developing Countries: Challenges and Prospects." *Journal AWWA*, 101(12), 57-68.

Mutikanga, H. E., Sharma, S. K., and Vairavamoorthy, K. (2011b). "Assessment of Apparent Losses in Urban Water Systems." *Water and Environment Journal*, 25(3), 327-335.

Mutikanga, H. E., Sharma, S. K., and Vairavamoorthy, K. (2011c). "Investigating water meter performance in developing countries: A case study of Kampala, Uganda." *Water SA*, 37(4), 567-574.

Mutikanga, H. E., Sharma, S. K., and Vairavamoorthy, K. (2011d). "Multi-criteria Decision Analysis: A strategic planning tool for water loss management." *Water Resources Management*, 25(14), 3947-3969.

Nazif, S., Karamouz, M., Tabesh, M., and Moridi, A. (2010). "Pressure Management model for Urban Water Distribution Networks." *Water Resources Management*, 24, 437-458.

Nicklow, J., Reed, P., Savic, D., Dessalegne, T., Harrel, L., Chan-Hilton, A., Karamouz, M., Minsker, B., Ostfeld, A., Singh, A., and Zechman, E. (2010). "State of the Art for Genetic Algorithms and Beyond in Water Resources Planning and Management." *Journal of Water Resources Planning and Management*, 136(4), 412-432.

Nicolini, M., Giacomello, C., and Deb, K. (2011). "Calibration and Optimal Leakage Management for a Real Water Distribution Network." *Journal of Water Resources Planning and Management*, 137(1), 134-142.

Nicolini, M., and Zovatto, L. (2009). "Optimal Location and Control of Pressure Reducing Valves in Water Networks." *Journal of Water Resources Planning and Management*, 135(3), 178-187.

Nixon, W., Ghidaoui, M. S., and Kolyshkin, A. A. (2006). "Range of validity of the transient damping leakage detection method." *Journal of Hydraulic Engineering*, 132(9), 944-957.

Noss, R. R., Newman, G. J., and Male, J. W. (1987). "Optimal Testing Frequency for Domestic Water Meters." *Journal of Water Resources Planning and Management*, 113(1), 1-14.

Obradovic, D. (2000). "Modelling of demand and losses in real-life water distribution systems." *Urban Water*, 2(2), 131-139.

OFWAT. (2006). "Security of Supply, leakage and water efficiency 2005-06 report. Retrieved January 2008 from www.ofwat.gov.uk." OFWAT, London.

OFWAT. (2010). "Service and delivery-performance of the water companies in England and Wales 2009-10 report ", OFWAT, UK.

Ong, A. N. C., and Rodil, M. E. H. (2012). "Trunk mains leak detection in Manila's West Zone." *Proceedings of the 7th IWA Water Loss Reduction Specialist Conference* Manila, Philippines, Feb 26-29, 2012.

Palau, C. V., Arregui, F. J., and Carlos, M. (2012). "Burst detection in water networks using principal component analysis." *Journal of Water Resources Planning and Management*, 138(1), 47-54.

Park, H. J. (2006). "A Study to develop strategies for Proactive Water Loss Management," Ph.D Thesis, Georgia State University, USA.

Pasanisi, A., and Parent, E. (2004). "Bayesian Modelling of water meters ageing by mixing classes of devices of different states of degradation." *Applied Statistics Review*, 52(1), 39-65 (**in French**).

Pearson, D., and Trow, S. (2008). "Identifying Economic Interventions against Water Losses." Water Loss Control Manual, J. Thornton, R. Sturm, and G. Kunkel, eds., McGraw-Hill, New York, 103-118.

Perez, R., Puig, V., Pascual, J., Peralta, A., Landeros, E., and Jordanas, L. (2009). "Pressure sensor distribution for leak detection in Barcelona water distribution network." *Water Science and Technology: Water Supply*, 9(6), 715-721.

Picazo-Tadeo, A., J, Saez-Fernandez, F. J., and Gonzalez-Gomez, F. (2008). "Does service quality matter in measuring the performance of water utilities?" *Utilities Policy*, 16, 30-38.

Picazo-Tadeo, A., J, Saez-Fernandez, F. J., and Gonzalez-Gomez, F. (2011). "Assessing performance in the management of the urban water cycle." *Water Policy*, 13, 782-796.

Pilcher, R., Hamilton, S., Chapman, H., Field, D., Ristovski, B., and Stapely, S. (2007). "Leak Location and Repair Guidance Notes, Version 1." IWA Publishing, London, UK.

Pilipovic, Z., and Taylor, R. (2003). "Pressure management in Waitakere City, New Zealand-a case study." *Water Science and Technology: Water Supply*, 3(1/2), 135-141.

Prescott, S. L., and Ulanicki, B. (2003). "Dynamic modelling of pressure reducing valves." *Journal of Hydraulic Engineering*, 129(10), 804-812.

Prescott, S. L., and Ulanicki, B. (2008). "Improved control of pressure reducing valves in water distribution networks." *Journal of Hydraulic Engineering*, 134(1), 56-65.

Price, R. K., and Vojinovic, Z. (2011). *Urban Hydroinformatics: Data, Models and Decision Support for Integrated Urban Water Management* IWA Publishing, London.

Pudar, R. S., and Liggett, J. A. (1992). "Leaks in pipe networks." *Journal of Hydraulic Engineering*, 118(7), 1031-1046.

Puust, R., Kapelan, Z., Savic, D. A., and Koppel, T. (2010). "A review of methods for leakage management in pipe networks." *Urban Water Journal*, 7(1), 25-45.

Reis, L. F. R., Porto, R. M., and Chaudhry, F. H. (1997). "Optimal Location of Control Valves in Pipe Networks by Genetic Algorithm." *Journal of Water Resources Planning and Management*, 123(6), 317-326.

Richards, G. L., Johnson, M. C., and Barfuss, S. L. (2010). "Apparent losses caused by water meter inaccuracies at ultralow flows." *Journal of American Water Works Association*, 105(5), 123-132.

Rizzo, A., M., V., Galea, S., Micallef, G., Riolo, S., and Pace, R. (2007). "Apparent Water Loss Control: The Way Forward." *IWA Water 21, August*

Romano, G., and Guerrini, A. (2011). "Measuring and comparing the efficiency of water utility companies: A data envelopment analysis approach." *Utilities Policy*, 19, 202-209.

Romano, M., Kapelan, Z., and Savic, D. A. (2009). "Bayesian-based online burst detection in water distribution systems." Integrating Water Systems, Boxall and Maksimovic, eds., Taylor and Francis Group, London, 331-337.

Rossman, L. A. (2000). *EPANET 2 users manual*, USEPA, Cincinnati.

Rossman, L. A. (2007). "Disscussion of "Solution for Water Distribution Systems under Pressure-Deficient Conditions" by Wah Khim Ang and Paul W. Jowitt." *J. Water Resour. Plann. and Manage.*, 133(6), 566-567.

Sattar, A. M., and Chaudhry, M. H. (2008). "Leak detection in pipelines by frequency response method." *Journal of Hydraulic Research*, 46(1), 138-151.

Sattary, J., Boam, D., Judeh, W. A., and Warren, S. (2002). "The Impact of Measurement Uncertainty on the Water Balance." *Water and Environment Journal*, 16(3), 218-222.

Savic, D. A., Boxall, J. B., Ulanicki, B., Kapelan, Z., Makropoulos, C., Fenner, R., Soga, K., Marshall, I. W., Maksimovic, C., Postlethwaite, I., Ashley, R., and Graham, N. (2008). "Project Neptune: Improved Operation of Water Distribution Networks." *Proceedings of the 10th Annual Water Distribution Systems Analysis Conference WDSA 2008, August 17-20*, Kruger National Park, South Africa, 543-558.

Savic, D. A., Kapelan, Z., and Jonkergouw, P. (2010). "Quo vadis water distribution model calibration." *Urban Water Journal*, 6(1), 3-22.

Savic, D. A., and Walters, G. A. (1995). "An Evolution Program for Optimal Pressure Regulation in Water Distribution Networks." *Engineering Optimization*, 24(3), 197-219.

Schouten, M., and Halim, R. D. (2010). "Resolving strategy paradoxes of water loss reduction: A synthesis in Jakarta." *Resources Conservation and Recycling*, 54, 1322-1330.

Seago, C., Bhagwan, J., and McKenzie, R. (2004). "Benchmarking leakage from water reticulation systems in South Africa." *Water SA*, 30(5), 25-32.

Sempewo, J., Pathirana, A., and Vairavamoorthy, K. (2008). "Spatial Analysis Tool for Development of Lekage Control Zones from the Analogy of Distributed Computing." *Proceedings of the 10th Annual Water Distribution System Analysis Conference (WSDA 2008), August 17-20*, Kruger National Park, South Africa, 676-690.

Sethaputra, S., Limanond, S., Wu, Z. Y., Thungkanapak, P., and Areekul, K. (2009). "Experiences using water network analysis modeling for leak localization." *Proceedings of the 5th IWA Water Loss Reduction Specialist Conference*, Cape Town, South Africa, 469-476.

Sharma, S. K., and Chinokoro, H. (2010). "Estimation of ELL and ELWL for Lusaka Water Distribution System." *Proceedings of the 6th IWA Water Loss Reduction Specialist Conference (CD-ROM)*, Sao Paulo, Brazil.

Sharma, S. K., and Vairavamoorthy, K. (2009). "Urban water demand management: prospects and challenges for the developing countries." *Water and Environment Journal*, 23, 210-218.

Shrestha, D. L. (2009). "Uncertainty Analysis in Rainfall-Runoff Modelling: Application of Machine Learning Techniques," PhD Thesis, UNESCO-IHE Institute for Water Education, Delft, The Netherlands.

Simonovic, S. P. (2009). *Managing Water Resources: Methods and Tools for a Systems Approach*, UNESCO Publishing, Paris.

Singh, M. R., Upadhyay, V., and Mittal, A. K. (2010). "Addressing sustainability in benchmarking framework for Indian urban water utilities." *Journal of Infrastructure Systems (ASCE)*, 16(1), 81-92.

Skipworth, P. J., Saul, A. J., and Machell, J. (1999). "The effect of regional factors on leakage levels and the role of performance indicators." *Water and Environment Journal*, 13(3), 184-188.

Smith, L. A., Fields, K. A., Chen, A. S. C., and Tafuri, A. N. (2000). *Leak and Break Detection and Repair of Drinking Water Systems*, Battelle Memorial Institute, Columbus, Ohio, USA.

Soares, A. K., Covas, D. I. C., and Reis, L. F. R. (2011). "Leak detection by inverse transient analysis in an experimental PVC pipe system." *Journal of Hydroinformatics*, 13(2), 153-166.

Speight, V., Khanal, N., Savic, D., Kapelan, Z., JonKergouw, P., and Agbodo, M. (2010). "Guidelines for Developing, Calibrating, and Using Hydraulic Models." Water Research Foundation, Denver, Colorado.

Stent, A. F., and Harwood, N. (2000). "Estimating Pipe Reticulation Losses in a Municipal Water Supply System." *Water and Environment Journal*, 14(4), 246-252.

Sullivan, J. P., and Speranza, E. M. (1992). "Proper Meter Sizing for Increased Accountability and Revenue." *Journal AWWA*, 84(7), 53-61.

Tabesh, M., Asadiyani, Y., and Burrows, R. (2009). "An Integrated Model to Evaluate Losses in Water Distribution Systems." *Water Resources Management*, 23(3), 477-492.

Tanyimboh, T. T., and Kalungi, P. (2008). "Holistic planning methodology for long-term design and capacity expansion of water networks." *Water Science and Technology: Water Supply*, 8(4), 481-488.

Thanassoulis, E. (2000). "The use of data envelopment analysis in the regulation of UK water utilities: Water distribution." *European Journal of Operational Research*, 126, 436-453.

Thanassoulis, E. (2001). *Introduction to the Theory and Application of Data Envelopment Analysis: A Foundation Text with Integrated Software*, Kluwer Academic Publishers, Boston, USA.

Thornton, J., Sturm, R., and Kunkel, G. (2008). *Water Loss Control*, McGraw-Hill, New York.

Tindiwensi, D. (2006). "An investigation into the performance of the Uganda construction industry," PhD Thesis, Makerere University, Kampala, Uganda.

Trifunovic, N., Sharma, S., and Pathirana, A. (2009). "Modelling Leakage in Distribution System using EPANET." *Proceedings of the 5th IWA Water Loss Reduction Specialist Conference*, Cape Town, South Africa, 482-489.

Trow, S. (2007). "Alternative Approaches to Setting Leakage Targets." *IWA Specialized Conference: Water Loss 2007*, Bucharest, Romania, 75-85.

Tucciarelli, T., Criminisi, A., and Termini, D. (1999). "Leak analysis in pipeline systems by means of optimal valve regulation." *Journal of Hydraulic Engineering*, 125(3), 277-285.

Ulanicki, B., Bounds, P. L. M., Rance, J. P., and Reynolds, L. (2000). "Open and Closed Loop Pressure Control for Leakage Reduction." *Urban Water*, 2, 105-114.

Vairavamoorthy, K., and Lumbers, J. (1998). "Leakage reduction in water distribution systems: Optimal valve control." *Journal of Hydraulic Engineering*, 124(11), 1146-1154.

van den Berg, C., and Danilenko, A. (2011). *The IBNET Water Supply and Sanitation Performance Blue Book: The International Benchmarking Network for Water and Sanitation Utilities Databook*, The World Bank, Washington, DC.

van der Linden, M. J. (1998). "Implementing a large meter replacement program." *AWWA*, 90(8), 50-56.

Van Zyl, J. E. (2011). "Introduction to integrated water meter management." Water Research Commission (WRC TT490/11), South Africa.

Van Zyl, J. E., and Cassa, A. M. (2011). "Linking the power and FAVAD equations for modelling the effect of pressure on leakage." *Proc. of the 11th Int. Conference on Computing and Control of the Water Industry (CCWI 2011) - Urban water management challenges and Opportunities, September 5-7*, Exeter, UK.

Vela, A., Perez, R., and Espert, V. (1991). "Incorporation of leakages in the mathematical model for a water distribution network." *Proceedings of the 2nd International Conference on Computer Methods in Water Resources*, Marrakesh, Morocco, 245-257.

Vitkovsky, J. P., Lambert, M. F., Simpson, A. R., and Liggett, J. A. (2007). "Experimental observation and analysis of inverse transients for pipeline leak detection." *Journal of Water Resources Planning and Management*, 133(6), 519-530.

Vitkovsky, J. P., Liggett, J. A., Simpson, A. R., and Lambert, M. F. (2003). "Optimal measurement site location for inverse transient analysis in pipe networks." *Journal of Water Resources Planning and Management*, 129(6), 480-492.

Vitkovsky, J. P., Simpson, A. R., and Lambert, M. F. (2000). "Leak detection and calibration using transients and genetic algorithms." *Journal of Water Resources Planning and Management*, 126(4), 262-265.

Waldron, T. (2008). "Expertise in water loss control applied to extreme problems of water distribution management." *Water Science and Technology: Water Supply*, 8(1), 107-112.

Wallace, L. P., and Wheadon, D. A. (1986). "An Optimal Meter Change-out Program for Water Utilities." *AWWA Annual Conference*, Denver, Colorado, 1035-1042.

Walski, T., Bezts, W., Posluszny, E. T., Weir, M., and Whitman, B. E. (2006). "Modeling Leakage Reduction through Pressure Control." *Journal American Water Works Association*, 98(4), 147-155.

Walter, M., Cullmann, A., von Hirschhausen, C., Wand, R., and Zschille, M. (2009). "Quo vadis efficiency analysis of water distribution? A comparative literature review." *Utilities Policy*, 17, 225-232.

Wang, X.-J., Lambert, M. F., Simpson, A. R., and Vitkovsky, J. P. (2001). "Leak Detection in Pipelines and Pipe Networks: A Review." *6th Conference on Hydraulics in Civil Engineering, The Institution of Engineers Australia*, Horbart, Australia, 391-400.

Wang, X. J., Lambert, M. F., Simpson, A. R., Liggett, J. A., and Vitkovsky, J. P. (2002). "Leak detection in pipelines using the damping of fluid transients." *Journal of Hydraulic Engineering*, 128(7), 697-711.

WSP. (2009). "Water Operators Partnerships: African Utility Performance Assessment." Water and Sanitation Program (WSP) - Africa, The World Bank, Nairobi, Kenya.

Wu, Z. Y., Farley, M., Turtle, D., Kapelan, Z., Boxall, J., Mounce, S., Dahasahasra, S., Mulay, M., and Kleiner, Y. (2011). *Water Loss Reduction*, Bentley Institute Press, Exton, Pennsylvania, USA.

Wu, Z. Y., Sage, P., and Turtle, D. (2010). "Pressure-dependent leak detection model and its application to a district water system." *Journal of Water Resources Planning and Management*, 136(1), 116-128.

Wyatt, A. S., and Romeo, K. J. (2010). "Application of a financial model for determining the optimal management of NRW in developing counrries." *Proceedings of the 3rd Regional Activity on NRW management: Solutions for drinking water loss reduction*, Rabat, Morocco, 107-120.

Yaniv, S. (2009). "Reduction of Apparent Losses Using the UFR (Unmeasured-Flow Reducer) - Case Studies." *Proceedings of the 5th IWA Specialist Conference on Efficient Water Use and Management (CD-ROM)*, Sydney, Australia.

Ye, G., and Fenner, R. A. (2011). "Kalman filtering of hydraulic measurements for burst detection in water distribution systems." *Journal of Pipeline Systems Engineering and Practice*, 2(1), 14-22.

Yee, M. D. (1999). "Economic Analysis for Replacing Residential Meters." *Journal American Water Works Association*, 91(7), 72-77.

Zhu, J. (2009). *Quantitative Models for Performance Evaluation and Benchmarking: Data Envelopment Analysis with Spreadsheets*, Springer, Boston, USA.

Chapter 3 - Water Distribution System Performance Evaluation and Benchmarking

Parts of this chapter are based on:

Mutikanga, H.E, Sharma, S.K, Vairavamoorthy, K., and Cabrera Jr., E. (2010). "Using Performance Indicators as a Water Loss Management Tool in Developing Countries". *Journal of Water Supply: Research and Technology-AQUA*, 59 (8), 471-481.

Mutikanga, H.E, Sharma, S.K., and Vairavamoorthy, K. (2010). "A Comprehensive Approach for Estimating Non-revenue Water in Urban Water Supply Systems". *Proceedings of the IWA World Water Congress and Exhibition*, September 19-24, Montreal, Canada, CD-ROM Edition.

Mutikanga, H.E, Sharma, S.K., and Vairavamoorthy, K. (2009). "Performance Indicators as a Tool for Water Loss Management in Developing Countries". *Proceedings of the 5th IWA Water Loss Reduction Specialist Conference*, Cape Town, South Africa, April 26-30, p 22 – 28, ISBN:978-1-920017-38-5.

Summary

Water distribution systems are probably the most costly and important assets of water utilities used to convey water from treatment plants to users. However, not all water that is put into the distribution system gets to its intended destination due to water losses. The amount of water lost is a measure of the efficiency of the distribution system. Well-devised methodologies for evaluating water distribution system efficiency are needed, in order to understand why, where and how much water is lost. In principle, asset condition should be directly taken into consideration. However, this is a challenging task because mains are located underground, not easily inspected and direct inspection methods are too costly. Indirect measures such as the water balance and PIs should be considered. The IWA and AWWA have developed a standard water balance methodology and an array of PIs. Whereas the IWA/AWWA water balance methodology and PIs are valuable tools, they cannot be directly applied and do not fully address the peculiar characteristics of water distribution systems in developing countries. In this study a performance assessment system for WLM in the developing countries has been developed based on the IWA/AWWA PI-concept. The developed system has been evaluated on real-developing countries water distribution systems in Uganda. The benefits of the developed PIs are further illustrated on some case study utilities in African cities. Apparent loss indices for developing and developed countries are also proposed. The uncertainty in the water balance input variables and uncertainty propagation in the NRW indicator is quantified for the Kampala water distribution system with aim of improving accuracy of the reported NRW figures. In order to improve water distribution system performance, a comparative efficiency study of 25 water utilities in Uganda was undertaken. Data Envelopment Analysis (DEA), a linear programming technique was applied to establish a *Pareto-efficient* frontier as a benchmark against which performance is evaluated and utility rankings established. A comparison of DEA and PIs as benchmarking tools was also made. The findings indicate high technical inefficiencies (36%) in these utilities with significant potential for water savings estimated at 42.6 ML/day. DEA and PI-based benchmarking methods were found to complement each other as decision-aid tools for improving water distribution system efficiency. The policy implications of the study for the Ugandan water sector are also highlighted. The water distribution utility rankings could serve as a catalyst for better stewardship of water resources. It is believed that the developed performance assessment system effectively addresses the problem of evaluating, quantifying and improving water distribution system efficiency.

3.1   Introduction

As highlighted in Chapter 1, one of the major challenges facing water utilities worldwide is the inefficiencies in operation of WDSs in terms of water and revenue losses. Performance assessment of WDSs is increasingly becoming a big issue in the water industry to ensure delivery of an acceptable level of service. However, measuring the technical performance of a WDS is not a straightforward task, given multiple factors involved and lack of a clear-cut definition of performance (Coelho 1997). Deb et al. (1995) recommends three WDS performance criteria: (i) *adequacy* – refers to the delivery of an acceptable quantity and quality of water to the customer, (ii) *dependability* – measures the ability of the WDS to consistently deliver an acceptable quantity and quality of water, and (iii) *efficiency* – reflects how well resources such as water are utilized. In this study, WDS performance is evaluated using the efficiency criteria. However, efficiency like other performance criteria is not directly measurable and water loss and/or NRW is used as an indirect performance measure for evaluating WDS efficiency.

Although the IWA/AWWA have developed a water balance methodology with an array of PIs for assessing WDS performance (Alegre et al. 2006; AWWA 2003), the methodology and PIs have some limitations to be directly applied in developing countries as discussed in Chapter 2. In addition, the IWA/AWWA-PI system provides a subjective confidence grading system to assess the quality of data (accuracy and reliability) used in the water balance calculation. Whereas the confidence of the generated PI depends on the confidence of the input data, the IWA/AWWA-PI system does not indicate how the uncertainty of the input data propagates and affects the uncertainty of the final PI result. Furthermore, the IWA/AWWA-PI system uses partial indicators (e.g. NRW (%), L/connection/d, L/km/d etc.) that have traditionally been the method of choice for performance assessment and will likely continue to be so in future. Whereas, it is possible to analyse different WDS aspects using partial indicators, they do not provide the overall efficiency measure and rakings required for effective benchmarking and performance improvement. Total methods that take into account multiple inputs and outputs are required for this task.

In this chapter, an integrated performance assessment system (PAS) for WDS performance assessment and improvement has been developed. The PAS is made up of three main components (Water balance model, PIs and benchmarking) as shown in Figure 3.1. The water balance model and PIs have been integrated into a water loss assessment tool (WLA-PI tool) to ease application in practice.

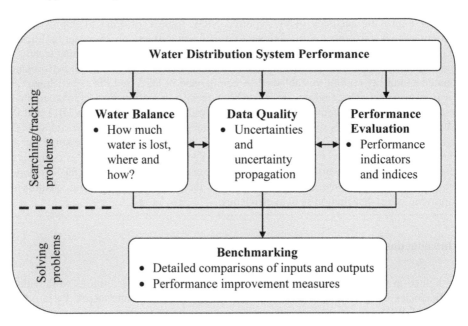

Figure 3.1 Performance assessment system for WLM

In order to select and develop appropriate PIs for water loss assessment (WLA), an eight-step participatory methodology is proposed as part of the system. The method is then applied to select and develop appropriate PIs which are tested and validated using various WDSs in Uganda. The method is generic and can be used to select and develop appropriate PIs for other WDSs in developing countries based on local conditions. The benefits of some of the PIs are highlighted using some African cities. The lessons learned in integrating a PI culture in developing countries using NWSC-Uganda are highlighted. Using the law of propagation of uncertainty, the impact of input uncertainty on the water balance NRW calculation is demonstrated using operational data for the KWDS. The usefulness of uncertainty analysis in the reported NRW figures is highlighted. A critique of two main performance indices used in WDS evaluation is also made and a new apparent loss index for developing countries is proposed. Lastly, the Data Envelopment Analysis (DEA) benchmarking methodology is applied to evaluate and improve WDS performance in some Ugandan water utilities.

The rest of the chapter is organized as follows. Section 3.2 presents the methodology for developing PIs, its application to develop appropriate PIs for developing countries and highlights the WLA-PI tool as well as the uncertainty estimation in the water balance model. Section 3.3 makes a critical discussion of two main performance indices used in WLM and proposes some improvements. Section 3.4 examines the technical efficiency in a sample of 25 Ugandan WDSs by application of the DEA benchmarking methodology. It identifies potential for water savings, compares DEA and PI-based methods and highlights policy implications of the study. Finally, section 3.5 draws some conclusions based on the study.

3.2   Methodology for PI Development, Definition and Selection

The eight-step participatory methodology followed in the development, definition and selection of PIs for water loss assessment (WLA) is presented in Figure 3.2. This methodology refers to the development of the PI system for general use and can be followed by any water supplier willing to implement a PI system. The methodology starts with defining objectives based on the utility strategic plan and ends with using PIs in daily operations. This methodology is consistent with the recommendations of Water Research Foundation (USA) and IWA of going from objectives to indicators (Alegre et al. 2006; Brueck 2005; Crotty 2004). It is also consistent with the PDCA approach (Plan Do Check Act) and the ISO 24500 standards (ISO, 2007c).

3.2.1   Establishing a PI system

The established WLA PI system is based on literature guidelines and the IWA/AWWA-PI concept. A list of relevant indicators was also selected from the existing PI systems mainly the AWWA/IWA-PI system. A total candidate list of 62 PIs (20 new ones and 42 selected) was assembled and made ready for review by the PI team. This was followed by appointing a PI team by the Managing Director (MD) of the water utility. The team comprised of 24 members of staff from the utility top management, senior managers, middle managers and operational staff. A PI team leader was appointed to coordinate the whole exercise. This was then followed by circulation of relevant literature to the PI team members to stimulate their thinking.

Three workshops were then conducted to provide forum for discussions, learning and articulating relevance of each indicator. To enhance speed, the team was split into 4 main working groups (physical losses, commercial losses, NRW and computational tool development). The final outputs were realised after 4 months with a final PI listing of 25 indicators, 51 variables and a PI computational tool.

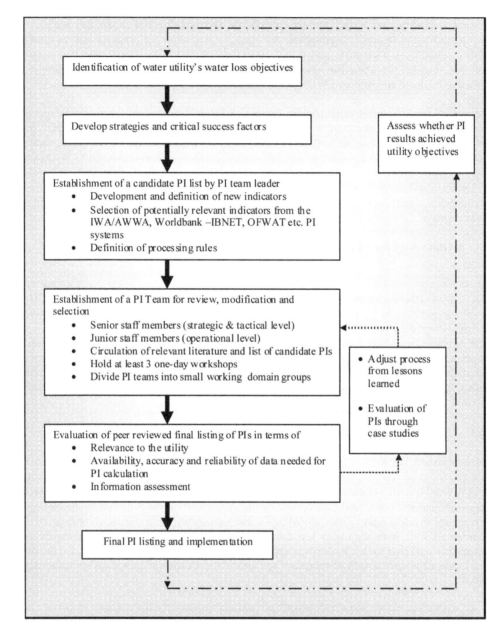

Figure 3.2 Methodology for PI development, selection and definition

Complimentary to the IWA/AWWA-PIs, there are 14 newly established PIs suitable for WLA in the water utilities of the developing countries with little or no relevance to water utilities in the developed countries (e.g. leakage handling efficiency, illegal use fines recovery efficiency and inactive accounts ratio). Detailed characterization of some selected PIs (objective, definition, processing rule, units of measurement, data required, results analysis, etc.) are presented in Tables 3.1, 3.2 and 3.3. The final step was pilot testing the feasibility of the adopted PIs and the developed computation tool.

This participatory methodology was preferred to draw continuing support from operational staff that will provide input data and the top decision making managers who are the users of the PI information in setting targets and assessing performance for continuous improvement. This approach of involving employees in the process creates ownership and ensures sustainability of the performance assessment system.

3.2.2   The PI system for water loss assessment

The PIs were structured and arranged into six main groups as shown in Figure 3.3. The purpose of each PI domain is discussed and the final user within utility is identified as follows.

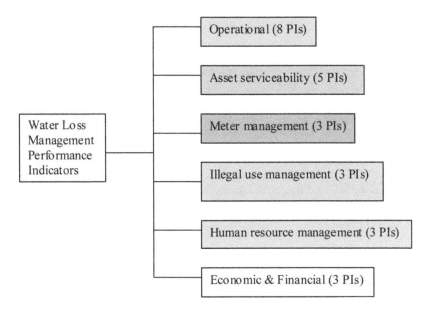

Figure 3.3 Performance indicators structure for WLM

Operational indicators (OpWL)

In this group, PIs are intended to assess the performance of the utility as regards operation and maintenance (O&M) activities. Utility managers need to pay much attention to O&M activities as the efficiency of the water utility can be lost or improved by these activities.

Asset serviceability indicators (AsWL)

In this group, PIs are intended to assess the long-term performance of the utility's assets to continue providing an acceptable level of service to its customers. There is normally a tendency for utilities to reduce asset renewing and operating cost to ensure short-term financial and economic sustainability especially in the developing countries.

Meter management indicators (MeWL)

In this group, PIs are intended to assess the utility's water meters functionality to accurately measure flows and safeguard against revenue losses. Choosing the right meters, keeping them in good operating mode or even the time to replace them presents big challenges to utilities in the developing countries and managerial efforts in this respect are assessed.

Illegal water use indicators (IlWL)

In this group, PIs are intended to assess the efficiency of the utility in handling illegal use of water and managerial efforts to promote proactive illegal use investigations and control. Currently there is very limited use of performance indicators to effectively assess illegal use of water that have become endemic in most cities of developing countries.

Human resource (personnel) indicators (PeWL)

In this group, PIs are intended to assess the efficiency and productivity of the utility's human resources and managerial efforts in recruitment and training of staff.

Economic and financial indicators (FiWL)

This group assess the utility's economic and financial sustainability. The basic (% NRW) indicator that is easy to compute though not very meaningful has been included as it is still widely used by many stakeholders.

In order to assess the selected PIs, utility data is required. The PI system input variables are structured in seven main domains, the same as those adopted in the IWA-PI system to ease usability of the system:

- A - Water volume data
- B – Personnel data
- C - Physical assets data
- D – Operational data
- E – Customer data
- F – Time data
- G - Economic and financial data

The final PI system input data totalling 51 variables have been defined. For meaningful diagnosis and decision-making, the PI system is supplemented by relevant utility context information and explanatory factors.

3.2.3 Selected PIs from the IWA/AWWA PI system

Some useful indicators selected from the IWA/AWWA menu of PIs (Alegre et al. 2006) for WLM included the following:

1. Real losses (m³/connection/day, when system is pressurised) as a proxy measure of water distribution network condition.
2. Mains break (number/km/year) – proxy measure for pipeline asset condition.
3. Apparent losses (m³/connection/day) – for benchmarking purposes.

High water losses could be a proxy for poor asset management. The breakdown of water loss into real and apparent loss components in the absence of reliable and accurate data is however subjective and debatable. The usefulness of these indicators in developing countries will heavily depend on acquisition of accurate data which is likely to be a challenging task amidst inadequate resources.

Table 3.1 Some of the developed operational indicators

OpWL1 – Leakage handling efficiency (%)
Concept: Number of reported leaks and bursts repaired within the target period / Total number of reported leaks and bursts
Processing rule: OpWL1 = (D2 / D1) x 100
Variables: D1 = Total number of leaks and bursts reported during the assessment period D2 = Total number of recorded leaks and bursts repaired within the target period
Objective of PI: To improve repair response time and minimize leakage run-times.
Comments: Very useful for developing countries with no zoned water distribution networks & DMAs but with administrative operational business units responsible for repairing leaks among others. Relevant for utilities with so many visible leaks that take days to be repaired.
OpWL5 - Apparent losses ratio (-)
Concept: [Apparent losses during the assessment period / Total amount of billed metered authorized consumption (including exported water) during the assessment period] x 100
Processing rule: OpWL5 = (A11 / A3) x 100
Variables: A11 = Apparent losses (m^3) A3 = Authorized billed metered consumption   (m^3)
Objective of PI: To quantify amount of commercial losses that do not generate revenue out of authorized consumption
Comments: In systems where all customers are metered, and water theft is low, most of the commercial losses will be due to customer metering errors.
OpWL6 – Inactive accounts ratio (-)
Concept: Total number of service connections off-supply, at the reference date / Total number of service connections, at the reference date
Processing rule: OpWL6 = C6 / C4
Variables: C4 = Service connections (No.) C6 = Inactive Service connections (No.)
Objective of PI: To quantify portion of service connections with high potential for illegal use of water.
Comments: Relevant to developing countries where water supply is turned off due to non-payment of water bills. The turned off customers often resort to stealing water in Kampala city through illegal reconnections.
OpWL7 – NRW per connection (m^3/connection/year)
Concept: (NRW during the assessment period x 365/assessment period) / number of service connections
Processing rule: OpWL7 = (A14 x 365 / F1) / C4
Variables: A14 = Non-revenue water (m^3) C4 = Service connections (No.) F1 = Assessment period (day)
Objective of PI: To compare performance among different utilities.
Comments: The indicator is easier to compute with reasonable accuracy. Helpful for developing countries with limited resources to collect data for reliable estimation of NRW components. Should not be assessed for periods less than one year, since it may lead to misleading conclusions. In case service connection density is less than 20 per km mains, then express OpWL7 as m^3/km water mains/year. This PI could also be expressed as m^3/conn./day depending on convenience.

Table 3.2 Some operational, asset management and metering indicators

OpWL8 – Transport availability (No. / 100 km)

Concept: Number of vehicles, motor cycles and bicycles available daily, on a permanent basis, in average, for field works in operations and maintenance activities / total mains length x 100

Processing rule: OpWL8 = D15 / C1 x 100

Variables: C1 = Mains length (km)
      D15 = Permanent transport (vehicles, motor cycles and bicycles) (No.)

Objective of PI: To assess speed capacity of utility regarding repairing of failures and for comparisons with other undertakings.

Comments: Relevant to developing countries where inadequate transport and frequent vehicle breakdowns are common and impact on response time to repair of leaks and bursts. Some areas (slums) in cities of developing countries can only be accessed by motor cycles and bicycles. They also come in handy during rush hours with congested traffic jams.

AsWL9 – Functional valve density (No. / 100 km)

Concept: Number of functional isolating valves / total transmission and distribution mains length, at the reference date x 100

Processing rule: AsWL9 = D7 / C1 x 100

Variables: C1 = Mains length (km)
      D7 = Functional isolating valves (No.)

Objective of PI: To assess the efficiency and effectiveness of isolating sections of the network to enable repairs to be done quickly and minimize water wastage.

Comments: Relevant to developing countries where a good number of network isolating valves are inaccessible and inoperable.

MeWL14 – Meter reading efficiency (%)

Concept: (Number of complying meter readings / total meter reading audits carried out during the assessment period) x 100

Processing rule: MeWL14 = (D9 / D8) x 100

Variables: D8 = Meter reading audits (No.)
      D9 = Complying meter readings (No.)

Objective of PI: To minimize apparent loses due to meter reading errors

Comments: Relevant for developing countries with low salaried field workers and integrity of meter readers is often compromised.

MeWL15 – Meter Failure (%/year)

Concept: (Number of customer water meters that are reported defective (stuck) during the assessment period

Processing rule: MeWL15 = (D10 / E1) x 100

Variables: D10 = Meter Failure (No.)
      E1 = Customer meters (No.)

Objective of PI: To assess utility's level of defective meters (stuck meters) at reference time.

Comments: Relevant for developing countries where frequently a good number of installed meters are defective (stuck) due to particulates in water, wrongly selected meter types etc.

Table 3.3 Some illegal use management and personnel indicators

IIWL17 – Illegal use detection efficiency (%)
Concept: (Number of illegal cases confirmed / total number of service connections investigated during the assessment period) x 100
Processing rule: IIWL17 = (D13 / D12) x 100
Variables: D12 = illegal use cases reported and investigated (No.)   D13 = illegal use cases confirmed (No.)
Objective of PI: To assess managerial efforts to proactively manage illegal water use.
Comments: Relevant for developing countries where illegal use of water is endemic. This PI is analogous to leaks found per property used to maximize efficiency of leakage detection and location in the UK.

IIWL18 – Fines recovery efficiency (%)
Concept: (Number of illegal use fines recovered / total number fines levied during the assessment period) x 100
Processing rule: IIWL18 = (G2 / G1) x 100
Variables: G1 = Illegal fines levied (No.)   G2 = Illegal fines paid (No.)
Objective of PI: To assess managerial efforts in enforcing laws and regulations to fight illegal use of water and improve financial sustainability.
Comments: Relevant for developing countries where illegal use of water is endemic.

IIWL 19 – Illegals found per investigator
Concept: (Total number of illegal cases found during the assessment/ total number illegal investigators during the assessment period)
Processing rule: IIWL19 = (D14 / C8)
Variables: D14 = Total number of illegal cases unearthed (No.)   C8 = Total number of illegal investigators(No.)
Objective of PI: To reduce apparent losses by maximize the productivity of illegal investigators
Comments: Relevant for developing countries where revenue protection units have been established to minimize unauthorized water use.

PeWL20 – Operations and maintenance staff (No. / 100 km)
Concept: Number of full time equivalent employees working in operations and maintenance of the water transmission, storage and distribution system / total mains length x 100
Processing rule: PeWL20 = B1 / C1 x 100
Variables: B1 = Operation and Maintenance personnel (No.)   C1 = Mains length (km)
Objective of PI: To assess the efficiency human resources deployment in handling water losses.
Comments: Essential in developing countries with high employee levels but poorly deployed for effective management of water losses. Useful for benchmarking. The indicator is assessed for a reference date.

3.2.4   The WLA- PI tool

As a part of this study, a computational MS Excel® spreadsheet tool has been developed to promote use of the PAS and assist utility managers in application of the standard water balance. The framework is user-friendly as highlighted on the homepage user interface (screenshot in Figure 3.4a); some computed PIs from input variables are depicted in the screenshot in Figure 3.4b.

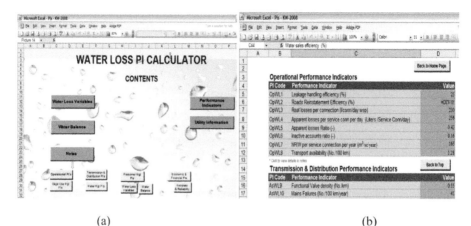

| (a) | (b) |

Figure 3.4 Screenshots of the PI computational tool for WLA (WL_PI_CALC.xls)

During the past decade, several efforts have been made to develop similar but more costly (typically priced from US $50,000 to $100,000) water balance software tools. Surely, these tools are beyond the reach of most cash-strapped water utilities in the developing countries and have hindered use of the IWA standardised water balance methodology. They include Aquafast software developed for AWWA's Water Research Foundation (Fanner et al. 2007), Aqualibre developed by Bristol Water Utility (UK) (Liemberger and McKenzie 2003) and SIGMA developed by Instituto Tecnologico del Agua (ITA) of the Universidad Politecnica de Valencia, Spain (Alegre et al. 2006).

3.2.5   Evaluating the effectiveness of the PAS

In order to test the robustness of the developed performance assessment system particularly the WLA-PI tool, a pilot implementation was carried out in five water utilities of NWSC-Uganda using preliminary data for 2008/09 financial year. The feasibility of the PIs was assessed using available data within the utility and additional easy to measure and less expensive data to collect but of much significance to water loss assessment. As a result 16 PIs out of 25 proposed PIs were successfully tested.

The water loss PI system was tested in 5 branch water utilities of NWSC- Uganda and extended to PI-based benchmarking in some African cities. The objectives of the pilot testing were:

- to confirm the relevance of the selected PIs;
- to assess, in practice, the feasibility of assessment;
- to set up and test guidance ranges or reference values;
- to test the computational tool
- to reveal the information gap and any problems of data quality and availability
- to do a preliminary performance benchmarking exercise and answer the question of : how well are we doing compared to others and how can we get better?
- to refine and modify the PIs

The results of the PI-based benchmarking using some of the newly developed PIs are presented in Table 3.4 for NWSC-Uganda branch utilities and Table 3.5 for some African cities.

Table 3.4 PI-based benchmarking in NWSC-Uganda (2008/09)

Town	Density (conns/km)	No. of service conns.	NRW (m^3/conn/day)	Leakage handling efficiency (%)	Mains replacement (%/yr)	Mains failure (No./100km/yr)	Service conns. failure (No/1000 conns/yr)
Kampala	63	133,198	0.45	25	0.49	41	133
Jinja	36	15,727	0.19	70	0.00	67	213
Entebbe	61	14,574	0.07	90	0.00	15	150
Mbale	32	6,885	0.05	94	0.93	19	43
Gulu	35	3,828	0.07	57	0.00	89	70

Table 3.5 PI-based benchmarking in African cities

City	Country	NRW (m^3/service connection/day)
Dar-es-salaam	Tanzania	1.00
Nairobi	Kenya	1.00
Khartoum	Sudan	1.00
Lagos	Nigeria	0.90
Lilongwe	Malawi	0.80
Johannesburg	South Africa	0.40
Cape Town	South Africa	0.20
Windhoek	Namibia	0.14

(Source: WSP, 2009)

The following observations can be made from Tables 3.4 and 3.5:

Asset management: the mains and service lines replacements are far from ideal for NWSC-Uganda distribution networks which are over 50 years old. For example, the average value observed for mains replacement was 0.3% with most utilities doing no replacement at all despite the high number of mains failures and leakage levels. Marques and Monteiro (2003), recommends a value of 2% as a reference value for water network rehabilitation of Portuguese water utilities despite the fact that the number of failures that exist in Portuguese water utilities is about a half of the average value that exists in NWSC-Ugandan utilities.

Leakage handling efficiency (LHE): the average value observed for LHE was 67.2% with a minimum value of 25% and a higher value of 94%, which means that most reported visible leaks take more than a day before they are repaired. The utility practices more of reactive maintenance than preventive maintenance. There is potential for improvement in the Kampala and Gulu branch utilities with respect to speed and quality of repairing leaks to bring it at par with other well performing utilities in Entebbe and Mbale. In general, the fire-fighting approach to control leaks should be reviewed and more proactive strategies adopted to minimize leak run-times and associated costs.

Network dependability: the high number of mains and service failures indicate poor structural performance of the water distribution systems in the NWSC-Ugandan utilities. These network deficiencies cause frequent service interruptions and consequent customer complaints. The main breaks in the NWSC utilities range from 15 to 89 breaks/km/year and this is much higher than the recommended goals for main breaks in American water utilities of 16 to 19 breaks/km/year (Deb et al. 1995). Pipeline systems having an average annual pipe break ratio per 100 km of less than 40 are considered to be in an acceptable state (Pelletier et al. 2003).

African cities: Out of 134 utilities that participated in the African water utilities benchmarking study, only 93 utilities were able to provide data for the basic NRW indicator (m^3/connection/day) (WSP 2009). The best practice performance benchmark was identified as 0.3 m^3/conn/day. Utilities operating at this benchmark or better are potential candidates for building capacity of other operators in improving water distribution efficiency under the water operator's partnerships (WOPs) program. Windhoek city in Namibia has the best water distribution efficiency probably due to the severe water scarcity conditions that exist and significant efforts have been made in water conservation and efficient water use.

However, it is important to note that, in order to meaningfully compare results to reference values one needs to look at the explanatory factors such as network age and materials, number and frequency of failures, meter management, cost of water etc. Nevertheless, the PIs developed seem to be appropriate for PI-based benchmarking in the developing countries and are likely to improve performance of WDSs provided that reliable data is generated and consistently used.

3.2.6 Analysis of Uncertainty in the Water Balance

The precise estimation of NRW figures from the water balance model depends on accurate measurement of the system input volume and revenue water. However, NRW figures are in practice derived from the water balance components that are themselves subject to potentially large estimation errors and uncertainties. Accurate measurement is not possible due to flow meter data inaccuracies and authorized unbilled unmetered consumption components. Consequently the NRW figure derived from a water balance is inherently uncertain. For meaningful reporting and performance benchmarking, it is imperative that uncertainty in NRW figures is recognized and properly accounted for. In this section, we demonstrate the estimation of how the uncertainty of the water balance input data propagates and affects the uncertainty of the generated NRW figures using the law of propagation of uncertainty and operational data for the KWDS.

The individual input measured volumes (SIV and RW) and the derived NRW volumes were statistically analyzed using monthly historical records for the last 5 years (2005-2009) and descriptive statistics techniques (Gottfried 2007). Thereafter standard uncertainties for individual input variables were computed and then combined to give the standard uncertainty in NRW using Equations 3.1 and 3.2. The uncertainties are all expressed at 95% confidence levels.

There are various methods for uncertainty analysis and no single method of uncertainty estimation can be claimed as being perfect in representing uncertainty. There is no methodology that can give accurate results from inaccurate data. Bargiela and Hainsworth (1989) used Monte Carlo simulation method to determine pressure and flow uncertainty in water systems. Mauris et al. (2001) used the fuzzy set theory to express uncertainty in measurement and concluded that the approach was compatible with the International Standards Organisation (ISO) Guide. Sattary et al. (2002) used the ISO guidelines for expressing uncertainty in measurements to evaluate uncertainty on the water balance calculation for Anglian Water company (UK). Herrero et al. (2003) also used the ISO methodology's law of propagation of uncertainty for assessing the impact of uncertainty in the input variables on the final PI. In this study, the ISO methodology that incorporates the law of propagation of uncertainty has been used as it is now widely accepted as the standardized method for estimating uncertainty in measurement (ISO/IEC 2008).

The combined uncertainty of a measurement is obtained combining the individual standard uncertainties of the input quantities using a first-order Taylor series approximation, termed as *law of propagation of uncertainty*:

$$u_C^2(y) = \sum_{i=1}^{N}\left(\frac{\partial f}{\partial x_i}\right)^2 u^2(x_i) + 2\sum_{i=1}^{N-1}\sum_{J=i+1}^{N}\frac{\partial f}{\partial x_i}\frac{df}{dx_j}u(x_i, x_j) \qquad (3.1)$$

where $u_C(y)$ is the combined standard uncertainty of the measurand Y (or *NRW* in this case)

$\quad u(x_i)$ is the standard uncertainty of input quantity X_i (or *SIV* and *RW* in this case)

$\quad f$ is the model that relates measurand and input quantities

$\quad u(x_i, x_j)$ is the covariance associated to X_i and X_j

$\quad \dfrac{\partial f}{dx_i}$ is the sensitivity coefficient of the measurand to the input quantity X_i

Covariance is a measure of mutual dependence between pairs of input quantities. The sensitivity coefficients, obtained as partial derivatives, quantify the influence of each input quantity on the result. For this reason, they are used in the propagation law for weighting the individual uncertainties.

Since the input quantities are not measured simultaneously but in different situations and with different assessment procedures, they are not correlated and the covariance in Equation 3.1 is zero and the second term vanishes. Assuming that SIV and RW influence the NRW volume equally (i.e. sensitivity coefficient of 1), then Equation 3.1 simplifies into Equation 3.2 as follows:

$$u_C(NRW) = \sqrt{(u_{SIV})^2 + (u_{RW})^2} \qquad (3.2)$$

3.2.6.1 Calculation of confidence limits for the input variables and NRW

In order to validate the degree of uncertainty, the standard approach of calculating variance (*VAR*) related to a certain volume of the water balance, based on 95% confidence limit is given by Equation 3.3 (Thornton et al. 2008):

$$VAR = (Volume\ in\ m^3\ x\ 95\%\ Confidence\ Limits/1.96)^2 \qquad (3.3)$$

Tables 3.6 and 3.7 summarize the results of uncertainty trends in the water balance and 95% confidence limits respectively. Uncertainties in Table 3.6 are shown in parentheses.

Table 3.6 Uncertainty trends in water balance inputs and NRW

Volume (m^3/month	2005	2006	2007	2008	2009
System input volume (SIV)	3,494,310 (117,874)	3,451,629 (133,023)	3,699,362 (331,107)	4,099,082 (199,735)	4,345,612 (180,778)
Revenue water (RW)	2,117,265 (83,805)	2,212,156 (144,543)	2,193,445 (61,673)	2,314,355 (75,558)	2,503,217 (127,194)
Non-revenue water (NRW)	1,377,045 (144,630)	1,239,473 (196,438)	1,505,917 (336,802)	1,784,727 (213,549)	1,842,395 (221,041)

Table 3.7 Confidence limits for water balance inputs and NRW

Period	Component	Variance (m^6)	Volume (m^3)	95% confidence limit
2005	SIV	13,894,379,063	3,494,310	6.6%
	RW	7,023,340,569	2,117,265	7.8%
	NRW	**20,917,719,632**	**1,377,045**	**20.6%**
2006	SIV	17,695,253,752	3,451,629	7.6%
	RW	20,892,681,141	2,212,156	12.8%
	NRW	**38,587,934,893**	**1,239,473**	**31.1%**
2007	SIV	1.09632E+11	3,699,362	17.5%
	RW	3,803,558,228	2,193,445	5.5%
	NRW	**1.13436E+11**	**1,505,917**	**43.8%**
2008	SIV	39,894,250,274	4,099,082	9.6%
	RW	5,708,953,318	2,314,355	6.4%
	NRW	**45,603,203,592**	**1,784,727**	**23.5%**
2009	SIV	32,680,876,063	4,345,612	8.2%
	RW	16,178,215,910	2,503,217	10.0%
	NRW	**48,859,091,973**	**1,842,395**	**23.2%**

3.2.6.2 Discussion of water balance uncertainty

The wider 95% confidence intervals (> 5%) indicated in Table 3.7 casts doubt in the quality of volumes used for each water balance component. Uncertainty in NRW is on the increase, from ±144,630 m^3/month in 2005 to ±221,041 m^3/month in 2009. This indicates potential to improve on the quality of measurements used for the water balance input and subsequently the uncertainty in the derived NRW. The high uncertainty in volume figures and wider intervals for the year 2007 could be attributed to the commissioning of a new water plant (Gaba III) that increased system input volume by about 70,000 m^3/d.

This initial analysis indicates that the uncertainty in the SIV is the dominant source of uncertainty in the water balance. The utility is now investigating SIV in more detail to identify the meters and other sources that contribute most to the uncertainty and develop cost-effective interventions for improvement. Thereafter, a more detailed analysis to identify sources of uncertainty in all the over 150,000 customer meters will be carried out until a point where no further improvement in the NRW uncertainty can be realised cost-effectively. Sattary et al. (2002) in their study on the impact of measurement uncertainty on the water balance of Anglian Water (UK) reported a reduction of uncertainty in the imbalance between water delivered and customer use from ±74 ML/d to ±56 ML/d through calibration of one single meter of SIV. They projected further reductions of upto ±48 ML/d by re-calibrating additional two SIV meters. At ±48 ML/d, the uncertainty in the imbalance represented about 4.2% of distribution input which still fell short of OFWAT's A1 band requirements of 2% (Sattary et al. 2002).

3.2.7 Challenges and lessons learned in introducing a PI culture in NWSC-Uganda

In the process of establishing and implementing a performance assessment system, the following challenges and key lessons have been learned:

I. *Mobilization and teamwork.* The main challenge was getting people from different towns (utilities) to work together on a project that demanded a lot of time and input from staff members over and above their routine work. The burden of additional paperwork and data collection, created fear and resistance among team members and this influenced decision on number of final PIs adopted. Careful selection of a PI team is critical as the developing of PIs require knowledge of the business, a lot of thought, patience and commitment. In addition, the process is lengthy and iterative. A PI team not exceeding 10 members is recommended for maximum productivity. Like the old adage of "too many cooks spoil the soup", likewise use of many PI members often encourages gossip rather than work. Instilling a PAS culture in the organization takes time and deliberate efforts from top management.

II. *Engagement of the organization's CEO.* Performance indicator systems are useless if results achieved are not used to support improvement measures within the organization (Alegre et al. 2009). The active participation of the MD who is the top most executive in the water utility confirmed the importance of the exercise and made more team members' buy-into the process. However, despite the involvement of the MD, about 50% of the PI team especially senior managers were not actively participating in the process in the last 2 workshops. Supportive organizational cultures that change the mindset of the utility managers is necessary to ensure PAS are embedded into all aspects of the utility's operations.

III. *Quality of data.* PIs are useless if data used to generate them is not reliable. The validity of the results, conclusions and decision-making are all based on the data quality. The establishment of a PAS helped identify institutional structures needed to ensure integrity in data collection and management via data quality assurance mechanisms. Although good progress has been made, the issue of data accuracy, uncertainty in measurements and establishment of effective data management systems remains a major challenge. Whereas decision makers do not have confidence in the reliability and accuracy of input data used for computation of NRW figures, they do not fully recognize the usefulness of uncertainty analysis in the water balance model input variables and the NRW figures in practice. The collection of good quality data is imperative for effective water loss reporting and capital investment planning and NWSC must allocate adequate resources needed to support the established PAS.

IV. *The number of PIs to be used.* Generating data required for calculating PIs is costly and the number of indicators must be kept as few as possible. Use of too many PIs and more analysis is not necessarily better. Only relevant indicators critical for WDS evaluation to support decision-making must be selected. A cost-benefit analysis of data collection, validation, archiving and processing must be carried out in establishing a PAS.

V. *Performance-based pay.* Performance measures and the feedback they provide are only as good as the database and the underlying analysis from which they are derived and are subject to manipulation by some utility managers especially if their pay is performance based. Managers need to be convinced that the generated PIs are to help them improve performance in service delivery and not for penalising them. Otherwise, utility managers will often corrupt data and produce numbers they are asked to deliver.

VI.   *Only best results welcome.* Traditionally, public reported has glossed over areas of underperformance and presented only the good results. Focusing only on good results without auditing how they were derived and the causes of poor performance could be counter-productive. A fault-positive culture in the utility where bad performance is looked at as an opportunity for potential improvement in the future must be cultivated within the organization (Alegre et al. 2009).

VII. *Consistency in PI reporting.* Though periodic reviews and refinements of PIs is recommended, there is need to balance continuous improvement with stable old PIs to support historical trends and analysis.

VIII. *High short-term expectations.* The "100-days" results syndrome of utility leaders is often very unrealistic especially for NRW improvement. It is not unusual to suggest a reduction of NRW in Kampala city by 10% in 100 days. There is need for utility leaders to appreciate the fact that certain improvement measures (e.g. asset renewal) take time to cause improvement in performance figures. Unless performance measures are looked at in the spirit of continual improvement, misunderstandings and wrong decisions are likely to arise.

IX.   *Balanced score card.* Having a balanced set of well defined PIs that are very well aligned to the utility strategic plans is very critical for the successful implementation of a PAS. For example, although NWSC-Kampala has excelled in commercial operations and customer care (Mugisha and Brown 2010), its technical efficiency in water distribution with NRW of about 40% of water delivered, remains one of the highest on the African continent. This is because NWSC's performance measures have paid less attention to technical aspects compared to commercial and financial operations.

X.   *No "one-size-fits-all" solution.* There is no cookie-cutter blueprint for a PAS; each utility must examine its own operational processes and practices, infrastructure, organizational culture, supporting data, technology and available resources and develop performance measures most appropriate for their local working environment.

The lessons learned in Uganda will benefit immensely other water utilities in developing countries trying to establish and implement PASs. The challenges encountered provide valuable information on the pitfalls to avoid. Further challenges and how to overcome some of the misunderstandings while implementing PAS can be found in Alegre et al. (2009).

3.3   Applicability of performance indices for WLM in developing countries

According to Alegre et al. (2009), performance indices are measures: (i) resulting from the combination of more disaggregated performance measures (e.g. weighted average of PIs), aiming at aggregating several perspectives into a single measure, and (ii) derived from analysis tools (e.g. simulation models, statistical tools and cost efficiency methods). There are currently two main indices used for assessing WDS efficiency: infrastructure leakage index (ILI) and apparent loss index (ALI). Although they have proved to be useful for performance benchmarking in the developed countries, their applicability in developing countries is doubtful as discussed in Chapter 2. There is need for customising them to suit local conditions.

3.3.1   Infrastructure leakage index (ILI)

The ILI is a technical indicator of how well a WDS is managed for the control of real losses at the current operating pressure. It is defined as the ratio of current annual volume real losses (CARL) to the unavoidable annual real losses (UARL) (Lambert et al. 1999). It is a technical concept rather than an economic concept. ILI higher than 1 may indicate that leakage reductions are technically feasible but may not necessarily be economically feasible. Being a ratio, it is useful for benchmarking and is widely used in the developed countries (McKenzie et al. 2007).

During the peer review, ILI was found inappropriate by the team for use in NWSC-Uganda for the following reasons: (i) it does not fulfil some of the basic requirements for PIs and (ii) it required detailed data that was difficult and costly to obtain. The UARL, which is the lowest technically achievable level of leakage that could be achieved at current operating pressures assuming: (i) no financial or economic constraints, (ii) well maintained infrastructure, (iii) intensive state-of-the-art active leakage control, and (iv) all detectable leaks and bursts are repaired quickly and efficiently. The convenience of ILI for the Ugandan case was dubious as all assumptions made in deriving the UARL are not valid and practical. This is likely to be the case in most WDSs of the developing countries where resources are limited, leakage control is reactive and infrastructure is in a deplorable state. The ILI could be used sequentially in future as resources become available and technology evolves in line with IWA's recommendation of a step-by-step implementation.

3.3.2 Apparent loss index (ALI)

Based on the analogy of ILI, Rizzo et al. (2007) proposed the ALI as the ratio of the current actual apparent loss volume (CAAL) to the unavoidable annual apparent loss volume (UAAL). In the absence of data, 5% of water sales was recommended as a reference volume for UAAL. According to Thornton et al. (2008), the 5% UAAL benchmark may be high for water utilities in developed countries, which typically have good customer meter management and where buildings do not have roof tanks which present an opportunity for very low flows that pass unregistered through many water meters. These utilities typically also have reasonable policies and safeguards that prevent exorbitant unauthorized consumption. They conclude that the 5% assumption may be high in developed countries, but reasonable for developing countries.

From a literature survey carried out (Tables 3.8 and 3.9) (AWWA 2009; Bidgoli 2009; Dimaano and Jamora 2010; Fanner et al. 2007; Garzon-Contreras et al. 2009; Guibentif et al. 2007; Kanakoudis and Tsitsifli 2010; Mutikanga et al. 2009; OFWAT 2010; Schouten and Halim 2010; Sharma and Chinokoro 2010), we propose the following benchmarks for the technically feasible UAAL at the current level of technology and operating practices and assuming: (i) universal metering of all utility customers, (ii) 75% reduction of the current mean AL value (Table 3.8) in the developing countries, and (iii) 50% reduction of the current mean AL value (Table 3.9) in the developed countries:

- developed countries, UAAL = 3% of water sales or revenue water (RW);

- developing countries, UAAL = 7% of water sales or revenue water (RW).

Table 3.8 Apparent losses as a proportion of revenue water in some developing countries

Work	Country	City/Utility	RW (million m³/yr)	AL (million m³/yr)	AL (% of RW)	ALI (UAAL= 7% of RW)
Schouten & Halim (2010)	Indonesia	Jarkata Palyja	130	47	36.1	5.2
Dimaano & Jamora (2010)	Phillipines	Manila Maynilad	370.84	98.19	26.5	3.8
Garzon-Contreras *et al* (2009)	Colombia	Medellin EPM	185.7	29.71	16.0	2.3
		Cali	121.71	46.11	37.9	5.4
		Bogota	287.55	72.5	25.2	3.6
Sharma & Chinokoro (2010)	Zambia	Lusaka LWSC	35.64	11.77	33.0	4.7
Mutikanga *et al* (2009)	Uganda	Kampala NWSC	26.63	9.32	35.0	5.0
Bidgoli (2009)	Iran	Tehran NWWEC	720.72	112.43	15.6	2.2
Mean					28.2	

The rationale behind these proposed benchmarks is that even new meters do not register 100% of water sales and errors of ±2% and ±5% are allowable depending on the flow rate. In addition, for developing countries, the influence of roof tanks and high cases of unauthorised use had an influence on the high proposed benchmark. Furthermore, apparent losses for most developed countries are not assessed on basis of field investigations but default values. Studies based on field investigations indicate high meter under-registrations e.g. 14% for Madrid city in Spain (Flores and Diaz 2009). The proposed ALI values could be revised as metering technology evolves, water supply conditions improve, techniques and resources improve.

Table 3.9 Apparent losses as a proportion of revenue water in some developed countries

Work	Country	City/Utility	RW (million m³)	AL (million m³)	AL (% of RW)	ALI (UAAL= 3% of RW)
AWWA (2009)	USA	Philadelphia	222.3	21.3	9.6	3.2
Fanner *et al*. (2007)	USA	Boston/BWSC	92.4	2.9	3.1	1.0
Fanner *et al*. (2007)	USA	Florida/Charlote County Utilities	10.1	0.2	2.1	0.7
Fanner *et al*. (2007)	Canada	Halifax	166.9	2.6	1.6	0.5
Guibentif *et al*. (2007)	Switzerland	Geneva/SIG	58.6	2.2	3.8	0.9
Kanakoudis & Tsitsifli (2010)	Greece	Larisa/DEYAL	11.7	1.1	17.5	5.8
OFWAT (2010)	England & Wales	22 Companies	4213.2	117.9	2.8	1.8
Mean					5.8	

In order to assist utilities assess their performance and trigger appropriate action to improve the ALI, we propose four different performance groups (A-D) as shown in Figure 3.5.

Region	Technical Performance Group	ALI	Remarks
Developed Countries	A	1 - 2	Acceptable performance. Further reduction may be uneconomical unless if the cost of water is very high.
	B	2 - 3	There is room for improvement.
	C	3 - 4	High revenue losses. Acceptable where cost of water is very low.
	D	> 4	Very inefficient with poor meter management practices and inadequate policies for revenue protection. Urgent action required to minimize revenue losses.
Developing Countries	A	1 - 2	Acceptable performance. Further reduction may be uneconomical unless if the cost of water is very high.
	B	2 - 4	There is room for improvement.
	C	4 - 6	High revenue losses. Acceptable where cost of water is very low.
	D	> 6	Very inefficient with poor meter management practices and inadequate policies for revenue protection. Urgent action required to minimize revenue losses.

Figure 3.5 Apparent loss index (ALI) performance bands

3.4 Benchmarking Using Data Envelopment Analysis (DEA)

The business of water distribution services involves multiple inputs and outputs with often no standards to define efficient and effective performance. Although PI-based benchmarking is a valuable tool, it cannot suffice as one single measure to diagnose inefficiency and potential improvements in WDSs taking into account resource inputs. Benchmarking models that are able to deal with multiple performance measures and provide an integrated benchmarking measure are needed (Cook et al. 2004). DEA has been demonstrated as a powerful benchmarking tool where multiple inputs and outputs need to be assessed to identify best-practices and improve efficiency in organizations (Cooper et al. 2000; Zhu 2009). DEA has been used for benchmarking in many services including electricity distribution (Giannakis et al. 2005), healthcare (Kontodimopoulos and Niakas 2005), banking (Sherman and Zhu 2006), and education (Beasley 1995).

In the water industry, DEA is the most applied non-parametric method for benchmarking (Berg and Marques 2011) and has been applied by various researchers: in the regulation of UK water distribution companies (Thanassoulis 2000a; 2000b), performance benchmarking of Indian urban water utilities (Singh et al. 2010), assessing relative efficiency of water supply systems in Palestinian territories (Alsharif et al. 2008), assessing efficiency of Brazilian water utilities (Tupper and Resende 2004), inclusion of service quality in measuring performance of Spanish water utilities (Picazo-Tadeo et al. 2008), measuring efficiency in Italian water companies (Romano and Guerrini 2011), and international benchmarking of the water distribution sector (De Witte and Marques 2010). In this study, DEA is applied to 25 Ugandan water distribution systems to assess technical efficiency in water delivery with aim of reducing water losses and improve operating efficiency.

DEA has several advantages over other frontier-based methods (discussed in Chapter 2) but some noteworthy weaknesses as well. The advantages of DEA include:

- It allows comparison of organizations that use multiple inputs to produce multiple outputs which is suitable for WDSs and has ability to incorporate them simultaneously;
- It allows to measure inputs and outputs in their natural units without converting resources used and outputs into monetary units;
- It allows calculation (not estimation) of relative efficiency rather than absolute efficiency;
- It does not require specification of a cost or production function for the frontier as opposed to typical econometric models;
- It is transparent and easy to understand;
- It can be used with Malmquist indices which measure productivity change between two points in time;
- It can also examine the effect of specific factors, often referred to as environmental variables that are beyond the control of the utilities, thus accounting for the differences in operating efficiencies;
- Inefficient firms are compared to actual firms rather than some statistical measure;
- It identifies a set of peer firms (efficient firms with similar input and output mixes) for each inefficient firm;
- It has capabilities of identifying actual good practice and to set utility specific efficiency targets and revenue caps required for incentive regulation.

However, DEA has some drawbacks as well that include:

- It utilises a limited amount of available information, namely frontier utilities, to derive the efficiency scores. Any departure from the frontier is measured as inefficiency. The measurement of comparative efficiency rests on the hypothesis that efficient units are genuinely efficient;
- As a data-driven deterministic technique, DEA results are highly sensitive to outlier observations and to measurement errors in the frontier and insensitive to statistical noise;
- The DEA optimisation algorithm has no in-built test checks to ascertain the appropriateness of inputs and output variables;
- Efficiency scores tend to be sensitive to the choice of input and output variables. As more variables are included in the models, the number of utilities on the frontier tends to increase, so it is important to examine the sensitivity of the efficiency scores and rank order of the firms to model specification;
- It can be biased with limited samples. Bias is greatly reduced if there are enough efficient (or near efficient) utilities to define the frontier adequately. The large number of required samples may not be available during the early years of regulation;
- Traditional hypothesis tests are not possible. Statistical inference requires elaborate and sensitive re-sampling methods like bootstrapping techniques (Simar and Wilson 2000);
- It can fail to discriminate on the performance of firms concerned in presence of large numbers of inputs and outputs due to its weight flexibility i.e. most firms attain the maximum or near maximum efficiency score;
- It offers no basis for setting performance targets for relatively efficient firms.

3.4.1 DEA Models

DEA is a piecewise linear programming-based methodology whose optimisation algorithm is applied to construct a non-parametric production efficient and/or best-practice frontier of a sample of firms. These firms could be operating units such as utility branches, bank branches, schools, hospitals, university departments etc. Each unit is usually referred to as a decision-making unit (DMU). The efficient frontier is the benchmark against which the relative performance of DMUs is measured. DMUs can be classified as either Pareto-efficient or inefficient DMUs. There are several measures of efficiency with the most applied measure being the *radial* measure of efficiency. The radial efficiency scores of the DMUs are calculated on a scale of 0 to 1, with the most efficient DMUs on the frontier receiving a score of 1. DEA provides a set of scalar measures of efficiency, namely input and output-oriented measures. Output-oriented models maximize output for a given amount of input factors while input-oriented models minimize the input factors required for a given level of output (Thanassoulis 2001).

Charnes et al. (1978) introduced the generic DEA model assuming constant returns to scale (CRS) that was later extended to variable returns to scale (VRS) (Banker et al. 1984). The CRS model is commonly referred to as the Cooper-Charnes-Rhodes (CCR) model while the VRS as Banker-Charnes-Cooper (BCC) model. The CCR model uses an oriented radial measure of efficiency. The CRS frontier allows smaller firms to be benchmarked against bigger firms and vice versa. Conversely, the BCC model is free from scale induced biases and allows firms of similar operational size to be benchmarked against each other. The ratio of CRS efficiency to VRS input efficiency is the scale efficiency (SE). SE measures part of inefficiency of a DMU attributed to its operating away from the most productive scale size. Many DMUs are likely to be labelled as efficient under VRS compared to the CRS case (Cubbin and Tzanidakis 1998).

The DEA models are all based on the concept of production function, where the best DMU(s) with the best efficiency in converting inputs $(x_1, x_2, ..., x_n)$ into outputs $(y_1, y_2, ..., y_m)$ is identified, and becomes a benchmark for all other DMUs.

For DMU j_0, the basic DEA CCR model is calculated as follows:

$$h_{j0} = \underset{u_r, v_i}{Max} \frac{\sum_{r=1}^{s} u_r y_{rj0}}{\sum_{i=1}^{m} v_i x_{ij0}}$$

subject to

$$(3.4)$$

$$\frac{\sum_{r=1}^{s} u_r y_{rj}}{\sum_{i=1}^{m} v_i x_{ij}} \leq 1$$

$$j = 1 ... j_0 ... N;\ u_r > 0,\ v_i > 0$$

where N is the number of DMUs, the jth one using input levels x_{ij}, $i = 1, ..., m$, to secure output levels y_{rj}, $r = 1, ..., s$. The evaluated observation is labelled by the subscript "o". The interpretation of u_r and v_i is that they are weights applied to outputs y_{rj} and inputs x_{ij} and are selected to maximize the efficiency score h_{j0} for DMU j_0. The constraint ensures that the efficiency score for any DMU is less than one. The efficiency frontier enveloping all data points in a convex hull is established. The DMUs located on the frontier represent an efficiency level of 1, and those located inside the frontier (production possibility set) are operating at efficiencies less than 1(Thanassoulis 2001).

Model (Equation 3.4) is manipulated as follows in order to solve it as a LP model:

$$h_{j0} = Max \sum_{r=1}^{s} u_r y_{rj0}$$

subject to (3.5)

$$\sum_{i=1}^{m} v_i x_{ij} = 1$$

$$\sum_{r=1}^{s} u_r y_{rj} - \sum_{i=1}^{m} v_i x_{ij} \leq 0, \qquad j = 1, \dots j0, \dots, N,$$

$$u_r, v_i, \geq 0$$

In a simpler notation, model (Equation 3.5) can be written as:

$$max (v,u) = uy_0$$

subject to
$$vx_0 = 1$$
$$uY - vX \leq 0$$
$$v \geq 0, u \geq 0 \qquad (3.6)$$

Finally, before solving, the linear program is converted to its dual for computational efficiency reasons:

$$min (\theta, \lambda) = \theta$$

subject to
$$\theta x_0 - X\lambda \leq 0$$
$$Y\lambda \geq y_0$$
$$\lambda \geq 0 \qquad (3.7)$$

With the addition of slack variables, the dual problem becomes:

$$min (\theta, \lambda) = \theta$$

subject to
$$X\lambda + s^- = \theta x_0$$
$$Y\lambda - s^+ = y_0 \qquad (3.8)$$
$$\lambda \geq 0, \; s^+ \geq 0, \; s^- \geq 0$$

The slack variables can be interpreted as the output shortfall and the input excesses compared to the efficient frontier. Radial measures of efficiency will normally offer less discrimination of efficiency as they ignore "slacks" which reflect further potential input reductions beyond the radial improvement to inputs (Podinovski and Thanassoulis 2007). The BCC model is basically the CCR model with an additional "convexity" constraint $e\lambda = 1$. This additional constraint gives the frontiers piecewise linear and concave characteristics. Further details on DEA-benchmarking methodologies and applications can be found in Cooper et al. 2000, Thanassoulis, 2001 and Zhu, 2009.

3.4.2   Data and model specifications

The selection of input-output variables is perhaps the most important step in applying DEA for calculating the relative efficiency of water utilities. DEA models are sensitive to the number of input and output variables as well as the sample size (De Witte and Marques 2010).

3.4.2.1 Choice of variables

There is no general consensus on the input and output variables used in DEA measurement for water distribution systems. In a survey of benchmarking studies carried out in the water sector using DEA (Table 3.10), the most used input variables include operating expenditure (OPEX), capital expenditure (CAPEX), water losses, mains length, number of employees and energy costs; while output variables include number of customers, mains length, water delivered or supplied, water losses and revenues (Alsharif et al. 2008; Cubbin and Tzanidakis 1998; De Witte and Marques 2010; Romano and Guerrini 2011; Singh et al. 2010; Thanassoulis 2000a; Thanassoulis 2000b). Some studies also provide additional explanatory factors such as water source and/or treatment, volume by consumer category (industrial, domestic, etc.), metered and non-metered volume, peak factors, water losses and customer density. More or less similar input and output variables have been reported for electricity distribution benchmarking studies (Jamasb and Pollit 2001).

Table 3.10 Benchmarking studies in water distribution using DEA

Study	Scope	Focus	Inputs	Outputs	DEA Model
Cubbin and Tzanidakis (1998)	29 water companies in the UK	Regression versus DEA in efficiency measurement	OPEX	number of properties, water delivered and network length, non-household water supply	CCR (with CRS)-input oriented and BCC (with VRS)-input oriented
Thanassoulis (2000a,b)	32 water companies in the UK	Regulation and potential cost saving	OPEX	number of properties, water delivered and network length	CCR (with CRS)-input oriented
Alsharif et al. (2008)	33 Palestinian municipalities	Evaluation of water supply systems efficiency	Water losses and OPEX	Total revenue	BCC (with VRS)-input oriented
De Witte and Marques (2010)	17 DMUs in Australia and 105 DMUs in Europe	International benchmarking of technical efficiency	Labour and capital	Water delivered & number of connections	BCC (with VRS)-input oriented
Singh et al. (2010)	Efficiency in 18 Indian urban water utilities	Sustainability of water supply systems	Total expenditure and UFW	Water produced, total no. of conns., length of network	BCC (with VRS)-input oriented
Romano & Guerrini (2011)	Efficiency in 43 Italian water utility companies	To measure global cost efficiency	Cost of labour, material services & leases	Water delivered & population served	CRS & VRS - input oriented

From Table 3.10, water losses are considered either as inputs, outputs or explanatory factors. In this study, we use OPEX and water losses and/or non-revenue water (NRW) as inputs since our goal is to quantify the water loss or NRW reduction potential. The amount of water loss is a good indicator of water distribution system inefficiency (Deb et al. 1995; Park 2006). NRW has been used in several studies as a proxy measure for service quality (Coelli et al. 2003; Picazo-Tadeo et al. 2011; Tupper and Resende, 2004). OPEX is used as a proxy of labour and mains repair costs. Whereas Mugisha (2007) used number of employees as an input variable for Ugandan water utilities, in this study, the variable was not directly considered due to the different levels of outsourcing in the utilities. Although total expenditure (TOTEX) has been widely used as an input, it was not possible to get data for the variable from the Uganda's water sector annual performance report (MWE 2010). This limitation could affect the outcomes as some water loss control interventions could be identified under TOTEX rather than OPEX.

The model output variables selected were water supplied total number of service connections, and distribution network length. The model input and output variables used are presented in Table 3.11. The number of customer connections and length of the network are used as proxy for population and network density. Water losses are likely to increase with network length and number of connections, thus influencing OPEX and NRW. They capture the scale size of the WDS and in this way, all WDSs with the same network length and connections will tend to be benchmarked against each other. Likewise, water supplied is a measure of effort made by the utilities in conveying water from the treatment works to customers. These variables are the major cost drivers for OPEX and highly influence the level of NRW. The high correlation coefficients indicated in Table 3.12 between input and output variables for the dataset support the choice of variables for the study and confirm the reliability of the well structured DEA model. The outputs are also highly and positively correlated with each other.

Table 3.11 Inputs and outputs of DEA Model applied to Ugandan water utilities

Inputs	Non-revenue water (NRW)	Operating expenditure (OPEX)	
Outputs	Water Supplied (WS)	Total number of service connections (TSC)	Length of pipe network (LPN)

Table 3.12 Correlation coefficients

	NRW	OPEX	WS	TSC
OPEX	0.993			
WS	0.997	0.998		
TSC	0.991	0.998	0.998	
LPN	0.981	0.993	0.988	0.989

There was very poor correlation between the inputs and outputs with NRW as a percentage, so it was not used and NRW expressed as volume was used. The use of pairwise correlation is a valuable tool for validating the selected inputs and outputs of the DEA model (Park 2006; Podinovski and Thanassoulis 2007; Thanassoulis 2000b).

3.4.2.2 Model specification

In this study, we use an input-oriented model for DEA as opposed to the output-oriented model as the optimization objective is to minimize the excess, if any, in inputs (NRW and OPEX) for a given set of outputs. Input-oriented models have been found to be more appropriate for electricity distribution utilities, as demand for their output is a derived demand (fixed) which is beyond the control of utilities and can be taken as given (Giannakis et al. 2005; Jamasb and Pollit 2001). Since water utilities are analogous to electricity utilities, the selected model seems appropriate for the study. The input orientation is applied using both the CCR (with CRS) and BCC (with VRS) models in order to assess the uncertainty of scale-induced biases and discrimination on the performance of DMUs. The sample size and variables was also verified against some rules of thumb. The sample size n should satisfy $n \geq$ max $[m \times s, 3(m+s)]$ (Banker et al. 1984; De Witte and Marques 2010) or $n \geq 2m \times s$ (Dyson et al. 2001); where m is the number of inputs and s the number of outputs used in the analysis. These rules of thumb are satisfied in our analysis. Instead of relying on rules of thumb, it is advisable to explore other options such as increasing the number of DMUs, reducing the number of inputs and outputs, applying weight restrictions, production trade-offs, unobserved DMUs and selective proportionality (Podinovski and Thanassoulis 2007).

3.4.2.3 Dataset

The dataset used for comparative performance was the most recent data used in the annual water and environment sector performance report for the financial year 2009-2010 by the Ministry of Water and Environment (MWE) in Uganda (MWE 2010). This was the most comprehensive reporting format compared to the previous performance reports and was deemed more accurate bearing in mind that the reporting culture in the sector is just taking root. The 25 WDSs analyzed fall under two categories: (i) 20 large towns under the management of NWSC (public utility) including Kampala city, and (ii) 5 small towns under local governments. Water services in NWSC are provided under Internally Delegated Area Management Contracts (IDAMCs) (Mugisha 2007) while in small towns local private operators are hired to provide services under management contract frameworks and public-private partnerships (PPPs) (MWE 2010). The difference in size between utilities is large and is dominated by Kampala city on all variables. For example, the annual minimum water supplied is 41,039 m^3 (Pallisa town) and the maximum is 49,965,795 m^3 (Kampala city) with an average of about 2.8 million m^3, median of 645,894 m^3 and standard deviation of about 9.9 million m^3. The highest customer density is 92 connections per km in Entebbe municipality and the smallest is 18 connections per km in Kisoro town council.

The regulatory framework in the water sector is weak compared to other sectors such as electricity and telecommunications and casts doubt on the quality of data used for reporting. Where data was missing, the NWSC annual report for the same reporting period was used. The Ministry and NWSC performance reports are freely accessed on their websites. Comparative trends in performance could be made in future studies as consistence in data reporting and reliability improves. However, the single-year data set still provides insight on the comparative efficiency of the different water distribution systems in Uganda.

3.4.3   Results and discussion of DEA-based benchmarking

This section presents and discusses the main findings of the DEA-based benchmarking study. The DEA analysis was carried out using the PIM-DEA software, version 3.0 developed at Aston University, Birmingham, UK. Table 3.13 summarises the technical efficiency (TE) scores and rankings based on the CCR model (assuming CRS) and the BCC model (assuming VRS) while potential reductions in NRW are shown in Table 3.14 and Figure 3.6 presents potential savings in OPEX. The rank of 1 is used for the best performing DMUs i.e. the higher the efficiency score, the lower the numerical value of the rank.

Table 3.13 Technical efficiency scores and utility rankings (2009-2010)

DMU	Utility Location	CCR Model	Ranking	BCC Model	Ranking
	City				
1	Kampala	0.618	13	1.000	1
	Large Towns				
2	Jinja	0.517	19	1.000	1
3	Entebbe	0.598	16	1.000	1
4	Mbarara	0.630	11	1.000	1
5	Mbale	0.750	6	1.000	1
6	Masaka	0.447	22	1.000	1
	Medium Sized Towns				
7	Lira	0.658	10	1.000	1
8	Fort Portal	0.505	20	0.854	21
9	Tororo	0.750	6	0.979	14
10	Gulu	0.670	8	0.870	19
11	Soroti	0.601	15	0.913	16
12	Kasese	0.551	18	0.887	18
13	Arua	0.624	12	0.988	13
14	Kabale	0.659	9	0.897	17
	NWSC Small Towns				
15	Hoima	0.380	23	0.864	20
16	Masindi	0.605	14	1.000	1
17	Iganga	0.582	17	0.786	22
18	Mubende	0.466	21	0.605	23
19	Bushenyi	0.347	25	0.428	24
20	Lugazi	0.376	24	0.408	25
	Other Small Towns				
21	Kisoro	1.000	1	1.000	1
22	Busia	0.962	4	0.974	15
23	Mityana	1.000	1	1.000	1
24	Kamuli	1.000	1	1.000	1
25	Pallisa	0.754	5	1.000	1
	Mean Score	0.642		0.898	

Table 3.14 Potential efficiency gains for utilities under CCR Model

DMU	Utility Location	Actual NRW (m^3/year)	Target NRW (m^3)	Potential NRW Savings (m^3)	Potential Efficiency Gains (%)
1	City				
	Kampala	19,577,095	5,811,619	13,765,476	70.3
	Large Towns				
2	Jinja	1,021,967	486,335	535,632	52.4
3	Entebbe	637,669	381,523	256,146	40.2
4	Mbarara	363,609	229,260	134,349	36.9
5	Mbale	115,551	86,635	28,916	25.0
6	Masaka	288,974	129,245	159,729	55.3
	Medium Sized Towns				
7	Lira	130,364	85,815	44,549	34.2
8	Fort Portal	156,857	79,223	77,634	49.5
9	Tororo	63,767	47,830	15,937	25.0
10	Gulu	73,027	48,932	24,095	33.0
11	Soroti	98,302	59,078	39,224	39.9
12	Kasese	131,323	72,326	58,997	44.9
13	Arua	81,837	51,030	30,807	37.6
14	Kabale	54,544	35,937	18,607	34.1
	NWSC Small Towns				
15	Hoima	135,643	51,559	84,084	62.0
16	Masindi	56,824	34,368	22,456	39.5
17	Iganga	124,007	43,745	80,262	64.7
18	Mubende	77,665	36,167	41,498	53.4
19	Bushenyi	86,144	29,897	56,247	65.3
20	Lugazi	54,650	20,543	34,107	62.4
	Other Small Towns				
21	Kisoro	19,480	19480	-	0.0
22	Busia	42,114	15482	26,632	63.2
23	Mityana	6,864	6864	-	0.0
24	Kamuli	11,178	11178	-	0.0
25	Pallisa	22,379	6366	16,013	71.6
	Total	23,431,834	7,880,437	15,551,397	66.4

3.4.3.1 Technical efficiency (TE) scores under the CCR Model

The results in Table 3.13 indicate a mean overall efficiency of 64.2% with 3 DMUs on the efficient frontier. If the 22 DMUs had adopted the best-practices exhibited by the 3 Pareto efficient DMUs, the current levels of mean outputs could have been realised with a 36% reduction of resources utilized. The highest inefficiency is exhibited by the NWSC small towns varying from 65.3% (DMU 19) to about 40% (DMU 16). Inefficiencies not only drain the public purse but also seriously undermine the capacity of utilities to expand services. Conversely, the model identifies the other smaller towns outside NWSC (DMUs 21 to 25) as the most efficient with TE scores ranging from 75% to 100%. The mean overall efficiency of NWSC towns was 56.7% compared to 94.3% for the small towns outside NWSC. The difference in performance is probably explained by the different management frameworks (public-public versus public-private partnerships) and the different utility priorities. In

NWSC, the emphasis has been more on increasing water production and service coverage through new connections rather than reducing NRW. Conversely, the small towns have put much emphasis on operations and maintenance of the WDSs that culminated into hiring private operators (POs). Introduction of POs in management of WDSs has been found to lead to improved operating efficiency in developing countries such as Senegal where NRW reduced by 10% between 1996 and 2005 (Ringskog et al. 2006) and the east zone of Metro Manila where NRW has been reduced from 63% to 11% in the past 14 years saving over 0.6 million m^3 of water per day (Luczon and Ramos, 2012). It is also likely that NWSC is good at managing big towns rather than small towns. The other reason could be the high OPEX generally exhibited by all DMUs under NWSC. Mugisha (2007), reports that NWSC provides incentives to its operating utilities for improving performance including reducing NRW. This study reveals that, it is likely that the incentives are inadequate or are poorly targeted for reducing the high levels of NRW particularly in Kampala city.

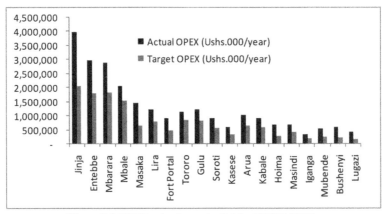

Figure 3.6 OPEX reduction potential for NWSC DMUs

3.4.3.2 Potential input savings under the CCR Model

Although the study was triggered by the desire to reduce water losses, the motivation for cost savings is strong. Table 3.14 shows potential NRW savings of about 15.5 million m^3 per annum and 88.5% of these savings are in Kampala city. In terms of money, this translates into about US$ 12 million annually which is substantial for the financially struggling water utilities in Uganda. Therefore, all efforts to save water losses in the sector should probably be focused on Kampala city. DMU 25 (Pallisa town) has the highest efficiency gains potential with respect to NRW, although the volumes to be recouped are significantly small. The NRW slacks were observed for DMUs 1, 2, 17, 22 and 25, indicating further potential for reduction beyond the calculated radial improvements.

The overall average reduction potential in OPEX is about 40% with significant cost-savings in the NWSC towns. The annual potential cost savings for NWSC towns is estimated at about US$ 0.5 million. The targets are shown in Figure 3.6 excluding Kampala which is too big to be represented graphically on the same scale. The highest cost savings (65%) have been predicted in DMU 19 (Bushenyi Town). Utilities that have decentralized or adopted private sector management have substantially lower hidden costs than those that have not (Banerjee and Morella 2011). The recent urban water sector reforms in Uganda that led to private sector management in small towns' water operations could have been the key driver in lowering inefficiency in OPEX.

3.4.3.3 Comparison between the CCR and BCC model results

As shown in Table 3.13, more utilities are ranked as efficient under the BCC model (with VRS) compared to the CCR model (with CRS). The number of utilities on the frontier increases from 3 utilities under CCR to 13 utilities under BCC. Likewise, the mean efficiency score increases from 64% to 90%. The reason for the increase in efficiency scores is that under BCC models, utilities are benchmarked against other comparable utilities in terms of size whereas in CCR models, economies of scale are not accounted for. CCR models assume that a small utility should be able to operate as efficiently as a large one (i.e. CRS). When competing explanatory factors or cost drivers have low correlation, BCC models are more likely to view each utility as being unique and therefore fully efficient especially when there are limited observations and the utility is an outlier. In such a case, BCC model efficiency scores tend to be equal to or very close to 1 for most utilities, which results in the loss of discriminatory power (Cubbin and Tzanidakis 1998). CRS models generally provide much better discrimination than VRS models (Podinovski and Thanassoulis 2007).

Efficiency under VRS is termed pure TE while efficiency under CRS is termed overall TE (Thanassoulis 2000a). The pure TE scores and rankings increased significantly in larger towns generally. The largest increase in efficiency score was 55.3% for DMU 6 (Masaka town) with a dramatic increase in ranking from number 22 to joint number 1. BCC models usually favour utilities at the extremes of the size range (Cubbin and Tzanidakis, 1998), but ironically, DMU6 is neither the largest or smallest water utility under the study. For example, the efficiency score of Kampala city changed from 61.8% under CCR model to 100% under the BCC model. The BCC model suggests that there is no room for improvement in Kampala which lacks merit based on corroborative evidence on the ground and the magnitude of input variables. The validity of the BCC model in analysing the dataset is therefore questionable from the perspective of assessing potential water loss reduction and improving water distribution efficiency. Although the CCR model may under estimate the utility's pure TE, lack of comparable utilities may put some inefficient utilities on the frontier under BCC models and thus produce misleading results. BCC models are likely to be generous to inefficient firms unless the dataset includes a sufficient number of comparable utilities in all size categories. In the Netherlands and Norway, where regulators use DEA for benchmarking electricity distribution, the issue of scale has not been found so compelling to necessitate the use of BCC models (Giannakis et al. 2005). In this benchmarking study, the results of the CCR DEA model that assumes CRS were preferred as they seem to best describe the relationships between inputs and outputs for the Ugandan water utilities. Nevertheless, both models under VRS and CRS do indicate that NWSC towns (large and small) are generally inefficient compared to small towns in delivering water services.

3.4.3.4 Comparison between the CCR model and PI-based benchmarking rankings

In order to gain more insight into the potential water and cost savings calculated with the CCR model, the rankings under the CCR DEA-based and PI-based benchmarking were examined. The rankings and their differences are presented in Table 3.15.

The ranking correlation coefficients of the CCR model with NRW was moderate ($R^2 = 0.423$), indicating that one benchmarking method can explain only 18% features of the other. Although weakly correlated [CCR with OPEX ($R^2 = 0.101$) and NRW with OPEX ($R^2 = 0.091$)], the DEA CCR model and the PI-based rankings can be interpreted as an indication of poor agreement between individual PIs and DEA CCR model on the performance of DMUs. However, the following observations can be made from the rankings shown in Table 3.15:

Table 3.15 CCR Model and PI-based rankings (2009-2010)

DMU	City/Town	Frontier-based A TE CCR-Model	PI-based B NRW L/conn/d	PI-based C OPEX (Ushs/100 km/yr)	Change (A - B)	Change (A -C)
1	Kampala	13	25	25	-12	-12
2	Jinja	19	24	23	-5	-4
3	Entebbe	16	16	24	0	-8
4	Mbarara	11	15	22	-4	-11
5	Mbale	6	5	11	1	-5
6	Masaka	22	21	9	1	13
7	Lira	10	11	17	-1	-7
8	Fort Portal	20	13	16	7	4
9	Tororo	6	8	18	-2	-12
10	Gulu	8	7	21	1	-13
11	Soroti	15	10	13	5	2
12	Kasese	18	12	14	6	4
13	Arua	12	6	20	6	-8
14	Kabale	9	4	12	5	-3
15	Hoima	23	18	7	5	16
16	Masindi	14	9	6	5	8
17	Iganga	17	23	15	-6	2
18	Mubende	21	14	8	7	13
19	Bushenyi	25	22	9	3	16
20	Lugazi	24	19	18	5	6
21	Kisoro	1	3	1	-2	0
22	Busia	4	20	4	-16	0
23	Mityana	1	1	5	0	-4
24	Kamuli	1	2	3	-1	-2
25	Pallisa	5	17	2	-12	3

- The Pareto efficient DMUs (21, 23 and 24) identified under the CCR model are also ranked in the first, second and third positions under the NRW PI-based benchmarking, indicating no dispute and confirming that probably the efficient DMUs are genuinely efficient. When these two methods agree on performance, the PIs can support the communication of DEA results to non-specialists;

- The DMUs with hardly any change in their poor rankings by the two methods (DMUs 3, 6, 7, and 10) are likely to have both water loss and cost savings potential.

- The rankings are relatively stable except for DMUs (1, 22 and 25) that are ranked worse under the NRW PI-based benchmarking approach. This is probably explained by the fact that DEA models are flexible enough (under unrestricted weights) and will try to represent each DMU in the best possible way if it has good performance on any one of the PIs. DEA considers all inputs and outputs simultaneously rather than partial assessment based on one input and one output level. The DMUs also had slacks in the NRW variable under the CCR model. The slacks coupled with the poor rankings under the NRW PI-based approach, indicates further potential for NRW reductions by the DMUs;

- The DEA CCR model disagrees with the PI-based methods on performance of most DMUs. This disagreement provides useful information upon careful scrutiny. For example, DMU 23 is evaluated as efficient by the CCR model and the NRW PI indicating no room for improvement. However, the OPEX PI indicates that it is not the best, thereby suggesting potential for cost savings. The reverse is true for DMU 15 as another example. In this regard, PIs could enhance target setting for DEA-efficient DMUs. The combined information provided by both methods, supported by judgement, seems to be useful in further strengthening performance of DMUs in different aspects of operations;

- The PIs themselves seem to disagree on performance of most DMUs as well, complicating decision-making on the DMUs overall performance. Use of a single PI to select the best DMU biases the assessment. In this respect DEA results that include multiple inputs and outputs are more objective.

All in all, the two methods are complimentary in measuring performance and where feasible PI values should always be supplemented by DEA measures of performance. This two-way approach ensures that advantages of both methods are achieved without facing the disadvantages. The findings of this study are in agreement with those of Thanassoulis et al. (1996) who compared DEA and PIs as tools for performance assessment in the District Health Authorities of England.

3.4.3.5 T-tests for the mean efficiencies of publicly and privately managed water utilities

A paired t-test was performed to test the significance of the differences of mean technical efficiencies observed in Table 3.13. An independent sample t-test was carried out to test the differences of mean efficiencies for the five smallest publicly managed (under NWSC) water utilities and the five privately managed (under PPPs) water utilities outside NWSC. The results are presented in Table 3.16. The research hypothesis in this case is that the mean efficiencies of the publicly and privately managed WDSs are significantly different while the null hypothesis is that there is no difference.

Table 3.16 Sample t-test for publicly and privately managed small towns WDSs

	Group statistics			t-test for equality of means					95% Confidence interval	
Operations management	N	Mean TE score	Variance	t	df	P-value (2-tailed)	Mean TE score difference	Standard deviation	Lower	Upper
Public	5	0.475	0.014	8.562	4	0.001	0.468	0.122	0.316	0.620
Private	5	0.943	0.116							

The information in Table 3.16 provides statistically significant evidence to reject the null hypothesis and support the research hypothesis. The mean difference TE score (TE=0.468, SD=0.122, N=5) was significantly greater than zero, t (4) = 8.562, two-tail p = 0.001, providing evidence that the performance difference in water distribution efficiency was significant. A 95% confidence limit about the mean TE score is (0.316, 0.620). Studies have shown that well-designed PPPs are a valid option to turn around poorly performing urban water utilities in developing countries (Marin 2009).

3.4.4 Policy implications of the DEA-benchmarking study

In order to improve WDS efficiency in the Ugandan water utilities, the following policy weaknesses and/or shortcomings need to be addressed.

3.4.4.1 Sector regulation

Regulation in Uganda's water sector is very weak compared to other sectors such as electricity and telecommunications. The technical inefficiency in the large towns probably explains the difference in tariffs between the NWSC large towns and small towns. The average water tariff for NWSC towns is 1,750 UShs/m^3 (or 0.76 US $/m^3) (NWSC 2010), whereas in the small towns it is 1,000 UShs/m^3 (or 0.43 US $/m^3) (MWE 2010). In a benchmarking study carried out in the USA using a dataset of over 100 water utilities, it was confirmed that utilities with more water losses in the distribution systems charge customers more than utilities with low levels of water losses (Park 2006). In other words inefficient water utilities transfer the costs to customers especially in countries where regulation is weak and public utilities get tariff approvals so easily. There is need to strengthen regulation in the water sector to address shortcomings in the tariffs and quality of service.

Mugisha and Brown (2010) report significant improvements in Kampala Water's commercial and financial performance that could probably be explained by the high tariffs rather than improved technical efficiency in water distribution systems. The incentive fee formula for Kampala Water reported by the same authors indicate less weight for NRW compared to other key performance indicators such as working ratio (WR). The incentive formula may actually be a disincentive with respect to reducing NRW as the operator may decide not to repair leakages to save on OPEX and have an improvement in WR thus maximizing the cash incentive. There is need to review the existing regulatory framework to exert more pressure on inefficient utilities to improve operating efficiency. The utility incentives for water loss could include rewards and penalties per m^3 of water saved or lost. Some good lessons on the use of DEA in regulation of water distribution utilities can be borrowed from the UK water industry (Thanassoulis 2000b). Water distribution efficiency in the UK has improved tremendously over the years with leakage levels dropping by over 35% from 5,112 ML/d in 1994/95 to now 3,281 ML/d (2009/10) (OFWAT 2010).

3.4.4.2 NRW and OPEX target setting

DEA calculates potential savings in inputs based on the established efficiency frontier. These savings could provide a basis for both the regulator and utility management to set targets on controllable inputs such as NRW and OPEX. For the 25 water utilities examined under this study, NRW and OPEX targets have been calculated (Table 3.12) that could be used in this aspect. The calculated NRW target for Kampala city would translate in NRW reduction from the current level of about 39% to 12%. This is in-line with well-functioning water utilities in other cities of the developing countries that have been reported to have NRW figures of less than 10% of water supplied e.g. Phnom Penh city in Cambodia (ADB 2010). In Phnom Penh city, NRW declined from 70% in 1993 to 6.2 % in 2008, which translates in a reduction rate of 4.3% per year. Similarly, Manila Water reduced NRW from 63% in 1997 to 11 % in 2011, which translates in a reduction rate of 3.7% per year (Luczon and Ramos, 2012). The average rate from these best practice cities for NRW reduction is 4.0% per year. Benchmarking against this efficiency frontier, Kampala would realize its DEA efficient NRW target in about 7 years. This is equivalent to saving 2 million m^3 per year (or US $1.5 million annually) which looks a rather attractive and realistic target to pursue. The reported NRW reduction rate of 2-3% per annum in NWSC-Uganda (Mugisha et al. 2007) is rather low and needs to be upped to match best practices in Asian water utilities.

3.4.4.3 Data quality and reporting

Effective benchmarking requires use of accurate, consistent and reliable data. Whereas progress has been made in data collection and reporting in the Ugandan water sector in the last decade, scrutiny of the water sector review performance reports (2006-2010) freely available on the MWE website, indicate that gaps still exist. The reporting formats vary a lot with more variables being added every year. Some data for particular towns appear this year and disappear next year, some figures do not look realistic (e.g. NRW of 100%, 3% or NRW reducing by over 70% in a single year). NWSC reported data differs from small towns' data and financial/technical quantities can be affected by different accounting conventions and policies. There is need for establishing standard reporting formats, auditing measurements and enforcing consistent reporting to allow for assessing trends in performance and meaningful benchmarking.

3.4.4.4 Model specifications

This study used the DEA method to calculate TE but other methods such as stochastic frontier analysis (SFA) and corrected ordinary least squares (COLS) exist. Each method has its own pros and cons, and there is need to agree on acceptable benchmarking methods by all water sector stakeholders to promote transparency and acceptable outcomes. For example, the DEA BCC (with VRS) model is likely to provide incentives for mergers in order to improve utility scores and rakings without necessarily improving TE of water distribution. Care needs to be exercised when using DEA models especially in specifying the sample size and number of variables to avoid spurious efficiency scores. Benchmarking promotes competition which ensures that: (i) utilities produce all outputs at the minimum cost (productive efficiency), and (ii) these outputs are available to consumers at prices which accurately reflect these minimum costs (allocative efficiency). This study was limited to TE and there is need to assess allocative efficiency, cost efficiency and incorporating quality in DEA models. Quality of service comes at a cost and if not well regulated, utilities driven by profit-making are likely to compromise on quality of service with adverse effect on supply reliability and price of water.

3.5   Conclusions and Recommendations

The conclusions and recommendations of this chapter are summarized in the following sections.

3.5.1   Conclusions

This Chapter has developed a performance assessment system for evaluating and improving WDS efficiency. The performance assessment system consists of a methodology for selecting and developing PIs, a water balance model with PIs (WLA-PI tool) and a DEA-benchmarking model. The system was developed based on the IWA/AWWA water balance and PI-concepts. Although the system was tested and validated using data from Uganda, it is generic and easily adaptable for application in other water utilities of the developing countries. The major findings of the study can be summarized as follows:

- In addition to IWA/AWWA water loss PIs, appropriate PIs for developing countries have been developed that include NRW (m^3/connection/day) and OPEX ($/km/yr).

These two indicators were found to be the most readily available in African water utilities and probably the most cost-effective to generate with limited resources. The reference standard for NRW for African cities' WDSs was established as 0.3 m^3/connection/day. In addition an ALI benchmark of 7% of water sales as the UAAL has been proposed for developing countries and 3% for developed countries;

- A procedure for estimating uncertainty in the water balance model input values and uncertainty propagation into the NRW figures has been established. The study also provides insight into the uncertainty in NRW in Kampala and it indicates that it is heavily influenced by the measurement errors in the SIV. Calibrating the SIV bulk meters is likely to reduce the uncertainty in NRW and improve confidence in the reported NRW figures as a proxy for WDS efficiency.

- Small towns owned by Local Governments and managed by POs have the highest efficient scores with both DEA models (VRS and CRS) compared to NWSC towns, implying that privately managed towns deliver water services more efficiently than publicly managed towns.

- The study reveals that DEA-benchmarking is a powerful tool for evaluating and improving performance in water utilities. The mean overall efficiency score for a sample of 25 water distribution utilities in Uganda was 64%, indicating 36% potential for water and cost savings. In volumetric terms, this is equivalent to annual potential NRW savings of about 15.5 million m^3. In financial terms, this translates into about US \$12 million annually. The overall average reduction potential in cost savings in the NWSC towns is estimated at about US \$0.5 million annually;

- DMUs with high NRW levels also exhibited high OPEX probably due to increased costs of repairs and energy pumping costs;

- Effective and comprehensive water distribution benchmarking and performance improvement requires application of both the frontier-based and PI-based methodologies as they do complement each other.

It can be concluded that, the developed PAS will be a valuable and cost-effective tool for evaluating and improving WDS efficiency. The DEA-benchmarking results provide basis for utility managers and regulators to develop intervention policies and strategies for improving operating efficiency in water utilities of the developing countries where the problem of water loss is more prominent. Lastly, the results from this study should be interpreted in the context of the limitations imposed by the available data quality and the methodology applied.

3.5.2 Recommendations

The following recommendations for improving water distribution evaluation and performance are proposed:

1. The decision-makers in the Uganda water sector should further examine the policy implications of this study highlighted in section (§3.4.4) and make appropriate interventions required for efficient water distribution.

2. There is need to extend benchmarking of water distribution to regional and international utilities to find comparable utilities particularly for Kampala city that

seemed an outlier on the national scene in terms of size. Good lessons could be learned from international peer groups like Phnom Penh city in Cambodia, Asia.

3. The aim of this study was to assess the TE of water distribution and identify potential improvements. For completeness of benchmarking, future studies should include allocative measures which reflect input prices and cost efficiency as technically efficient utilities may not necessarily be cost efficient. In addition, to safeguard against asset stripping and ensure asset serviceability, it is important that future models integrate total costs and quality of service in benchmarking and incentive regulation of WDSs. Future evaluation using a data set for more than one year would provide more insight into the evolution of performance of the water utilities.

3.6 References

ADB. (2010). "Every Drop Counts: Learning from Good Practices in Eight Asian Cities." Asian Development Bank, Manila.

Alegre, H., Baptista, J. M., Cabrera, E. J., Cubillo, F., Hirner, W., Merkel, W., and Parena, R. (2006). *Performance Indicators for Water Supply Services, IWA Manual of Best Practice*, IWA Publishing.

Alegre, H., Cabrera Jr, E., and Merkel, W. (2009). "Performance Assessment of Urban Utilities: the case of water supply, wastewater and solid waste." *Journal of water supply: Research and Technology-AQUA*, 58(5), 305-315.

Alsharif, K., Feroz, E. H., Klemer, A., and Raab, R. (2008). "Governance of water supply systems in the Palestinian Territories: A data envelopment analysis approach to the management of water resources." *Journal of Environmental Management*, 87, 80-94.

AWWA. (2003). "Committee report: Applying worldwide BMPs in water loss control." *Journal AWWA*, 95(8), 65-79.

AWWA. (2009). "Water Audits and Loss Control Programs: AWWA Manual M36." American Water Works Association, Denver, USA.

Banerjee, S. G., and Morella, E. (2011). "Africa's Water and Sanitation Infrastructure." The World Bank, Washington, DC, USA.

Banker, R. D., Charnes, A., and Cooper, W. W. (1984). "Some models for estimating technical and scale inefficiencies in data envelopment analysis." *Management Science*, 30(9), 1078-1092.

Bargiela, A., and Hainsworth, G. D. (1989). "Pressure and Flow Uncertainity in Water Systems." *Journal of Water Resources Planning and Management*, 115(2), 212-229.

Beasley, J. E. (1995). "Determining Teaching and Research Efficiencies." *Journal of Operational Research Society*, 46, 441-452.

Berg, S., and Marques, R. (2011). "Quantitative studies of water and sanitation utilities: a benchmarking literature survey." *Water Policy*, 13, 591-606.

Bidgoli, A. M. (2009). "Water losses reduction programme in Iran." *Proceedings of international workshop on drinking water loss reduction: Developing capacity for applying solutions*, UN Campus, Bonn, Germany, 80-87.

Brueck, T. M. (2005). *Developing and Implementing a Performance Measurement System* IWA Publishing, London.

Charnes, A., Cooper, W. W., and Rhodes, E. (1978). "Measuring efficiency of decision-making units." *European Journal of Operational Research*, 2(6), 428-449.

Coelho, T. (1997). "Performance Indicators in Water Supply—A Systems Approach." Series Water Engineering and management systems, Research Studies Press, John Wiley and Sons, UK.

Coelli, T., Estache, A., Perelaman, S., and Trujillo, L. (2003). *A Primer on Efficient Measurement for Utilities and Transport Regulators*, World Bank Institute, Washington, D.C, USA.

Cook, W. D., Seiford, L. M., and Zhu, J. (2004). "Models for performance benchmarking: measuring the effect of e-business activities on banking performance." *Omega: The International Journal of Management Science*, 32, 313-322.

Cooper, W. W., Seiford, L. M., and Tone, K. (2000). *Data Envelopment Analysis: A Comprehensive Text with Models, Applications, References and DEA-Solver Software.*, Kluwer Academic Publishers, Boston.

Crotty, P. (2004). *Selection and Definition of Performance Indicators for Water and Wastewater Utilities*, Awwa Research Foundation, A merican Water Works Association (AWWA), Denver, Colarado, USA.

Cubbin, J., and Tzanidakis, G. (1998). "Regression versus data envelopment analysis for efficiency measurement: an application to the England and Wales regulated water industry." *Utilities Policy*, 7, 75-85.

De Witte, K., and Marques, R. C. (2010). "Designing performance incentives, an international benchmark study in the water sector." *Central European Journal of Operational Research*, 18, 189-220.

Deb, A. K., Hasit, Y. J., and Grablutz, F. M. (1995). *Distribution System Performance Evaluation*, AWWA Research Foundation and AWWA, Denver, USA.

Dimaano, I., and Jamora, R. (2010). "Embarking on the World's Largest NRW Management Project." *Proceedings of the 6th IWA Water Loss Reduction Specialist Conference*, Sao Paulo, Brazil, CD-ROM.

Dyson, R. G., Allen, R., Camanho, A. S., Podinovski, V. V., Sarrico, C., and Shale, E. A. (2001). "Pitfalls and protocols in DEA." *European Journal of Operational Research*, 132(2), 245-259.

Fanner, P., Thornton, J., Liemberger, R., and Sturm, R. (2007). *Evaluating Water Loss and Planning Loss Reduction Strategies*, Awwa Research Foundation, AWWA, Denver, USA

Flores, and Diaz. (2009). "Meter assessment in Madrid." *Proceedings of the 5th IWA Specialist Conference on Efficient Water Use and Management*, Sydney, Australia, October 26-28 CD-ROM.

Garzon-Contreras, F., Uribe-Preciado, A., Yepes-Enriquez, L., and Agredo-Perdomo, A. (2009). "Unauthorized consumption: The key component of Apparent Losses in Colombia's major cities." *Proceedings of the 5th IWA Water Loss Reduction Specialist Conference*, Cape Town, South Africa, 29-35.

Giannakis, D., Jamasb, T., and Pollit, M. (2005). "Benchmarking and incentive regulation of quality of service: an application to the UK electricity distribution networks." *Energy Policy*, 33, 2256-2271.

Gottfried, B. S. (2007). *Spreadsheet Tools for Engineers Using Excel* McGraw Hill, New York.

Guibentif, H., Rufenacht, H. P., Rapillard, P., and Ruetschi, M. (2007). "Acceptable level of water losses in Geneva." *Proceedings of the 4th IWA Water Loss Reduction Specialist Conference*, Bucharest, Romania, 138-147.

Herrero, M., Cabrera Jr, E., and Valero, F. J. (2003). "A New Approach to Assess Performance Indicators' Data Quality." Pumps, Electromechanical Devices and Systems Applied to Urban Water Management Systems, E. Cabrera and E. Cabrera Jr., eds., Swets & Zeitlinger, Lisse, The Netherlands, 69-78.

ISO. (2007c). "ISO 24512:2007-Service Activities Relating to Drinking Water and Wastewater - Guidelines for the Management of Drinking Water Utilities and for the Assessment of Drinking Water Services. Technical Committee TC 224, International Organization for Standardization." Geneva.

ISO/IEC. (2008). "Uncertainty of Measurement - Part 3: Guide to expression of uncertainty in measurement (GUM:1995)." International Organization for Standardization(ISO)/International Electrotechnical Commission (IEC), Geneva, Switzerland.

Jamasb, T., and Pollit, M. (2001). "Benchmarking and regulation: international electricity experience." *Utilities Policy*, 9, 107-130.

Kanakoudis, V., and Tsitsifli. (2010). "Results of an urban water distribution network performance evaluation attempt in Greece." *Urban Water Journal*, 7(5), 267-285.

Kontodimopoulos, N., and Niakas, D. (2005). "Efficiency measurement of hemodialysis units in Greece with data envelopment analysis." *Health Policy*, 71, 195-204.

Lambert, A. O., Brown, T. G., Takizawa, M., and Weimer, D. (1999). "A review of performance indicators for real losses from water supply systems." *Aqua- Journal of Water Services Research and Technology*, 48(6), 227-237.

Liemberger, R., and McKenzie, R. (2003). "Aqualibre: A New Innovative Water Balance Software." IWA Efficient 2003 Conference, IWA, Tenerife.

Luczon, L. C., and Ramos, G. (2012). "Sustaining the NRW reduction strategy: The Manila Water Company Territory Management Concept and Monitoring Tools." *Proceedings of the 7th IWA Water Loss Reduction Specialist Conference*, Manila, Philippines, Feb 26-29, 2012.

Marin, P. (2009). "Public-Private Partnerships for Urban Water Utilities: A Review of Experiences in Developing Countries." The World Bank, Washington DC.

Marques, R. C., and Monteiro, A. J. (2003). "Application of performance indicators to control losses-results from the Portuguese water sector." *Water Science and Technology: Water Supply*, 3(1/2), 127-133.

Mauris, G., Lasserre, V., and Foulloy, L. (2001). "A fuzzy approach for expression of uncertainity in measurement." *Measurement*, 29, 165-177.

McKenzie, R., Seago, C., and Liemberger, R. (2007). "Benchmarking of Losses from Potable Water Reticulation Systems - Results from IWA Task Team." *Proc. of the 4th IWA Specialised Water Loss Reduction Conference, September 23-26*, Bucharest, Romania, 161-175.

Mugisha, S. (2007). "Effects of incentive applications on technical efficiencies: Empirical evidence from Ugandan water utilities." *Utilities Policy*, 15, 225-233.

Mugisha, S., Berg, S. V., and Muhairwe, W. T. (2007). "Using Internal Incentive Contracts to Improve Water Utility Performance: The Case of Uganda's NWSC." *Water Policy*, 9, 271-284.

Mugisha, S., and Brown, A. (2010). "Patience and action pays: a comparative analysis of WSS reforms in three East African Cities." *Water Policy*, 12, 654-674.

Mutikanga, H. E., Sharma, S., and Vairavamoorthy, K. (2009). "Water Loss Management in Developing Countries: Challenges and Prospects." *Journal AWWA*, 101(12), 57-68.

MWE. (2010). "Water and Environment Sector Performance Report." Ministry of Water and Environment (MWE), Kampala, Uganda.

NWSC. (2010). "NWSC Annual Performance Report for Financial Year 2009-2010." National Water and Sewerage Corporation, Kampala, Uganda.

OFWAT. (2010). "Service and delivery-performance of the water companies in England and Wales 2009-10 report ", OFWAT, UK.

Park, H. J. (2006). "A Study to develop strategies for Proactive Water Loss Management," Ph.D Thesis, Georgia State University, USA.

Pelletier, G., Mailhot, A., and Villeneuve, J. P. (2003). "Modeling water pipe breaks-three case studies." *Journal of Water Resources Planning and Management*, 129(2), 115-123.

Picazo-Tadeo, A., J, Saez-Fernandez, F. J., and Gonzalez-Gomez, F. (2008). "Does service quality matter in measuring the performance of water utilities?" *Utilities Policy*, 16, 30-38.

Picazo-Tadeo, A., J, Saez-Fernandez, F. J., and Gonzalez-Gomez, F. (2011). "Assessing performance in the management of the urban water cycle." *Water Policy*, 13, 782-796.

Podinovski, V. V., and Thanassoulis, E. (2007). "Improving discrimination in data envelopment analysis: some practical suggestions." *Journal of Productivity Analysis*, 28, 117-126.

Ringskog, K., Hammond, M. E., and Locussol, A. (2006). "Using management and lease-affermage contracts for water supply." Gridlines Note No. 12, September 2006, PPIAF/WorldBank, Washington, DC.

Rizzo, A., M., V., Galea, S., Micallef, G., Riolo, S., and Pace, R. (2007). "Apparent Water Loss Control: The Way Forward." *IWA Water 21, August*

Romano, G., and Guerrini, A. (2011). "Measuring and comparing the efficiency of water utility companies: A data envelopment analysis approach." *Utilities Policy*, 19, 202-209.

Sattary, J., Boam, D., Judeh, W. A., and Warren, S. (2002). "The Impact of Measurement Uncertainty on the Water Balance." *Water and Environment Journal*, 16(3), 218-222.

Schouten, M., and Halim, R. D. (2010). "Resolving strategy paradoxes of water loss reduction: A synthesis in Jakarta." *Resources Conservation and Recycling*, 54, 1322-1330.

Sharma, S. K., and Chinokoro, H. (2010). "Estimation of ELL and ELWL for Lusaka Water Distribution System." *Proceedings of the 6th IWA Water Loss Reduction Specialist Conference (CD-ROM)*, Sao Paulo, Brazil.

Sherman, H. D., and Zhu, J. (2006). "Benchmarking with quality-adjusted DEA (Q-DEA) to seek lower-cost high-quality service: Evidence from a U.S. bank application." *Annals of Operations Research*, 145(1), 301-319.

Singh, M. R., Upadhyay, V., and Mittal, A. K. (2010). "Addressing sustainability in benchmarking framework for Indian urban water utilities." *Journal of Infrastructure Systems (ASCE)*, 16(1), 81-92.

Thanassoulis, E. (2000a). "DEA and its use in the regulation of water companies." *European Journal of Operational Research*, 127, 1-13.

Thanassoulis, E. (2000b). "The use of data envelopment analysis in the regulation of UK water utilities: Water distribution." *European Journal of Operational Research*, 126, 436-453.

Thanassoulis, E. (2001). *Introduction to the Theory and Application of Data Envelopment Analysis: A Foundation Text with Integrated Software*, Kluwer Academic Publishers, Boston, USA.

Thanassoulis, E., Boussofiane, A., and Dyson, R. G. (1996). "A comparison of Data Envelopment Analysis and Ratio Analysis as Tools for Performance Assessment." *Omega:The International Journal of Management Science*, 24(3), 229-244.

Thornton, J., Sturm, R., and Kunkel, G. (2008). *Water Loss Control*, McGraw-Hill, New York.

Tupper, H. C., and Resende, M. (2004). "Efficiency and regulatory issues in the Brazilian water and sewage sector: an empirical study." *Utilities Policy*, 12, 29-40.

WSP. (2009). "Water Operators Partnerships: African Utility Performance Assessment." Water and Sanitation Program (WSP) - Africa, The World Bank, Nairobi, Kenya.

Zhu, J. (2009). *Quantitative Models for Performance Evaluation and Benchmarking: Data Envelopment Analysis with Spreadsheets*, Springer, Boston, USA.

Chapter 4 - Water Meter Management for Reduction of Revenue Losses

Publications based on this chapter:

Mutikanga, H.E, Sharma, S.K., and Vairavamoorthy, K., 2011. "Investigating Water Meter Performance in Developing Countries: A Case Study of Kampala, Uganda". *Water SA*, 37(4), 567-574.

Mutikanga, H.E, Vairavamoorthy, K., Kizito, F., and Sharma, S.K., and (2011). "Decision Support Tool for Optimal Water Meter Replacement". *Proceedings of the Second International Conference on Advances in Engineering and Technology (AET 2011),* Entebbe, Uganda, Jan 30-Feb 1, pp. 649-655, ISBN 978-9970-214-00-7.

Mutikanga, H.E, Nantongo, O., Wozei,E., Sharma,S.K., and Vairavamoorthy,K., (2009). "Investigating the Impact of Utility Sub-metering on Revenue Water". *Proceedings of the Second International Conference on Advances in Engineering and Technology (AET 2011)*, Entebbe, Uganda, Jan 30-Feb 1, pp.633-639, ISBN 978-9970-214-00-7.

Mutikanga, H.E, Sharma, S.K., and Vairavamoorthy,K., (2010). "Customer Demand Profiling for Apparent Water Loss Reduction". *Proceedings of the 6th IWA Water Loss Reduction Specialist Conference*, June 6-9, Sao Paulo, Brazil, CD-ROM Edition.

Mutikanga, H.E, Nantongo, O., Wozei,E., Sharma,S.K., and Vairavamoorthy,K., (2009). "Assessing Meter Accuracy for Reduction of Non-revenue Water". *Proceedings of the IWA's Efficient Water Use Conference*, October 25-28, Sydney, Australia, CD-ROM Edition.

Summary

The water meter is an essential utility tool for effective utility revenue generation, equitable customer billing, water demand management, and generating data for network planning and management. When metering is inefficient, all benefits associated with metering are lost. Revenue losses caused by metering inefficiencies can be reduced by assessing water meter performance, identifying the main causes of losses and applying appropriate intervention measures. Using data from the Kampala water distribution system, this study examined water meter performance and developed an integrated water meter management (IWMM) framework to help water utilities minimize revenue losses due to metering inaccuracies. The framework covers all aspects of meter management from meter selection to optimal meter replacement. The influence of customer water use patterns, private elevated storage tanks, and sub-metering on meter accuracy were also investigated. From this platform of knowledge and meter testing results, optimal meter sizing and renewal decisions are made. Guidelines for quantifying water loss due to metering errors and failure are also established. Specifically customer meters of size 15 mm which constitute a large proportion of utility meters are the main focus of this study. The main findings of the study were high metering errors (-21.5%) and high meter failure (6.6%/year) in the Kampala water distribution system. The metering errors are magnified by sub-metering and the ball-valve effect that induce very low flow rates through the meter. The high meter failure was due to use of inappropriate metering technology, inadequate system operation and maintenance particularly poor repair practices, irregular supply and inadequate rehabilitation measures. The water losses due to the DN 15 mm metering inaccuracies and failure were estimated to be 38.2% of the global NRW (or 8,160 ML/year). The findings suggest several policy implications and recommendations for the utility to help address the need for better water meter management in developing countries.

4.1 Introduction

As discussed in the previous chapter, even when water is delivered to metered customers not all of it is measured due to metering errors. This chapter addresses the problem of customer metering inaccuracies which is one of the water loss components presented in Fig. 2.2 of Chapter 2. Apparent losses resulting from meter inaccuracies and poor water meter management can be reduced by assessing meters performance and identifying the main causes of mal-functioning.

Mechanical water meters' metrology become more and more inaccurate during their operating life due to "wear and tear" of the measuring components (Arregui et al. 2006b; Male et al. 1985). However, most studies carried out on water metering (chapter 2) have been reported in water utilities of the developed countries with well-managed distribution systems notably in the USA (Barfuss et al. 2011), Australia (Egbars and Tennakoon 2005), Spain (Flores and Diaz 2009), France (Pasanisi and Parent 2004), and Italy (Criminisi et al. 2009). Water meter performance in water systems of the developing countries, often with poorly managed networks and relatively lower water quality in the distribution system is not very well understood. This study attempts to close the knowledge gap by investigating water utility metering problems in developing countries, using a case study of KWDS.

There are four main drivers for universal metering namely: (i) equity, (ii) water use efficiency, (iii) economic benefits, and (iv) system management (Van Zyl 2011). However, universal water metering and sub-metering has been widely used as a tool to promote water conservation mainly in the USA. Water use reduction in the range of 10 to 30% has been reported (AWWA 2000). Nevertheless, water usage and wastage remain high in most cities of the developing countries despite universal and sub-metering metering. The term "sub-metering" refers to any metering that occurs downstream a water utility's master meter to measure individual resident water usage in apartments, condominiums, mobile home parks, and small mixed commercial properties (AWWA 2000). Universal metering and sub-metering, however, have not brought the much anticipated benefits to water utilities in developing countries most likely due to metering inefficiencies.

Although the water utility in Kampala has had a universal metering policy since 1990, there is no complementary policy on meter management (selection, installation, sub-metering, maintenance, replacement etc.) and this has lead to the installation of a variety of meter models which have degraded in accuracy over the years. There is an increase of customer billing based on sub-metering although its impact from the utility perspective of increasing revenue water is not very well understood. Most water utilities if not all supply water irregularly in the developing countries. As a result storage tanks are improvised by customers to bridge the gap between supply and demand. In Kampala city, over 80% of the customers have elevated storage tanks with regulating ball-valves. The influence of these tanks on meter performance is not well understood. There are no guidelines on how to install meters and sizing decisions are based on rules of thumb. With increasing revenue demand for service expansion and infrastructure rehabilitation coupled with water scarcity, optimum measures are required for accurate water measurement and accountability. In order to develop proactive measures for water meter management, it is essential to identify key factors that influence water meter accuracy and failure.

The main objective of this study was to develop an integrated water meter management (IWWM) model to help water utilities understand how different aspects of water metering can be integrated to maximize utility revenue water. The specific objectives of the study were:

1. To establish customer water use patterns for optimal metering decisions
2. To analyze the customer meter performance of size DN 15 mm
3. To investigate the effect of private elevated storage tanks on meter accuracy
4. To investigate the effect of sub-metering on revenue water
5. To establish guidelines for estimating water losses due to metering inaccuracies of working meters and meter failure (stuck meters)
6. To determine the most suitable meter model for the KWDS
7. To demonstrate optimal meter sizing based on demand profiling techniques
8. To develop a decision support tool for optimal meter replacement frequency

The conceptual model for IWMM for maximising utility metering benefits is illustrated in Figure 4.1.

The rest of the chapter is organized as follows. Section 4.2 outlines the materials and methods used in the study. Section 4.3 presents and discusses the meter performance results. The estimation of water loss due to working and failed meters is presented in section 4.4. Optimal meter sizing and selection is discussed in section 4.5. The optimal meter replacement frequency model is presented in section 4.6. Finally, section 4.7 concludes the research findings and discusses policy implications of the research and suggests some recommendations for effective sub-metering.

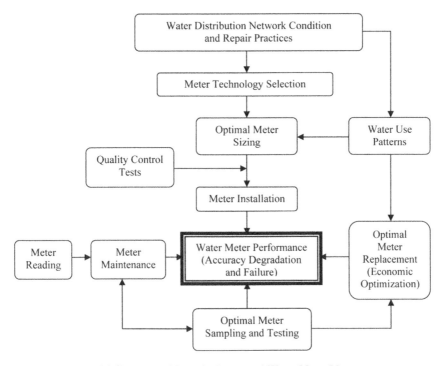

Figure 4.1 Conceptual Model for Integrated Water Meter Management

4.2   Materials and Methods

The main activities to assess water meter management included: (i) sampling of meters and properties, (ii) in-situ data logging, (iii) analysis of effect of sub-metering, and (iv) laboratory studies. The methodology for each of these activities is elaborated in the following paragraphs.

The methodology used in this study for sampling and meter testing has been used by many researchers (Allander 1996; Arregui et al. 2006a; Male et al. 1985; Newman and Noss 1982; Rizzo and Cilia 2005; Tao 1982; Yee 1999). The method for estimating weighted meter accuracy was adapted from Arregui et al. (2006a) and is summarized in Figure 4.2. This methodology is used for small customer meters and is also applicable to large customers with each user's demand profile assessed separately.

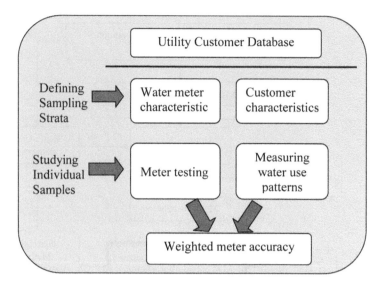

Figure 4.2 Methodology for determining weighted meter accuracy
[Source: Adapted from Arregui et al. (2006a)]

4.2.1 Sampling meters and properties

There are about 150,000 installed customer meters and over 1.5 million users in Kampala city. Testing all of these meters and measuring consumption patterns can only be realized by defining and establishing representative samples.

Meter sampling for accuracy estimation

In this study only meters of size 15 mm were considered as they constitute about 92% of all customer meters installed in the KWDS. For this size of meters, there are two types of in-service meter models in the KWDS: velocity type meters (multi-jet and single-jet) and volumetric or displacement (oscillating-piston type) meters. Three different meter models were examined; namely, two types of volumetric meter models (model 1 and 2) from two different manufacturers and the multi-jet meter type (model 3). Models 1 and 2 were the most dominant in the network and make up 76% of all the small meters of size 15 mm; 24% are of the velocity type. Out of the total meters of size DN 15 mm, model 1 makes up 43%, model 2, 41% and 16% is model 3. The average age of the meters is about 10 years with some meters being more than 20 years old. Single-jet meters were not considered for this study as they have only recently been introduced (< 1 year prior to the time of this study).

Statistical sampling techniques (stratified random and cluster sampling) were applied in determining the accuracy of meters. Meters were grouped based on the billing index or throughput. The billing index is the meter odometer reading when the meter is removed from the field. The expected meter life for a half-inch meter is when the odometer reading clocks 8000 m^3 (Egbars and Tennakoon 2005). Meters were grouped to enable assessing accuracy of each meter at 25, 50, 75 and 100 percent of its meter life, which is five sub-groups (0-2000 m^3, 2000-4000m^3, 4000-6000m^3, 6000-8000 m^3 and more than 8000 m^3). Although most previous researchers grouped meters by age (Criminisi et al. 2009; Yee 1999), this was not possible in the case of this study due to inconsistent and unreliable data on when meters were first installed in the field. It was preferred to use volumes which could be easily verified. In any case it is usage and not age that affects meter accuracy and it may be prudent to measure the life of a meter by total consumption rather than by time (AWWA 1999; Hill and Davis 2005; Wallace and Wheadon 1986).

The larger the sample, the higher the cost of sampling, but the more reliable the statistical inferences derived from the sampling. A trade-off had to be made between the numbers of samples that would yield meaningful data versus the number that could be realistically tested within the available resource envelope. In this case two sources of data were identified: (i) the utility water meter management database and (ii) removing additional in-service water meters and examining them. Meters were selected to ensure fair representation of the entire population based on meter type, usage and other system characteristics (water quality, pressure, regular or irregular supply etc.). To do this, the samples were picked from different parts of the network to create a homogeneous population. The more homogeneous, or well-mixed the entire sampling population is, the fewer the samples needed to obtain a good representation.

Water meter accuracy tests that were carried out between 2007 and 2009 were obtained from the utility's meter laboratory database for analysis. The database that was analysed consisted of 2698 meters of size 15 mm. In addition 100 in-service meters were pulled and tested making a total number of 2798 meters examined during the study.

Of these, 980 (35%) were Model 1 meters, 839 (30%) were Model 2 and 895 (32%) were Model 3 meters. Although the samples do not fully represent the meter distribution in the network, they still provide insight into the meter performance in the KWDS. The other 3% were other meter models of same technology but different manufacturers and were not considered for the study.

Upon stratification and screening only 515 meters were selected for the study (Table 4.1). Screening of the data in each stratum involved eliminating of accuracy tests not having all the test ranges. Accuracy tests reading errors higher than -90% and 2% were outliers and therefore assumed to be faulty readings and were purged from the sample. Tests not having the cumulative volume through the meter were also eliminated. This method of screening left only complete and reasonable results in each of the meter groups.

Table 4.1 Sample description by total registered volume

Billing Index (m^3)	Model 1	Model 2	Model 3
0 - 2000	115	55	72
2000 - 4000	85	10	45
4000 - 6000	42	3	20
6000 - 8000	28	4	15
> 8000	9	2	10
Total	279	74	162

Property (Users) Sampling

In order to minimize cost of data acquisition while minimizing uncertainties in sampling users' consumption patterns, classification of users by type was done. Four types of users were identified: (i) public standpipes or yard taps with direct supply, (ii) domestic users with elevated storage tanks, (iii) institutions with elevated storage tanks and (iv) commercial users with elevated storage tanks. To ensure a statistically representative sample per group, a total of 100 users' (25 per group) water consumption profiles were targeted for the study.

4.2.2 In-situ measurements

Data logging

The customer water use profiles were measured using high precision master meters of excellent metrology and data loggers that continuously recorded the readings from the meter. The key to obtaining accurate flow data is generating a sufficient number of pulses per time interval (AWWA 2004).

During the study, the following equipment was used for measurement of the various parameters:

- Very accurate "Volumetric positive displacement" type master meters with a low start up flow (Class D, Q_n = 1.0 to 1.5 m^3/h size 15 mm with starting flow rate of 1 L/h). These meters are equipped with pulse emitters with minimum resolution of 0.1 L/pulse. Meter Class (A,B,C,D) refers to the ISO classification for water meters and indicates the ability of the meter to measure low flows. Class D meters have the greatest ability to measure low flows and Class A have the least ability.
- Sensus Cosmos Data loggers (CDL-4U) with extended memory capacity and high resolution sensors designed for water metering analysis. One-week data loggings were carried out via pulse count on a 10-second interval.
- The Sensus free software (CDL WIN 3.5) was used for data retrieval from data loggers to the computer for analysis.
- The Hydreka Vistaplus pressure data loggers (OCTC511LF/30).

The actual data logging arrangement in the field for a single-family residential is shown in Figure 4.3.

Figure 4.3 Installation set-up of logging equipment

Private elevated storage tanks

In order to assess their impact on meter accuracy, data loggers with master meters of high accuracy were placed in series at the tank inlet and outlets as shown in Figure 4.4.

First, the households (HHs) were metered in two different positions (B and C) using new class D meters of size 15 mm. The two meters were data logged. The first meter (meter A) was the in-service revenue meter and measured the inflow of water into the complete HH. The second meter (meter B) measured the inflow at the roof tank inlet and the third meter (meter C) measured the flows at the roof tank outlet. The tank sizes for both HHs were each 1000 litres.

Daily readings of the two meters were recorded for a period of 7 days and the volumes through the two meters B and C were obtained. The difference between the volumes measured by the two meters was expressed as a percentage of the volume measured by outlet meter C. This was taken as the metering error due to the ball-valve effect of the storage roof tanks. The process was repeated using a 15-year-old meter with billing index of 9,056 m^3 (Class C volumetric meter, DN = 15 mm, Q_n = 1.5 m^3/hr) at the tank inlet (position B) to assess impact as the meter loses accuracy with usage over time. The water levels in the tank before each meter reading cycle was measured using a calibrated dip-stick to ensure the volume was the same.

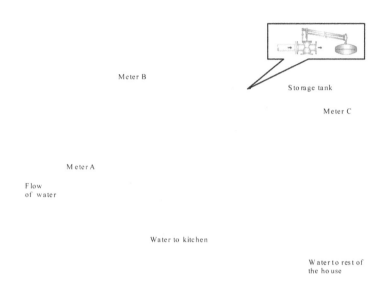

Figure 4.4 Experimental set-up for household storage tanks

Sub-metering

A sample of eight sub-metered properties was selected from four Operational Branches in different parts of the KWDS to ensure a well-mixed homogeneous population that is a representative of all sub-metered properties in the network. The properties were of varying sizes with the smallest having six sub-meters and the largest 34 sub-meters (Table 4.7). The total number of sub-meters for the eight properties was 137, comprised of 119 residential sub-meters and 18 sub-meters for the single commercial property (mainly supermarket and office blocks). The property references (Table 4.7) indicate actual network block maps where the properties are exactly located. The sub-meters examined in this study were of size DN 15 mm while the master meters were of mixed small sizes of DN 15 mm and DN 20 mm (Table 4.7). All of the in-service sub-meters investigated were of Class C and B multi-jet and positive displacement (piston-type meters). The distance between the master meter and the sub-meters varied from 0.5 m to 15 m depending on the premises set-up.

To assess the impact of sub-metering on meter accuracy, new monitoring master meters were installed (where non-existent) in series with property sub-meters for individual tenants as indicated in Figure 4.5.

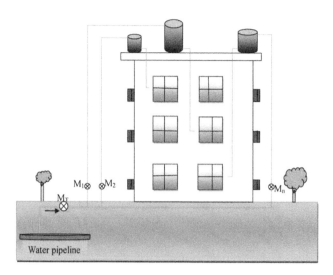

Figure 4.5 Sub-metering Illustration

where M_T is the master meter and M_1, M_2, and M_n are the in-service sub-meters for the individual property units from 1 to n. Weekly and monthly meter readings were taken for the master meter and the sub-meters for a period of four months. This was done after thorough investigations to eliminate any leaks, illegal use and meter by-pass between the master meter and the sub-meters. The difference between the initial and final readings for each of the meters was obtained and taken as the volume of water consumption through the meter. The total water consumption of all the sub-meters for each of the properties was obtained and compared to the master meter consumption for the same time period. The total difference for all of the properties was obtained and expressed as a percentage of the master meter total volume; this was taken as the metering error due to sub-metering.

In order to understand the cause of the difference in registered consumptions, the smallest property with six sub-meters was examined in details. Meter accuracy tests for the master meter and sub-meters were carried out to determine their level of accuracy using the utility's meter test-benches. The volumes that were registered by these meters were then corrected using the calculated individual meter errors in accordance with the water balance equation, Eq. (4.1). Thereafter the master meter and all the sub-meters were replaced by highly accurate meters (Class D, $Q_n = 1.0$ to 1.5 m^3/h, size DN 15 mm) and the experiment was repeated to eliminate any influence of the meter age on the metering error.

$$Q_{\text{master meter}} + Errors_{\text{master meter}} = \sum_{i=1}^{n} Q_{\text{consumer meter}} + \sum_{i=1}^{n} Errors_{\text{consumer meter}} \qquad (4.1)$$

where, $Q_{\text{master meter}}$ = Volume measured by the monitoring master meter, $Errors_{\text{master meter}}$ = Errors of the master meter, $Q_{\text{consumer meter}}$ = Volume measured by the separated meters, and $Errors_{\text{consumer meter}}$ = Errors of the sub-meters.

Simultaneous customer demand profiling for the master meter and sub-meters was carried out to investigate whether customer demand patterns had an influence on the metering error due to sub-metering.

Pressure measurements were also carried out at the properties to investigate whether pressure differences also have an impact on the metering error due to sub-metering. To determine the effect of pressure on the metering error, the set-up used to determine the metering error due to individual sub-meters (using Class D meters) was repeated under different network residual pressures at the metering points (5 m and 20 m) in collaboration with the utility's water supply department.

4.2.3   Laboratory studies

Water meter testing

The metrological requirements (water meter accuracy) for any type of water meter are given by the mean of four flow rates (EN-14154-1:2005+A1 2007; ISO-4064-1 2005). These are indicated on the standard meter error curve in Figure 4.6, clearly indicating the minimum flow rate (Q1), transitional flow rate (Q2), permanent flow rate (Q3) or nominal flow rate (ISO 4064:1993) and the overload flow rate (Q4) or maximum flow rate (ISO 4064:1993) . The transitional flow rate occurs between the permanent and minimum flow rates, and divides the flow rate range into two zones: the upper zone with maximum permissible error of ±2% and the lower zone with maximum permissible error of ±5%.

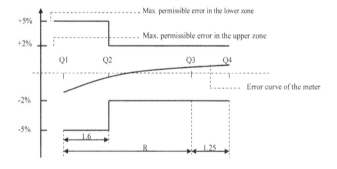

Figure 4.6 Standard water meter error curve (ISO 4064-1, 2005)

The measurement characteristics of a water meter are defined by:

- Q3: continuous flow rate
- the dynamic range "R": Q3/Q1 ratio (where Q1 is minimum flow rate)

with secondary characteristics of:

- Q2: transition flow rate defined by Q2/Q1=1.6
- Q4: overload flow rate defined by Q4/Q3 =1.25

The dynamic range "R" is a useful parameter for evaluating water meter efficiency. However, for proper evaluation of meter efficiency, it should be used in combination with the meter error curve and user consumption profile. A meter with a lower "R" could have higher or comparable efficiency with a meter with a higher "R" depending on the shape of the error curve and the consumption profile.

Meter testing was carried out using a volumetric calibration meter test-bench (Figure 4.7) in accordance with ISO 4064-3 (1993) test methods and equipment.

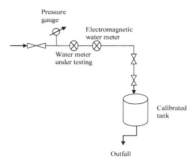

Figure 4.7 A schematic diagram of a water meter test-bench

The test-bench consists of a water supply feed tank, pipe works, a calibrated tank for precise volume measurement, a pressure gauge, and a reference electromagnetic water meter capable of measuring very low flow rates (< 5 L/h). The meter placement in the test section is consistent with ISO 4064-2 (2005) and EN 14154-2:2005 + A1 (2007). The mandatory ISO 4064:1993 flow rates used for testing the different metrological classes are indicated in Table 4.2. Meters with equal capacity (Q3) will have different minimum and transitional flow rates depending on the metrological class. It is important to note that since we are testing old in-service meters, the old version of the ISO standard for water meter testing is applicable to meter models approved before October 2006 until 2016 (Arregui et al. 2006b).

Table 4.2 Flow test rates (ISO-4064-3, 1993).

Class	Low (L/h)	Medium (L/h)	High (L/h)
B	31.5	126	2850
C	15.8	23.6	2850

The problem with the ISO and EN standards is that they are defined for new meters only and they do not define the starting flow of meters. In this study, the ISO and EN standards were modified to include small flow rates that are vital for studying the performance of old meters. Metrological tests for a few meter samples were performed at 11 different flow rates (3.75, 7.5, 15, 22.5, 30, 120, 185, 375, 750, 1500 and 3000 L/h) to depict as closely as possible the customer profiles, including the meter starting flow rate (Q_s). The number of samples subjected to these 11 flow tests was limited due to the long flow times required to measure the extremely small flow rates and minimize uncertainty in the measurements.

The error ε, expressed as a percentage, is computed using Equation 4.2.

$$\varepsilon = \frac{V_m - V_a}{V_a} * 100 \qquad (4.2)$$

where, V_m is the measured (registered) volume and V_a is the actual volume.

Analysis of meter failure data

Meter failure records in the laboratory were also analyzed. However, no records were kept on meter failure until 2007. The 2007-2009 failure data was lacking in detail necessary for a meaningful analysis. The data details required included:

1. The meter number and make.
2. The date of meter installation and date of removal from service.
3. The billing index at time of removal from service.
4. The type and cause of failure (e.g. stuck piston).

As a result, the data used for this study was the most recent data i.e. that for the 2010 calendar year.

4.2.4   Weighted meter accuracy

The weighted meter accuracy (*WMA*) is computed by integrating the accuracy of a meter at each flow range with amount of water used at that flow range as shown in Equation 4.3.

$$WMA = x\%.A(Q1) + y\%.A(Q2) + z\%.A(Q3) \qquad (4.3)$$

where, x, y and z are the fractions of the volumes consumed at flow rate ranges of $Q1$, $Q2$ and $Q3$ respectively. The meter accuracies $A(Q1)$, $A(Q2)$ and $A(Q3)$ are the meter accuracies at minimum flow test ($Q1$), medium flow test ($Q2$) and high flow test ($Q3$) respectively. It is important to note that ($x+y+z$) may not be equal to 1 because meters do not register at some flows.

In order to obtain the final average accuracy of the installed meters, the multiple subgroups of water consumption patterns and water meter classes should be weighted appropriately. The weighting coefficients should be calculated as follows (Arregui et al. 2006b):

- percentage of meters in the group over the total number of meters;
- percentage of volume registered by meters in the group over total registered volume.

4.2.5   Data analysis

Sampling and flow measurements are all subject to uncertainties which must be computed in order to draw conclusions about the results. Data generated in this study for field and laboratory measurements was analyzed using MS/Excel statistical data analysis tools (Gottfried 2007) and the ISO/IEC (2008) guide to expression of uncertainty in measurement (ISO/IEC 2008). Meter tests were repeated twice at high and medium flow rates and three times at low flow rates where variability and thus uncertainty is much higher. The commonly used data set parameters (mean, variance, standard deviation and standard error) were examined in order to draw conclusions.

4.3   Results and Discussions

The results and discussions of field investigations for integrated water meter management activities are described in the following sections.

4.3.1   Demand profiling results

Ninety six customer data loggings were obtained in eight months. However, only 78 were analyzed as 18 were incomplete and un-usable. Fifty user's service connections had meters of size DN 15 mm, 20 were DN 20mm and the rest were for large customers. The two common user patterns for users with DN 15 mm service connections, mainly domestic users are presented in Table 4.3. The measurements were classified as follows:

Household Type I: Single family household with elevated storage tank; this is graphically represented in Figure 4.8.

Household Type II: Apartment block of 9 units (4 floors) fed from elevated storage tanks at roof level. The average use per apartment was 528 L/d (or 4,756 L/d for the 9 apartments).

Table 4.3 Customer water use patterns

Group	Flow Range (L/h)	% of Customer Usage	
		Type I	Type II
1	0-13	10.3	0.0
2	13-21	5.8	1.0
3	21-35	7.0	2.0
4	35-96	16.4	9.0
5	96-158	10.3	10.0
6	158-711	40.0	78.0
7	711-1173	9.6	0.0
8	1173-1935	0.6	0.0
9	1935-3000	0.2	0.0
	Approximate Average Usage (L/day)	783	4,756

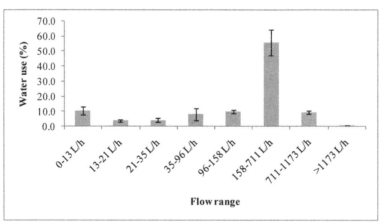

Figure 4.8 Type I usage pattern

From Table 4.3, more than 15% of water used by single residential households in Kampala occurs at flow rates below 21L/h. Most of this water will not be measured by new Class B meters of size DN 15 mm (Qn = 1.5 m³/h) whose minimum flow rate is 30 L/h. The uncertainty for the average water use pattern is low (Figure 4.8) particularly in the low flow ranges that are vital for accurate computation of the weighted meter accuracy. Water use patterns provide data essential for decision making including sizing meters, evaluating meter performance, meter renewal, water audit/balance and engineering design studies. Of interest to this study is use of water use patterns for meter sizing, evaluation and renewal.

4.3.2   Weighted meter accuracy results

The weighted meter accuracy (Eq. (4.2)) was calculated using the established metering accuracies at different flow rates for the different meter groups and the household type I (Figure 4.1) consumption pattern that makes up about 80% of the total households in Kampala city. The results are shown in Table 4.4.

Global Weighted Metering Accuracy

The global weighted meter accuracy (WMA_{global}) was computed using the weighting coefficient based on number of meters in each group as shown in Eq. (4.4).

$$WMA_{global} = (WAMA_1 * PS_1) + (WAMA_2 * PS_2) + (WAMA_3 * PS_3) \qquad (4.4)$$

where $WAMA_1$ is the weighted average meter accuracy of all tested Model 1 meters, $WAMA_2$ is the weighted average meter accuracy of all tested Model 2 meters and $WAMA_3$ is the weighted average meter accuracy of all tested Model 3 meters, PS_1 is the percentage of Model 1 meters in the system, PS_2 is the percentage of Model 2 meters in the system, and PS_3 is the percentage of Model 3 meters in the system.

$$WMA_{global} = (73.7 * 0.43) + (83.1 * 0.41) + (79.9 * 0.16) = 78.5\% \pm 0.9\%$$

The weighted average meter accuracy of DN 15 mm meters in the system was therefore computed at 78.5% and thus it can be concluded that the meter weighted error for the meters in the KWDS with an average age of 10 years was -21.5% ± 0.9%.

Table 4.4 Calculated weighted average accuracy of the meter models

Model	Flowrate	% of Use	% Avg. Accuracy	Weighted Avg. Accuracy (%)
1	Low	16.0 ± 2.1%	62.3 ± 1.2%	
	Medium	74.0 ± 1.7%	74.3 ± 0.1%	73.7 ± 0.6%
	High	10.0 ± 6.1%	87.9 ± 0.1%	
2	Low	16.0 ± 2.1%	80.1 ± 1.4%	
	Medium	74.0 ± 1.7%	83.4 ± 0.1%	83.1 ± 0.6%
	High	10.0 ± 6.1%	85.3 ± 0.2%	
3	Low	23.0 ± 3.4%	61.0 ± 0.9%	
	Medium	67.0 ± 4.2%	85.0 ± 0.3%	79.9 ± 0.3%
	High	10.0 ± 3.1%	89.2 ± 0.1%	

In a study carried out by the University of Massachusetts (USA) to estimate metering errors for two water utilities of Westchester Joint Water Works (WJWW) in Mamaroneck, New York and Taunton Water Works (TWW) in Taunton, Massachusetts, metering errors of -11.2% (WJWW) and -13.7% (TWW) were reported for meters of sizes DN 15 mm and average age of 14 and 10 years respectively (Newman and Noss 1982). In another study in Canal de Isabel II water utility in Madrid, Spain, an average weighted metering error of -14% was reported for old meter sizes of DN 13 mm to DN 40 mm based on a sample size of 2000 meters and 226 customer consumption profiles (Flores and Diaz 2009). The difference in metering errors across continents could be attributed to the following differences:

- *Domestic water use profiles.* Water used at flow ranges below 113 L/h (0.5 gpm) was about 40% in Kampala, 8% in Madrid, and in the USA 15% and 4% for the two different profiles used in the study. This partly explains the high metering inaccuracy for Kampala as meters are least efficient at measuring ultralow flow rates (Richards et al. 2010). The importance of accurate demand profiling to assess water meter accuracy for small domestic meters cannot be over emphasized.
- *Meter testing standards and procedures.* The recommended test flow rates for domestic small meters by the American Water Works Association (AWWA) at minimum, intermediate and maximum rates are 0.25 gpm (56.7 L/h), 2 gpm (453.6 L/h) and 15 gpm (3402 L/h) respectively (AWWA, 1999). The test flow rates (minimum, transitional and maximum) used to calculate the weighted error in Kampala are those recommended by the ISO (Table 4.2). AWWA test flow rates at low and medium ranges are much higher than ISO flow rates which certainly will yield different results for the same meter under test. In addition, the way meters are handled after pulling from the field in Kampala allows meters to run-dry as no plugs are inserted at both ends of the meter and this could have affected the results.
- *Water distribution systems.* Meter performance is greatly influenced by the water system in which it is installed (Wallace and Wheadon 1986). The different system characteristics (e.g. water quality, regular or intermittent supply, installed meter models and quality of meters etc.) all have influence on meter performance.
- *Inherent errors in sampling and measurements of different magnitude.*

4.3.3 Influence of private elevated storage tanks

The results of the influence of ball valves in elevated storage tanks on meter accuracy for two households (HH 1 & HH 2) with an old meter of billing index of 9,056 m^3 (about 15 years old) is indicated in Table 4.5. The limitation with this study was the few numbers of households sampled. There was resistance from customers due to water supply inconveniences during set-up of the study equipment that necessitated temporary closure of supply.

The average meter under-registration due to the ball valve effect in Kampala was about 67.2% ± 0.1% using an old meter. The influence was less when highly accurate new meters were used and was around 4.0% ± 0.1% of the total household consumption (Table 4.6). Although the sample may not have been representative, the results still provide insight of apparent losses due to the ball-valve effect of the elevated storage tanks.

Table 4.5 Influence of ball-valve on meter accuracy (old meter)

HH 1	Inlet (m^3)	Outlet(m^3)	Difference	HH 2	Inlet (m^3)	Outlet(m^3)	Difference
Day 1	0.220	0.505	0.285	Day 1	0.151	0.505	0.354
Day 2	0.203	0.501	0.298	Day 2	0.180	0.501	0.321
Day 3	0.059	0.464	0.405	Day 3	0.045	0.464	0.419
Day 4	0.077	0.398	0.321	Day 4	0.109	0.398	0.289
Day 5	0.136	0.445	0.309	Day 5	0.485	1.868	1.383
Day 6	0.152	0.506	0.354	Day 6	0.321	0.767	0.446
Day 7	0.195	0.467	0.272	Day 7	0.532	0.878	0.346
Total	1.042	3.286	2.244		1.823	5.381	3.558
Under-registration (%)			68.3				66.1

Table 4.6 Influence of ball-valve on meter accuracy (new meter)

HH 1	Inlet (m^3)	Outlet(m^3)	Difference	HH 2	Inlet (m^3)	Outlet(m^3)	Difference
Day 1	0.464	0.467	0.003	Day 1	0.635	0.646	0.011
Day 2	0.191	0.203	0.012	Day 2	0.434	0.44	0.006
Day 3	0.114	0.116	0.002	Day 3	0.508	0.517	0.009
Day 4	0.055	0.057	0.002	Day 4	0.257	0.305	0.048
Day 5	0.345	0.354	0.009	Day 5	0.332	0.353	0.021
Day 6	0.512	0.542	0.03	Day 6	0.655	0.695	0.04
Day 7	0.504	0.521	0.017	Day 7	0.514	0.545	0.031
Total	2.185	2.26	0.075		3.335	3.501	0.166
Under-registration (%)			3.3				4.7

In their study on under-registration caused by the ball valves of roof tanks in Malta, Rizzo and Cilia (2005), reported meter under-registration of about 6% of the total household consumption using similar new meters, and 92% under-registration using a 5 year old test meter. In more recent studies carried out in Palermo (Italy) and Spain, metering errors due to the ball valve effect were reported to be 49% with 11 years old meter (Criminisi et al. 2009) and 40% using a 14 year old meter (Cobacho et al. 2008) respectively. These findings are more or less similar to the Kampala study but should in no case be generalized as they are system specific and each utility should try to compute its own figures.

The differences observed could be explained by the different water use patterns, the size of tank and the degree of water supply reliability. When there is no water, the tank empties and when water supply returns, the tank fills at high flow rates and the ball valve effect has little influence. The larger the surface area of the tank the more the effect of the ball valve on meter under-registration.

4.3.4 Impact of sub-metering on meter accuracy

The results of measured revenue water for the eight investigated properties with sub-meters are presented in Table 4.7. Properties PR 1331, PR1327, PR1619, PR1720, PR3016 and PR3414 were residential apartments (2 to 6 floors); PR 3120 was a single-family residential property with semi-detached houses and PR 2027 was a commercial complex (5 floors).

Table 4.7 Water under-registration due to sub-metering

Sub-meters (No.)	Property Reference (PR)	Size of Master Meter (DN mm)	Total Sub-meters (m^3)	Master Meter (m^3)	Difference (m^3)	Error (%)
6	1331	15	133	152	-19	-12.5
8	1327	15	213	203	10	4.9
9	1619	15	714	676	38	5.6
12	1720	20	625	898	-273	-30.4
18	3120	15	485	504	-19	-3.8
18	2027	15	4276	5960	-1684	-28.3
32	3016	20	1138	1312	-174	-13.3
34	3414	20	1363	1692	-329	-19.4
Mean (excluding PRs 1327 & 1619 with booster pumps)						-18.0

From Table 4.7, it is clear that there was meter under-registration on all properties apart from two (PR 1327 and PR 1619). The two exceptions had one common collection tank on the ground floor and a booster pump for pumping water to the various elevated storage tanks that feed different apartments. Since the booster pump was before the sub-meters meters, it is likely to have increased flow rates to within the higher ranges of meter performance. The under-registration (metering errors) of revenue water due to the six sub-metered properties in Kampala city is estimated at 18.0%. The high variability in error values could be due to the inherent uncertainties in user flow rates and random errors during meter readings. Increasing the sample size of sub-metered properties and measurement precision could reduce the spread in error values. Despite these limitations, the results do provide insight into the effects of sub-metering on revenue water.

Influence of individual metering errors:

The experiment to determine the individual metering error due to sub-meters was carried out for a 1-month period and the results are summarized in Table 4.8. The estimated individual meter errors were a total of 14% (53% - 39%) and after correction the new error due to sub-metering was 39%. The error due to age of meters was 19% and after correction; the new error was 20%. This therefore implies that the error due to the individual sub-meters was 33% (53% - 20%) for these premises.

Influence of pressure: By increasing residual pressure at the property master meter from 5 m to 20 m, the water under-registration (metering error) dropped to -4% implying that the sub-meters were actually over registering by 4%. This confirmed the earlier assumption of over-registration for properties with booster pumps.

Since this error value was below 5%, it was concluded that after taking care of both errors due to individual meters and errors caused by low flows due to low network pressures, the volume measured by the monitoring master meter was equal to the sum of the sub-meter volumes (Eq. (4.1)).

Table 4.8 Influence of the quality of meters and meter age

In-Service Meters (average pressure at MM = 0.5 bar, private tank elevations = 7 m)				Metering using Class D meters
	Consumption (m3)	Weighted meter error (%)	Corrected consumption (m3)	Consumption (m3)
Master Meter (MM)	19.0	-4	19.76	8.3495
Sub-meters				
1	0.3	-89.8	0.57	0.4499
2	0.6	-0.7	0.60	0.2132
3	2.0	-89.8	3.80	0.6864
4	4.9	-0.5	4.93	1.8009
5	0.7	-88.9	1.32	0.0006
6	0.2	-89.8	0.89	6.6632
Total	9.0		12.11	6.6632
Difference	10.0		7.65	1.6860
Error	53%		39%	20%

Influence of demand profile: To assess the influence of users' profile on metering errors, an analysis of the demand profiles logged simultaneously at all the meters was carried out before the pressure was increased and the consumptions taking place at a rate of below 15L/h (minimum flow for a 15 mm size, Class C meter, $Q_n = 1.5$ m^3/h). The results indicated that about 26% of the total apartment usage was taking place below Q_{min} and thus were not recorded by the meters. It is important to note that the percentage of the consumption that was below Q_{min} (26%) is quite close to the metering error that was eliminated (20%) when the pressure in the system was increased. It is likely that other factors such as the 'ball valve effect' of the private service tanks could have amplified the metering errors of the sub-meters.

Sub-metered properties are more vulnerable to illegal use through meter by-pass especially when the distance between the master meter and sub-meters is more than 5 m. In this study, three meter by passes were discovered on PRs 3414, 3016 and 2027. These properties had the highest number of sub-meters and it's likely that the more the number of sub-meters, the more likelihood of meter by-pass. Another observed scenario was the wrong meter installations on sub-metered properties. The fifth meter (from left to right) in Figure 4.9 is a velocity-type meter that is very sensitive to the way it is installed i.e. the velocity profile. For example installing meters 15 m above ground (observed at PR 3016) in different inclinations is likely to affect their accuracy and allows meters to run dry whenever pressure dropped. Meters are designed to be installed in a horizontal position and any orientation affects their performance. A study carried out to determine the influence of orientation during meter installation on in-service meters of class B (6-9 years) reported an increase in the starting flow rate of 10 L/h (from 24 to 34 L/h) with a 45^0 inclination (Arregui et al. 2006b).

Figure 4.9 Incorrect meter installation in Kampala

In Kampala service area, about 10% of total service connections (or 15,000 connections) are billed directly by sub-metering. Using an average annual consumption per service connection of 240 m^3 and average tariff (July-June 2010) of 1,800 Ugandan Shillings (UShs) per m^3 (or 0.82 US $ per m^3), the annual financial loss to the utility as a result of sub-metering is conservatively estimated at UShs 778 million (US $0.35 million). This excludes the sewerage charge component for apartments with a sewerage connection, cost of sub-meters and metering installations, cost of meter reading and billing administrative costs. Consequently, the findings of this study have policy implications for sub-metering and recommendations have been made with aim of promoting water use efficiency and maximizing utility revenues.

4.3.5   Meter failure analysis

Meter failures were grouped, based on billing index, into six sub-groups (0-500 m^3, 500-1,000 m^3, 1,000-2,000 m^3, 2,000-5,000 m^3, 5,000-10,000 m^3 and more than 10,000 m^3). The total number of meter failures that were registered in the laboratory in the year 2010 was 10,235. Out of these, 9,416 were of size DN 15 mm. The number of meter failures was then used with the total number of meters (DN 15 mm) in the system to estimate the meter failure rate for the KWDS for the DN 15mm size group. The estimated failure rate for the KWDS (DN 15 mm) meters was 0.066 failures per meter per year.

However, not all reported defective meters ended up in the meter laboratory as records from the customer billing database indicated an average of about 6,000 defective meters per month for the year 2010. The technical plumbers who service and replace defective meters are often engaged in other revenue collection related tasks. Often meters are also serviced on site and information not sent to the meter laboratory for updating the database; thus the meter failure rate estimated is rather conservative. The problem of data from the billing database is that it does not lend itself to easy analysis. Apart from numbers, it does not provide meter failure data in cohorts of size, make, type and cause of failure. It is a billing database not designed for water meter management.

The meter failure was categorized according to type and cause of failure, as shown in Figures 4.10 and 4.11, which was observed on disassembling the meters and from laboratory records.

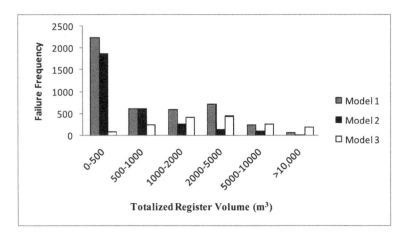

Figure 4.10 Meter failure frequency by model

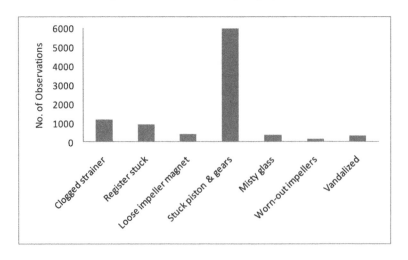

Figure 4.11 Number of observed defects

From Figure 4.10, it is evident that Meter Models 1 and 2 have a high frequency of failure and most are actually failing 2 to 3 months after installation and well before they register 500 m^3. The most common cause of meter failure is related to water quality (Figure 4.11). Due to high frequency of bursts and leaks coupled with poor pipe repair practices and inadequate mains flushing, a lot of silt and other suspended particulates enter the water system and get lodged in the meter drive mechanisms (reducing gear trains and oscillating piston) halting them from moving. In addition, meter movement is halted by deliberate meter vandalism by customers who would like to use water free of charge. Stone pebbles are put in between meter impellers with often help of utility plumbers or former employees to stop the meter from moving (Figure 4.12).

Figure 4.12 Deliberate meter impeller vandalism

A system with corroded pipes operating under intermittent conditions is likely to re-suspend a lot of deposited particulates in the pipe with devastating consequences for water quality and water meter performance (Van Zyl 2011; Vreeburg and Boxall 2007). Over 75% of failures were observed in oscillating-piston type meters (Models 1 and 2). One of the disadvantages of oscillating-piston meters is their sensitivity to suspended solids in water (Arregui et al. 2006b). Although 15 mm displacement meters (nutating disc and oscillating piston) are unrivalled for accuracy (Richards et al. 2010) and widely used in the developed countries, they are not suitable for the KWDS, and, most likely other developing countries with similar water system characteristics. Generally, systems with a high number of pipe repairs are not suitable for positive displacement meters or meters with wet dial counters (Arregui et al. 2012). Inferential (or velocity) water meters are less affected because only part of the turbine is in the stream and does not get blocked up with particulates in water. The advantages and disadvantages of different meter technologies can be found in Arregui et al. (2006b) and van Zyl (2011).

Lund (1988) reported meter failures of 1% for newer types of meters in Seattle Water Department. The meter failure rate in Kampala is about 7 times that in Seattle despite advancement in metering technology over the years. This is probably due to the different water system characteristics. Meter failure and accuracy are greatly influenced by the water supply system in which they are installed (Wallace and Wheadon 1986). Research has shown that meters in developing countries are more likely to malfunction due to intermittent supply (Criminisi et al. 2009). A study carried out to evaluate residential water meter performance in the USA by the Water Research Foundation indicated that meters subjected to pulsed flows decreased in accuracy at every flow rate to a greater degree than at constant-flow operation

(Bowen et al. 1991). Gokhale (2000), as cited in Butler and Memon (2006), highlights meter problems associated with intermittent supply systems as:

- Sudden variations in flow rates subject the meter mechanisms to undesirable strain
- Alternate drying and wetting of meter parts coming into contact with water is detrimental to continued satisfactory performance of meters
- Air that enters the distribution system is forced out through the meter at the start of supply causing meters to run at excessive speeds and thus increasing "wear and tear" of meter parts.

4.4 Estimation for Water Loss due to Metering Inaccuracy and Meter Failure

In order to manage water meters better, it is important to quantify water loss due to metering errors and meter failure. Until now, the procedures for estimating these losses are not yet well established. In this section, a procedure for estimating these losses is presented.

4.4.1 Procedure for estimating water losses due to metering

The procedure for estimating water loss is adapted from Male et al. (1985) and is shown in Figure 4.13.

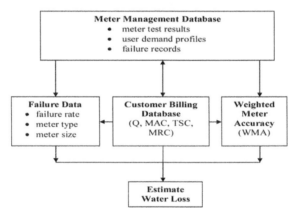

Figure 4.13 Flowchart for estimating water loss due to metering

The proposed method is summarized in the following eight steps for a meter size of interest:

1. From the meter management database (test results and customer demand profiles), estimate weighted meter accuracy (WMA).
2. Estimate meter failure rate (MFR) by dividing the total number of meter failures per year with the total number of in-service meters.
3. Calculate the average annual water used (Q) through the meter from the customer billing database.
4. Establish the total number of metered active connections (*MAC*) from the customer billing database. If all service connections are metered and on supply, then *MAC* is equal to the total number of service connections (*TSC*).
5. Estimate the average detection-replacement/service time (*DRT*) for failed meters based on utility meter reading cycles (*MRC*). If *MRC* is say one month, meter failures would be expected to exist for 15 days on the average and 2 weeks would be a

reasonable time within which to service or replace the meter. DRT in this case would be about 1-month.

6. Estimate total unregistered water due to meter failure (UWF m^3/yr) from Eq. 4.5.

$$UWF = MFR \times MAC \times Q \times (DRT/12) \qquad (4.5)$$

7. Estimate total unregistered water due to working meter errors (UWW m^3/yr) from Eq. 4.6.

$$UWW = [(100/WMA) - 1] \times Q \times MAC \qquad (4.6)$$

8. Estimate total water loss due to meter failure and working meters as the sum of UWF and UWW.

4.4.2   Estimating water losses for case study due to meter failure and errors

Using the developed procedure, water used but unmeasured by DN 15 mm meters due to metering errors and failures in Kampala has been estimated and is summarized in Table 4.9.

Table 4.9 Estimates of water loss due to metering errors and failure

Description	Unit	Value
Total service connections with meter sizes of DN 15 mm	No.	141,674
Connection efficiency (Active/Total)	%	85.9
Active service connections with meter sizes of 15 mm (MAC)	No.	121,698
Average water use for DN 15 mm service connection (Q)	m^3/year	240
Estimated average age of DN 15 mm meters	years	10
DN 15 mm meter failure rate (MFR)	failure/meter-year	0.066
Detection-replacement/service time to failed meters (DRT)	months	1
Weighted meter accuracy at average age (WMA)	%	78.5
Estimated unregistered water due to failure (UWF)	m^3/year	160,641
Estimated unregistered water due to working meters (UWW)	m^3/year	7,999,510
Total unregistered metered water (working & failed)-TWL	m^3/year	8,160,151
Unregistered water delivered per meter per year	m^3/meter-year	58
System Input Volume (SIV)	m^3/year	52,237,838
Non-revenue water (NRW)	m^3/year	21,346,351
Non-revenue water (NRW)	%	40.9
% of SIV delivered but unregistered by DN 15 mm meters	%	15.6
% of NRW due to DN 15 mm meters	%	38.2

From Table 4.9, the unmeasured water by DN 15 mm meters in Kampala makes up about 15% of annual water supplied and 38% of NRW. The amount of water unmeasured by an average meter is 58 m^3/meter-yr. These figures are rather high compared to estimated figures of 11% and 4% of NRW for Westchester Joint Water Works (WJWW) in Mamaroneck, New York and Taunton Water Works (TWW) in Taunton, Massachusetts respectively. The amount of water unmeasured by an average meter for both utilities was 43 m^3/meter-yr (Newman and Noss, 1982).

Although a lot of resources are put in by the utility to service defective metes, the impact of these meters on global NRW is minimal (about 1%) and more proactive water meter management strategies are needed. Arregui et al. (2012) have proposed nine steps for better water meter management that are likely to be of benefit to water utilities if adapted and implemented.

4.5   Optimal Meter Sizing and Selection

Apparent losses caused by metering inaccuracies can be reduced by properly selecting and sizing customer meters. Oversized meters can result in reduced revenues due to meter under-recording while undersized meters are likely to be damaged by high flow rates through the meter causing rapid degradation of meter accuracy. To get maximum benefit from a water meter, it is critical that the meter is properly selected and correctly sized. Optimal meter sizing and resizing is achieved by combining water demand profile data with the meter manufacturer's specifications to ensure that the meter selected accurately records the anticipated range of flows without damaging or accelerating wear of the meter during the brief peak demands (AWWA 2004).

In addition to charging for water used, water utilities often have a new service connection charge based on the size of the customer service line. Sizing decisions were often based on rules of thumb and the drive to maximize revenue from new connection fees resulting in oversized service lines. When customer metering was introduced, meters were installed to match the size of existing service lines often resulting in oversized customer meters with consequences of reduced revenue due to meter under-registration.

Whereas sizing of domestic water meters is straightforward, the selection of the right meter size for large consumers is not trivial and requires demand profiling techniques and good judgement. Large customers are often few in numbers compared to domestic consumers, but they contribute a substantial amount in terms of revenue. In Kampala there are about 400 large customers who use over 500 m^3 of water per month. These customers account for only 0.3% of the total number of customer connections but represent about 30% in terms of total water sales, and roughly 50% of revenue generation including sewerage charges. It is important therefore that they are correctly sized following systematic procedures for maximum utility revenues.

According to van Zyl (2011), a good guide for the right water meter selection should provide answers to the following questions: (i) what is the purpose of the meter?, (ii) does the meter comply with the required standards and policies?, (iii) is the meter rated for the expected flow rates and operating conditions?, and (iv) which is the most economical meter to use? Some examples on meter sizing and selection using demand profiling techniques are provided in the next sections.

4.5.1   Example of water meter optimal sizing

In this section, a hospital in-service meter is resized to illustrate use of demand profiling techniques. With users other than single-family residential, each user generates a unique demand profile, and the meter should be sized accordingly. The demand profile for one of the large users in Kampala city (Rubaga Hospital) is indicated in Table 4.10 and flow rate data in Figure 4.14.

Table 4.10 Demand profile for a large customer

Group	Flowrate range (L/h)	Used volume (L)	% Usage
1	0-158	30	0
2	158-711	260	0
3	711-1173	200	0
4	1173-1935	350	0.01
5	1935-3193	3,460	0.35
6	3193-5268	233,420	27.7
7	5268-8692	553,980	57.39
8	8692-14342	124,960	14.51
9	>14342	0	0

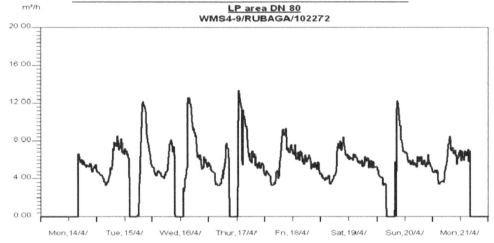

Figure 4.14 Typical weekly consumption flow rates of the Hospital

From Figure 4.14, it can be observed that flow rates through the meter are usually below 12 m^3/h, corresponding to the maximum flow rate of a 32 mm water meter. From the water consumption pattern (Table 4.10) and taking into account the average consumption of 123 m^3/day recorded in seven days, the most adequate meter would have been a single jet Class C meter, of size 50 mm (Q_n = 15 m^3/h) or Woltmann 50 mm (Q_n = 25 m^3/h). The choice was the single jet Class C meter of 50 mm (Q_n=15 m^3/h) since 99.96% of the customer's flow is below 15 m^3/h. Although the customer experiences brief demand spikes for about 7 seconds once a week of up to 20 m^3/h, it is prudent to size the meter to accurately collect the 99.96% of the flow as the head loss during the spike is within acceptable limits (6 m at Q_{max} = 30 m^3/h).The in-service meter was a Woltmann meter, Class B of size 80 mm, with a nominal flow rate of 60 m^3/h, which was clearly oversized.

As a result of this meter resizing from DN 80 mm to DN 50 mm, the billing database records indicate that the average monthly water consumption in 5 months increased by 1,354 m^3 (from 5,238 m^3 to 6,593 m^3), which in monetary terms translates to revenue increase of US $15,200 per annum. This increase is attributed to both resizing and improvement of the error curve by replacing the old meter with a new one. Adopting the same approach, plans are under way to optimally resize and replace over 400 large customer in-service meters that were sized based on "rules of thumb". This is anticipated to increase utility revenues by about US $6 million per annum and reduce the apparent water loss component due to metering by about 12,000 m^3/day.

Demand profiling techniques for meter resizing are being applied in most efficient water utilities worldwide. In Boston, Massachusetts, the Boston Water and Sewer Commission reported revenue water recovery of over 593,924 m³ (156, 915,320 gal) per year by downsizing over 400 meters $1^{1}/_{2}$ inch and larger. In addition to reducing non-revenue water, the meter resizing effort could generate over 700,000 US $ annually (Sullivan Jr and Speranza 2008).

4.5.2   Example of a single-family water meter optimal selection

In this section, a domestic water meter is selected with focus on the class of a meter using demand profiling techniques. Although selecting the meter size for a single-family residential may be straight forward (usually DN 15 mm), the decision on what class of meter to choose is not all that trivial and often puzzles most utility managers. The problem is partly answered by demand profiling techniques (Figure 4.15) coupled with incremental cost-benefit analyses of selecting one option over the other.

A Class C single-jet meter of size 15 mm and $Q_n = 1.5$ m³/h has a starting flow rate of 6 L/h. Replacing a Class B volumetric (piston-type) meter (minimum flow rate of 30L/h) with a single-jet meter (starting flow rate of 6L/h) results in additional registration of revenue water due to increased ability of a Class C meter to measure low flows. Likewise, a Class D positive displacement meter (piston-type) with a starting flow rate of 1 L/h would even register more water usage but as discussed earlier, this metering technology is expensive and unsuitable for the KWDS conditions. There is need to find a trade-off between ability of meters to measure low flows and costs of frequent meter failures in the decision process of selecting a Meter Class.

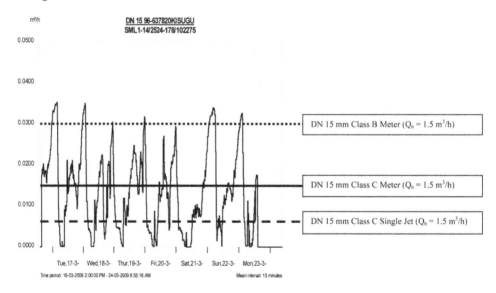

Figure 4.15 Demand profiling for optimal selection of a meter class

4.6   Optimal Meter Replacement Frequency Model

Once meters have been installed, the problem of replacement arises. The question of how long to leave a meter in service has for long troubled utility managers (Arregui et al. 2003;

Williams 1976). This study contributes to resolving this problem for the case study water utility.

Most previously developed models for meter replacement used the simple methodology of average cumulative net present value (NPV) cost per year which limits the lifetime of all residential water meters to one single optimum replacement period. This is not realistic in practice due to different user patterns. According to Arregui et al. (2006b), under no circumstances should a simple NPV calculation be used to provide the least cost option as it leads to miscalculations.

The individual water meter replacement model, referred to here as "I-WAMRM" is developed to determine the optimal meter replacement period based on totalized registered volume through the meter. The optimal meter replacement period has been defined as the time when revenue loss due to a drop in accuracy equals cost of replacing a meter (Wallace and Wheadon 1986). The condition driving the decision to replace the ith asset has been defined as (NAMS 2004; Ugarelli and Di Federico 2010):

$$CT_i(n) - [IN_i(n) + D_i(n)] \geq 0 \qquad (4.7)$$

At stage n (Eq. (4.7)), if the costs to maintain the existing ith asset (CT_i) are greater than the cost of investing in a new asset (IN_i), including the eventual depreciation charge of the existing asset (D_i), the asset should be replaced. In developing this model, the objective was to minimize the sum of these costs over an infinite number of succession replacement periods as in classical regeneration problems (Arregui et al. 2011; Eagle and Kiefer 2004; Lund 1988) shown in Figure 4.16. This is referred to as the minimum net present value cost of the replacement chain ($MNPVC_n$).

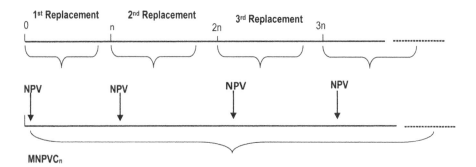

Figure 4.16 Minimum NPV costs of the replacement chain

4.6.1 Framework of I-WAMRM

The model guiding framework for the optimal and decision-making process is shown in Figure 4.17.

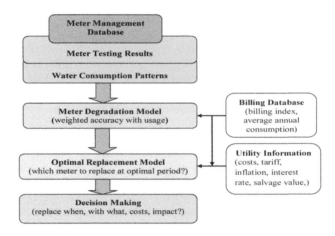

Figure 4.17 Model flow chart

4.6.2   NPV of the life cycle costs

The life cycle costs (LCC) seeks to optimize the costs of acquiring, owning and operating physical assets over their useful lives by attempting to identify and   quantify all the significant costs involved in that life, using the present value technique. LCC enables the trade-off between costs and benefits during the asset life to be studied to ensure optimal selection (Woodward 1997).

The optimal replacement period (ORP) is found by minimizing the total annual costs of replacement defined as (Male et al. 1985):

1. Cost of replacement policy (CRP): the cost of removing, testing, repairing, replacing and disposing meters.
2. Cost of water lost through failed meters (CWLF): water used but not measured after failure and before repair.
3. Cost of water lost through inaccurate meters (CWLI): derived from accuracy versus usage relationships.

The total annual cost (TAC) is calculated as follows:

$$TAC = CRP + CWLF + CWLI \qquad\qquad (4.8)$$

The elements of LCC for a water meter have been identified as: (i) initial costs of meters; (ii) cost of meter replacement; (iii) administrative (information and feedback) costs; (iv) meter under-registration costs; and (v) disposal costs. Since the costs of water lost due to failure for the selected small water meter model (multi-jet velocity type meters) are negligible, Eq. (4.8) becomes:

$$TAC = CRP + CWLI \qquad\qquad (4.9)$$

Where $CRP = [(C_{IN,} + C_{INST,} + C_{Admin}) - C_{SV}]$ and $CWLI = \sum_{t=1}^{n} \dfrac{P_W Q_t \varepsilon_t}{(1+r')^{t-1}}$

and n is the number of years for the meter replacement period, r' is the real discount rate, $C_{IN}, C_{INST}, C_{Admin}$ is the cost of meter purchase, installation and initial administrative costs, P_w is the price of water ($/m^3$) and assumed to be constant throughout the analysis period, Q_t is the average volume of water consumed by the user in the year t, c_t is the weighted meter error , C_{SV} is the salvage value of the meter often sold as scrap at disposal time and is a function of meter material (plastic or Bronze). Meters with bronze housings are sold as raw materials for steel industries but plastic meter bodies are hardly bought and their salvage value could be neglected.

Finally, the minimum present value cost of this infinite series of replacement is given by:

$$MNPVC_n = [CRP + CWLI] \left[\frac{(1+r')^n}{(1+r')^n - 1} \right]$$

(4.10)

Eq. (4.10) is the main engine of the optimal meter replacement model in selecting a period n that minimizes the total costs (or maximizes the revenues). The model calculates the NPV of the costs of infinite replacements conducted at fixed time steps while taking care of inflation. The period n is constant over the present and future replacement periods provided real costs and interest rates remain constant.

The Real Discount Rate (r')

The real discount rate can be calculated by Equation 4.11 (Arregui et al. 2006b).

$$r' = \frac{1 + i_R}{1 + I} - 1$$

(4.11)

where, i_R is the interest rate and I is the general inflation. As the life cycle costs are discounted to their present value, selection of a suitable discount rate is a crucial decision in LCC analysis. In the case of water meters, since the investment is reasonably risk free, the discount rate used will be quite close to the ones set by each country as risk free rates or state bonds (Arregui et al. 2006b). The appropriate discount rate should be determined by the utility's corporate planning department rather than mere arbitrary selection.

4.6.3 Predicting water meter accuracy

Most problems in operations research and engineering involve establishing the relationship between two or more variables. Regression analysis is the statistical technique that is often used for such types of problems (Montgomery and Runger 2007). An important aspect of predictive models is to be able to predict how condition will deteriorate over time. Water meter accuracy degradation is a function of many variables and it's not easy to predict meter accuracy degradation rate with certainty. However, it is important to understand the meter accuracy degradation process in every metering strategy.

Many researchers have assumed a linear relationship between accuracy and age or cumulative volume through the meter for domestic small meters (Arregui et al. 2006b; Hill and Davis 2005; Noss et al. 1987). In this study a regression analysis has been used assuming a linear relationship to predict meter accuracy degradation rate due to its simplicity.

In general, the dependent variable Y may be related to k independent variables by a multiple linear regression model. The form of the regression model is:

$$Y = \beta_0 + \beta_1 x_1 + \beta_2 x_2 + \dots + \beta_k x_k + \varepsilon \qquad (4.12)$$

where β_j $(j = 0, 1, \dots, k)$ are the regression coefficients and ε is the random error term. Assuming that regression models are performed for specific meter models (same manufacturer, same meter size, same metering technology) and including other variables implicitly apart from the totalized registered volume (usage) which is explicitly included, then the model takes the form:

$$Y = \beta_0 + \beta_1 x_1 + \varepsilon \qquad (4.13)$$

where x_1 is the totalized registered volume through the meter (proxy for meter age). The term "ε" value is used as an adjustment factor to account for different system characteristics and how they impact on meter accuracy.

Based on statistical random sampling techniques and meter testing records from a recently established database in the Kampala meter testing laboratory, data for a total of 122 meters (multi-jet type of size DN 15 mm) was analyzed. Out of the 122 data sets, only 83 were finally used after data filtering of suspicious outliers. The resulting regression model is presented in Figure 4.18.

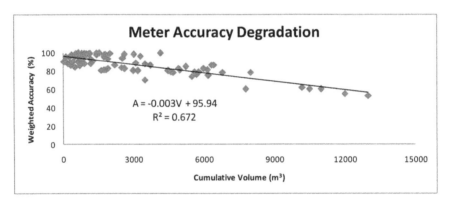

Figure 4.18 Accuracy degradation rate for a multi-jet water meter type

The fitted simple linear regression model for the in-service multi-jet velocity meter model in the KWDS that relates water meter accuracy to volume is estimated as:

$$A = -0.003v + 95.94 \qquad (4.14)$$

where, A is meter accuracy (%) and v is the totalized registered volume through the meter (m^3). The maximum allowable meter accuracy degradation (Eq. (4.14)) for a half-inch multi-jet meter in Kampala is 72% when the meter's totalized register volume equals 8,000 m^3.

The goodness of fit of the regression line which is measured using the coefficient of determination (R^2 = 67.2%) is rather low probably due to the limited dataset used and uncertainties of the input data. In addition, other key factors such as water quality characteristics were not included in the model due to inadequate reliable data. The utility management is working towards establishing a more accurate and comprehensive meter management database to improve future prediction models. A rich database of both customer use demand profiles and accurate meter testing results will improve the value of R. However, the regression model (Eq. (4.13)) can be used to fairly predict average weighted water meter accuracy based on meter condition (totalized registered volume at any time *t* as an indicator).

4.6.4   Model application to the case study water utility

The I-WAMRM algorithm was implemented using MS Excel® spreadsheet application. The tool has a friendly user interface (Figure 4.19) for data input and analysis. The required utility model input data is summarized in Table 4.11. The multi-jet meter model is the subject of analysis; however any other meter could be used as long as its degradation rate of weighted meter accuracy can be accurately established.

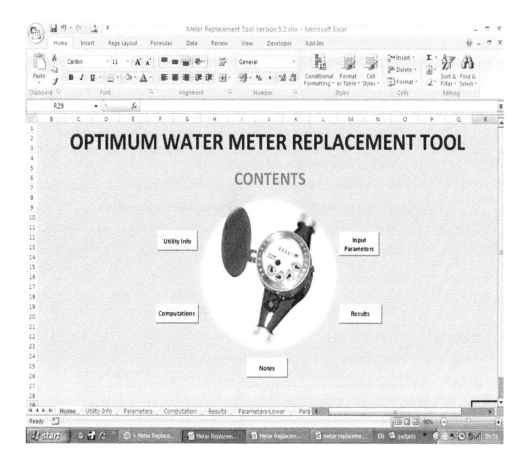

Figure 4.19 Screenshot of tool user interface

Table 4.11 Kampala Water utility parameters

Parameter	Unit	Value
Water tariff	US$/m^3	1
Interest rate	%	10
Inflation rate	%	6
Real discount rate	%	3.8
Average annual usage	m^3/year	Variable
Multi-jet meter (Class C, Q_n=1.5 m^3/h)	mm	DN 15
Retail price for new multi-jet meter	US$	30
Removal and installation labour costs	US$	10
Administrative costs	US$	3
Salvage meter value	US$	3
Initial meter weighted error	%	-4
Meter accuracy degradation rate	%/year	-0.3

4.6.5 Numerical results and discussions

The details of model computations for an individual meter (Meter No: 96-712461) and property reference (PR: 321577) are shown in Table 4.12. The results of the model for the selected two individual meters are summarized in Table 4.13.

Table 4.12 Summary of I-WAMRM predicted optimal metering conditions

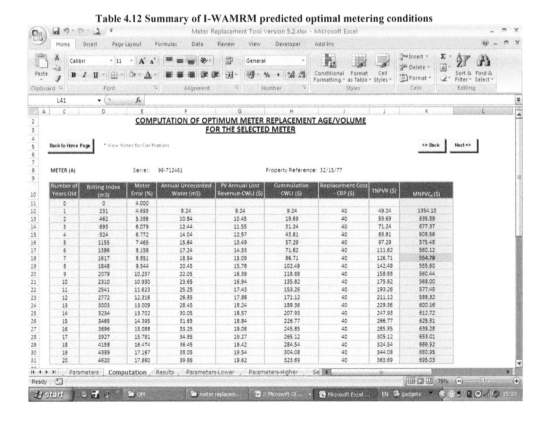

Table 4.13 Summary of I-WAMRM of two meter replacements

Parameter	Unit	Meter Identification Number	
		96-712461 (A)	96-638391 (B)
Average annual usage	m^3/year	231	579
Current billing index (register total)	m^3	1,741	10,912
Billing index at optimality	m^3	1,617	1,158
Average accuracy at optimality	%	91.8	92.5
Optimal replacement period (ORP)	years	7	3
Total cost at optimality	US$	555	1,278

From Table 4.13, meter A is within the optimality replacement volume range (1,617-2,079 m^3) or a frequency replacement period of every 7 to 9 years and needs to be replaced now before the cost of water loss due to meter inaccuracy exceeds replacement costs. At 9 years, the billing index is 2,079 m^3 and the total cost is US$ 560. So it does not make much difference whether meter A is replaced after 7, 8 or 9 years as the annual incremental cost from 7 to 9 years is minimal. Meter B has far exceeded the optimality billing index and should be a priority for replacement. At a billing index of 10,912 m^3, the total cost to the utility is US$ 2,879 which is more than twice the cost at optimality of US$ 1,278. For both, meters, about 30% of the total cost is due to meter replacement activities and 70% is revenue lost as a result of meter under-registration.

4.6.6 Sensitivity analysis

In both the LCC and regression analysis models, uncertainties are not accounted for in an explicit manner. However, the disadvantage is partially compensated for by a sensitivity analysis. Sensitivity analysis was performed with respect to (i) meter degradation rate, and (ii) cost of water. The sensitivity results for meter A are presented in Figures 4.20 and 4.21.

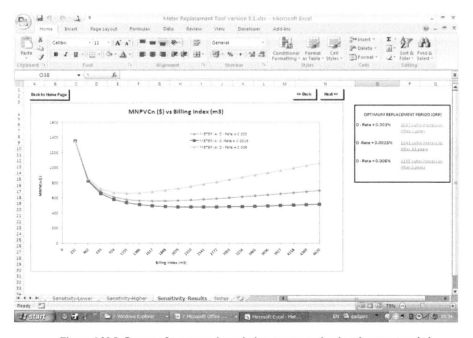

Figure 4.20 Influence of accuracy degradation rate on optimal replacement period

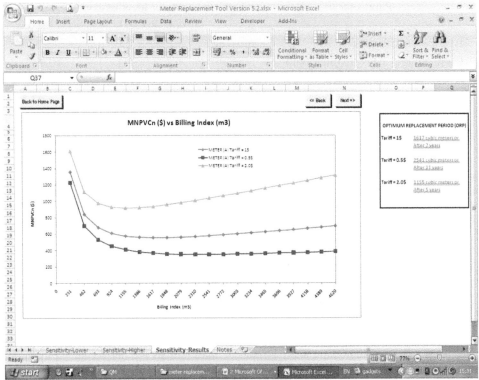

Figure 4.21 Influence of tariff on optimal replacement period

As depicted in Figure 4.20, the meter accuracy degradation rate has a major influence on the ORP. By changing the degradation rate from 0.3% per year to 0.15% per year increases ORP by 4 years i.e. from 7 years to 11 years. Therefore it is important to ensure that input data (customer demand profiles and test bench results) used to generate the meter accuracy degradation rate is very accurate, reliable and representative for all the different cohorts of meters under consideration.

The price of water has a significant influence on the ORP as indicated in Figure 4.21. Reducing the cost of water by half from US $1 to US $0.5 per m^3, extends ORP by 4 years. The result is straight forward, when the cost of water increases, the meter ORP happens sooner to minimize costs.

4.6.7   Limitations of the Model

Although the model is a valuable decision support tool, it has the following limitations:

1. The model is developed specifically for optimal replacement of multi-jet velocity meter types with minimal failure frequency. However, when dealing with the positive displacement (piston-type) meters with high failure rates, the failure costs must be incorporated in the model using appropriate methodologies (Lund 1988; Noss et al. 1987).

2. The model requires costly establishment of consumption pattern and meters' error curves standardised databases which may be out of reach for many small water

utilities in the developing countries with often limited resources. To address this a more simplified tool based on comparison of monthly billed volumes should be considered (Arregui et al. 2003).

3. For the model to work effectively, it requires integration of the meter management and customer billing databases. The utility is currently in the process of integrating the two databases to start utilizing the tool.

4. The tool does not eliminate the continued use of meter readers for early detection of failed meters and subsequent service or replacement. It is a supplementary tool.

4.7 Conclusions and Recommendations

4.7.1 Conclusions

This Chapter evaluated water meter performance in the KWDS and developed an integrated water meter management model. The model will help water utility managers to make informed decisions in addressing metering challenges and thus maximizing utility revenues. Based on the analysis of performance of water meters in the KWDS, the following conclusions can be drawn:

1. Global metering error is high and has been estimated at an average of $21.5\% \pm 0.9\%$.
2. Metering errors due to household elevated storage tanks are estimated at $4.0\% \pm 0.1\%$ for new meters and $67.2\% \pm 0.1\%$ for old meters (10-15 years).
3. The revenue water loss in KWDS attributed to sub-metering has been estimated at 18.0%. This under-registration of sub-meters is due to individual ageing of sub-meters and low network pressures.
4. Water demand profiling data provides a powerful tool for optimal sizing decisions of in-service customer meters.
5. For single-family households with storage elevated tanks, about 25% of use occurs at flow rates between 0 and 35 L/h. The high percentage of water use at very low flows, where meters are least efficient contributes significantly to metering errors. Elevated water storage tanks with ball valves have influence on meter sizing, meter class selection and starting flow required for accurate measurements.
6. It is critical that water meters are correctly sized to minimize life cycle costs to the utility. Guidelines for meter selection and optimal sizing based on demand profiling and economic optimization techniques have been established. In addition, a procedure for estimating water losses due to metering inaccuracies and meter failure is proposed.
7. A decision-aid model has been developed from routine meter testing data and readily available utility data. The optimal replacement frequency is derived by minimizing the value of unmeasured water due to metering inaccuracies and the cost of metering.
8. The sensitivity analyses indicated that the optimum replacement period is very sensitive to the price of water and the meter accuracy degradation rate. On the contrary, it was less sensitive to the average annual water used through the DN 15 mm meter.
9. The volumetric (piston-type) meters are not suitable for the KWDS due to the observed high failure frequency.

4.7.2 Recommendations

In light of the key findings of this study, the following recommendations and guidelines are proposed to address the problems of billing systems based on sub-metering:

i.   Utilities must promote and support sub-metering as a policy in the context of promoting water conservation, and efficient and equitable use of water.

ii. From the utility perspective of increasing the proportion of revenue vs. NRW and improving cost savings, billing of multi-family apartments and commercial complexes must be based on the master meter, as is the practice in other States and cities such as in California, Arizona, Maryland and many others in the USA (AWWA 2000), and in Paris, France (Nguyen 2010).

iii. The property owner or his agent should pass on the charges to the tenants based on actual consumption billing via the sub-meters and an agreed upon allocation formula.

iv. For condominium properties, formation of home owners associations (HOA) to manage water billing payments (master meter) and allocations (sub-meters) is recommended.

v.   Development of national guidelines and policies is necessary to regulate the rapidly growing business of sub-metering to protect all stakeholders.

vi. Use of Unmeasured Flow Reducers (UFRs) should be promoted to enhance accuracy of sub-meters. UFRs have been reported to recover about 94% and 14% of water flows below start-up flow rate and at Q_{min} for domestic meters (1-7 years) of Class C turbine meters and with a Q_{min} of 25 ℓ/h (Fantozzi 2009).

vii. As a long-term goal, where feasible water utilities in developing countries should strive to provide 24 hour water supply of good quality and at sufficient pressures to minimize metering errors due to sub-metering.

viii. Further studies should be carried out on more sub-metered properties, to minimize uncertainties due to a small sample size and thus to enable more confidence in the results.

Finally, the most appropriate meter recommended for the KWDS is the simple but robust velocity meter types (single-jet and multi-jet) that are magnetically driven and equipped with dry dial sealed registers. The advantage of dry dial meters is that their counter gears do not come in contact with the water, thus minimizing failure rate due to particulates in water (van Zyl 2011). In a recent research study on the accuracy of in-service water meters at low and high flow rates, undertaken at Utah State University for the Water Research Foundation (USA), the multi-jet meter was found to out-perform the single-jet meter in withstanding sand particulates in water and accuracy degradation using clean water over the meter's full life (Barfuss et al. 2011). This is probably explained by the way water flow impacts on the meter turbines. Unlike the single-jet where there is direct impact on the turbine from the flow of water, in a multi-jet meter there are several points at which the water rotates the turbine. This means that the forces on the turbines are better balanced, thus reducing wear on the moving parts and providing a much longer life. In addition, the multi-jet water meter retains accuracy over a longer period as the load is evenly distributed across the turbines. This makes the multi-jet water meter more attractive for developing countries. However, this should not be generalised, as laboratory conditions are quite different from real network conditions. In addition, meter performance varies significantly between meter manufacturers and models. Another disadvantage of multi-jet meters is that they are not very sensitive to low flow rates, and the starting flow rate can deteriorate significantly with time (van Zyl 2011). The decision as to which meter is best depends on local network conditions and other policy factors involved in optimal meter replacement (Arregui et al. 2011). It is advisable that each water utility undertakes in-situ pilot studies with different meter technologies to track the evolution of the meter metrological performance over time. It is by only doing so, that a water utility can develop its own meter selection policy based on which meter performs best in its water distribution network.

4.8 References

Allander, H. D. (1996). "Determining the economical optimum life of residential water meters." *Water Engineering and Management*, 143(9), 20-24.

Arregui, F., Cabrera, E., Cobacho, R., and Garcia-Serra, J. (2006a). "Reducing Apparent Losses Caused by Meters Inacuracies." *Water Practice and Technology*, 1(4), doi:10.2166/WPT.2006093.

Arregui, F., Cabrera Jr, E., Cobacho, R., and Palau, V. (2003). "Management strategies for optimum meter selection and replacement." *Water Science and Technology: Water Supply*, 3(1/2), 143-152.

Arregui, F., Jr., C. E., and Cobacho, R. (2006b). *Integrated Water Meter Management* IWA Publishing, London.

Arregui, F. J., Cobacho, R., Cabrera Jr, E., and Espert, V. (2011). "Graphical Method to Calculate the Optimum Replacement Period for Water Meters." *Journal of Water Resources Planning and Management*, 137(1), 143-146.

Arregui, F. J., Soriano, J., Cabrera Jr, E., and Cobacho, R. (2012). "Nine steps towards a better water meter management." *Water Science and Technology*, 65(7), 1273-1280.

AWWA. (1999). "Water Meters - Selection, Installation, Testing, and Maintenance: Manual of Water Supply Practices (M6) ", American Water Works Association, Denver.

AWWA. (2000). "Water Submetering and Billing Allocation: Adiscussion of issues and Recommended Industry Guidelines. An AWWA Draft White Paper." American Water Works Association, Denver, Colorado.

AWWA. (2004). *Sizing Water Service Lines and Meters: Manual M22*, American Water Works Association, Denver, Colorado.

Barfuss, S. L., Johnson, M. C., and Neilsen. (2011). "Accuracy of In-Service Water Meters at Low and High Flow Rates." Water Research Foundation, Denver, CO, USA.

Bowen, P. T., Harp, J. F., Entwistle Jr, J. M., and Shoeleh, M. (1991). *Evaluating Residential Water Meter Performance*, AWWA Research Foundation., Denver, Colorado, USA.

Butler, D., and Memon, F. A. (2006). *Water Demand Management*, IWA Publishing, London.

Cobacho, R., Arregui, F., Cabrera, E., and Cabrera Jr, E. (2008). "Private Water Storage Tanks: Evaluating their Inefficiencies." *Water Practice and Technology*, 3(1), doi:10.2166/WPT.200825.

Criminisi, A., Fontanazza, C. M., Freni, G., and La Loggia, G. (2009). "Evaluation of the apparent losses caused by water meter under-registration in intermittent water supply." *Water Science and Technology:WST*, 60(9), 2373-2382.

Eagle, D., and Kiefer, D. B. (2004). "Compairing Capital Projects With Unequal Lives: Inflation and Technology Issues." *The Journal of Accounting and Finance Research*, 12(7).

Egbars, C., and Tennakoon, J. (2005). "Ipswich Water's Meter Replacement Strategy." Water Asset Management International, 19-21.

EN-14154-1:2005+A1. (2007). "Water Meters-Part 1: General Requirements."

EN-14154-2:2005+A1. (2007). "Water Meters-Part 2: Installation and Conditions of Use."

Fantozzi, M. (2009). "Reduction of customer meters under-registration by optimal economical replacement based on meter accuracy testing programme and unmeasured flow reducers." *Proceedings of the 5th IWA Water Loss Reduction Specialist Conference*, Cape Town, South Africa, 233-239.

Flores, and Diaz. (2009). "Meter assessment in Madrid." *Proceedings of the 5th IWA Specialist Conference on Efficient Water Use and Management*, Sydney, Australia, October 26-28 CD-ROM.

Gottfried, B. S. (2007). *Spreadsheet Tools for Engineers Using Excel* McGraw Hill, New York.

Hill, C., and Davis, S. E. (2005). "Economics of Domestic Residential Water Meter Replacement Based on Cumulative Volume." *AWWA Annual Conference*, San Francisco, California, CD ROM.

ISO-4064-1. (2005). "Measurement of water flow in a fully charged closed conduits-meters for cold potable water and hot water. Part 1 Specifications."

ISO-4064-2. (2005). "Measurement of water flow in fully charged closed conduits-meters for cold potable water and hot water. Part 2: Installation Requirements."

ISO-4064-3. (1993). "Measurement of water flow in closed conduits-meters for cold potable water. Part 3: Test methods and equipment."

ISO/IEC. (2008). "Uncertainty of Measurement - Part 3: Guide to expression of uncertainty in measurement (GUM:1995)." International Organization for Standardization(ISO)/International Electrotechnical Commission (IEC), Geneva, Switzerland.

Lund, J. R. (1988). "Metering Utility Services: Evaluation and Maintenance." *Water Resources Research*, 24(6), 802-816.

Male, J. W., Noss, R. R., and Moore, I. C. (1985). *Identifying and Reducing Losses in Water Distribution Systems*, Noyes Publications, New Jersey.

Montgomery, D. C., and Runger, G. C. (2007). *Applied Statistics and Probability for Engineers, Fourth Edition*, John Wiley and Sons, Inc, Arizona State University, USA.

NAMS. (2004). *Optimised Decision Making Guidelines*, NZ National Asset Management Steering Group, Auckland, New Zealand.

Newman, G. J., and Noss, R. R. (1982). "Domestic 5/8 Inch Meter Accuracy and Testing, Repair, Replacement Programs." *AWWA Annual Conference*, Miami Beach, Florida, 341-352.

Nguyen, B. (2010). "Paris enters a new automated era of water metering." *Water Utility Management International*, 5(1), 21-23.

Noss, R. R., Newman, G. J., and Male, J. W. (1987). "Optimal Testing Frequency for Domestic Water Meters." *Journal of Water Resources Planning and Management*, 113(1), 1-14.

Pasanisi, A., and Parent, E. (2004). "Bayesian Modelling of water meters ageing by mixing classes of devices of different states of degradation." *Applied Statistics Review*, 52(1), 39-65 (**in French**).

Richards, G. L., Johnson, M. C., and Barfuss, S. L. (2010). "Apparent losses caused by water meter inaccuracies at ultralow flows." *Journal of American Water Works Association*, 105(5), 123-132.

Rizzo, A., and Cilia, J. (2005). "Quantifying Meter Under-registration caused by the Ball Valves of Roof Tanks (for indirect plumbing systems)." *Leakage 2005*, Halifax, Canada.

Sullivan Jr, J. P., and Speranza, E. M. (2008). "Proper Meter Sizing for Increased Accountability and Revenues." Water Loss Control, J. Thornton, R. Sturm, and G. Kunkel, eds., McGraw Hill, New York.

Tao, P. (1982). "Statistical Sampling Technique for Controlling the Accuracy of Small Maters." *Journal American Water Works Association*, 74(6), 296-304.

Thornton, J., Sturm, R., and Kunkel, G. (2008). *Water Loss Control*, McGraw-Hill, New York.

Ugarelli, R., and Di Federico, V. (2010). "Optimal Scheduling of Replacement and Rehabilitation in Wastewater Pipeline Networks." *J. of Water Resources Planning and Management*, 136(3), 348-356.

Van Zyl, J. E. (2011). "Introduction to integrated water meter management." Water Research Commission (WRC TT490/11), South Africa.

Vreeburg, J. H. G., and Boxall, J. B. (2007). "Discolouration in potable water distribution systems: a review." *Water Research*, 41, 519-529.

Wallace, L. P., and Wheadon, D. A. (1986). "An Optimal Meter Change-out Program for Water Utilities." *AWWA Annual Conference*, Denver, Colorado, 1035-1042.

Williams. (1976). "Water Meter Maintenance." *AWWA 96th Annual Conference Proceedings*, New Orleans, La.

Woodward, D. G. (1997). "Life cycle costing - theory, information acquisition and application." *International Journal of Project Management*, 15(6), 335-344.

Yee, M. D. (1999). "Economic Analysis for Replacing Residential Meters." *Journal American Water Works Association*, 91(7), 72-77.

Chapter 5 - Assessment of Apparent Losses in Water Distribution Systems

Parts of this chapter are based on:

Mutikanga, H.E., Sharma, S.K., and Vairavamoorthy, K. (2011). "Assessment of Apparent Losses in Urban Water Systems". *Water and Environment Journal*, 25(3), 327-335.

Mutikanga, H.E, Sharma, S.K., and Vairavamoorthy, K. (2009). "Apparent Water Losses Assessment: The case of Kampala City, Uganda". Proceedings of the 5th IWA Water Loss Reduction Specialist Conference, Cape Town, South Africa, April 26-30, p 36 – 42, ISBN:978-1-920017-38-5.

Summary

Apparent losses (AL) are the non-physical loss component of water losses in the distribution system. They include all types of inaccuracies associated with customer metering, data handling errors (meter reading and billing) and unauthorised consumption (theft or illegal use). Most research carried out in the last decade has mainly focused on physical losses (leakage). Until now, there are no set procedures and guidelines for assessment of AL. In this study a methodology for assessing different components of AL has been developed. The methodology was then applied to the KWDS to estimate different apparent loss components. Guidelines for assessment of AL in water utilities of the developing countries with insufficient resources and data limitations to carryout in-depth assessment are also established. The major apparent loss components for Kampala were found to be high metering inaccuracies (-22%) and illegal use (-10%) expressed as a percentage of revenue water. Meter reading errors (-1.4%) and data handling and billing errors (-3.5%) were low. The influence of ultralow flows on the AL caused by meter inaccuracies for the different meter models in the KWDS was also examined. The findings reveal that AL caused by meter inaccuracies at low flow rates are influenced by meter type, manufacturer and usage (or age) of the meter. Although perceived to be a problem of the developing countries, AL in developed countries has also been found to be significant and seem to be steadily increasing. Apparent loss in urban WDSs is still a research area and more effort is still required in developing apparent loss interventions to match leakage interventions.

5.1 Introduction

One of the major challenges facing water utilities is the high levels of apparent water losses especially in developing countries. In this chapter, we tackle the problem of AL in line with the third objective of this study (Chapter 1) and research gaps identified in Chapter 2. Although AL are often smaller in volume, they are usually significant in monetary terms. According to a World Bank report (Kingdom et al. 2006), about 20% of NRW (or 2.4 billion m^3/year) and 40% (or 10.6 billion m^3/year) of water is AL in developed and developing countries respectively. Conversely, the same report indicates that the cost of AL is about 45% of the total cost of NRW (about 5.5 billion US $/year) for both developed and developing countries. This is probably due to the fact that water is more expensive in the developed countries than in the developing countries. Although perceived to be a problem of the developing countries, with about US$ 2.4 billion per year of AL in the developed countries, clearly, much work remains to be done as well to minimize the losses. However, most of the NRW components are based on estimates often using imprecise data and these figures should be cautiously treated.

AL result into appreciable revenue loss for water utilities and distort the integrity of consumption data required for various management decisions and engineering studies. This problem is more pronounced in the water utilities of the developing countries. As highlighted in Chapter 2, most research that has been done in the last decade in Europe, Australia and North America focused mainly on real loss (leakage) component of water losses (Fanner et al. 2007; Farley and Trow 2003; Puust et al. 2010; Thornton et al. 2008). In England and Wales for instance, household metering is still low and was reported to be 37% in 2009-2010 (OFWAT 2010) and this probably explains why research has focused less on apparent water loss. The UK water industry has not yet embraced the IWA Standard Water Balance Methodology and terminologies such as "apparent losses" are not used. For example, metering inaccuracies are accounted for as part of billed metered authorized consumption thus under-declaring NRW levels (Farley and Trow 2003; Lambert 1994; OFWAT 2010).

Apparent loss control in water supply systems is in its infancy, and much work remains to be done to bring it to par with available real loss interventions (AWWA 2003). Until now there are no set procedures and guidelines for assessment and control of apparent losses in the water distribution systems. Considerable efforts are being made by the IWA Water Loss Task Force (WLTF) in assessing components of AL and some initial results have been presented by Thornton and Rizzo (2002). However, most research studies have only focused on assessing the metering inaccuracies component of apparent losses (Arregui et al. 2006a; Criminisi et al. 2009; Richards et al. 2010; Rizzo et al. 2007a).

In the absence of adequate data and a proper methodology, most developed countries use default values or rules of thumb (e.g. unauthorised consumption is computed as 0.5% of total system input volume (SIV) and domestic meter under-registration as 2% of metered consumption) which tend to be lowest values for well managed water systems, for computation of apparent loss components (Seago and Mckenzie 2007). In the BENCHLEAK Model for calculating components of NRW, a lump sum default value of 20% of total water losses is used to compute the AL component (McKenzie et al. 2002). In Australia, AL are computed as 1% to 3% of SIV, 9% of SIV in Malaysia and 9.2% of SIV in Korea (Lambert 2002). In cases where illegal use is not excessive, Thornton et al. (2008) recommends use of 0.25% of water supplied to compute unauthorized consumption. These default values may not be appropriate for most developing countries where illegal use of water is rather high and water meter management policies are either non-existent or ineffective.

Attempts to assess apparent loss components have been made by several researchers. Rizzo et al. (2007b) proposed an apparent water loss audit based on a pilot zone or district metered area (DMA) approach. They proposed to first remove all leakages in the DMA or calculate the leakage component using the minimum night flow of the zone. They also propose that the first component of apparent loss to be analyzed is water theft and the other two components are analyzed later using automated meter reading (AMR). This approach is unrealistic, costly and is very difficult to apply in water distribution systems of the developing countries with intermittent water supply and widespread water theft. Tabesh et al. (2009) in their study of evaluating water losses in one of the Iranian towns attempted to assess apparent losses but did not clearly clarify how to verify meter reading errors. They compared meter readings and consumptions, which is erroneous as you cannot compare "apples" with "mangoes". They also introduced new errors of operational and management nature which they did not clearly define and compute. These operational and management errors are likely to cause confusion as they are different from those defined by IWA and accepted internationally.

In this chapter a methodology for assessing apparent water loss components is presented. The method is applied to assess the AL in the KWDS. The method was found to be effective in estimating the different components of AL. Furthermore, the apparent loss trends in some developing and developed countries are also analyzed. Guidelines for estimating AL in urban WDSs of the developing countries with often data-poor networks are established. Lastly, AL caused by water meter inaccuracies at ultralow flows in the KWDS is examined. Four different domestic meter models of size 15 mm are analyzed at low flow rates (below Q_{min}). These flows are not considered under the ISO standard but are very important in understanding the evolution of in-service water meter performance. Guidelines for estimating revenue losses due to metering inaccuracies are also established. The rest of the chapter is organized in the following way. Section 5.2 outlines the research methodology used in the study in developing a procedure for assessing AL. Section 5.3 presents the apparent loss components for the KWDS assessed using the developed methodology. Section 5.4 examines some case studies in developing and developed countries. Guidelines for estimating the apparent loss components for water utilities in developing countries with insufficient resources for detailed field investigations are presented in section 5.5. The AL caused by metering inaccuracies at very low flow rates in the KWDS are analyzed in section 5.6. Factors that influence the level of AL and strategies for minimizing the losses are discussed in section 5.7. Finally, section 5.8 draws some conclusions and makes recommendations for further research.

5.2   Research Methodology

A water balance was carried out for Kampala city's distribution system using the standard IWA methodology (Lambert and Hirner 2000; Thornton 2002). The only difference is that real losses were computed first based on operational data and well established procedures using the burst and background estimate (BABE) methodology and FAVAD theory (Fanner et al. 2007; Lambert 2002; May 1994; Thornton et al. 2008). These methods are suitable for systems with regular water supply. The Kampala central business district (CBD) with regular supply was used for the assessment of real losses. The quantified real losses were then assumed to represent the entire network. This assumption is realistic as the CBD is the oldest part of the network with a mixture of pipe materials and relatively high pressures. AL were then computed as the difference between systems input volume and real losses. Different approaches were then used to assess the different components of AL (metering inaccuracies, meter reading errors, billing and data handling errors and unauthorized use of water). The different approaches used for estimating different components apart from metering inaccuracies that were presented in the previous chapter are outlined. The only addition to metering was the inclusion of customer meter sizes of DN 20 to 40 mm in the assessment.

5.2.1   Assessment of meter reading errors

Customer meter reading in KWDS is carried out using the traditional approach whereby meter readers visit individual meters and collect monthly readings manually. This approach of reading meters is prone to human errors particularly where readings are taken hurriedly to meet meter reading targets. In Kampala, each meter reader on average reads about 150 meters per day.

Meter reading audits based on random sampling were carried out (in different zones of the network) in a day during the months of November 2008 and March 2009 to verify accuracy of meter readings submitted by meter readers for billing purposes. The readings of the auditors were then compared with the readings submitted by meter readers. The readings that show un-realistic variances were regarded as erroneous readings. The billed consumption volume based on erroneous meter readings was then summed up (z m^3) and expressed as a percentage of volume of water sold (y m^3) for the total audited accounts. The result is the meter reading error (z/y X 100) and is expected to be representative of the entire system. The auditors were utility employees from all departments including, the executive director, senior managers and lower cadre staff. The total numbers of meters audited was 12,000 (or about 10% of total customer meters).

5.2.2   Assessment of data handling and billing errors

These errors arise in the process of transmitting or capturing data from the meter reading sheets into the customer billing database. Gaining access to some customer meters located inside customer premises is difficult due to increasing number of working couples leaving no one at home or at times leaving guard dogs. In addition, there is an increase in the number of defective customer meters and a lag of new connections update in GIS that complicates tracing installed meters. For these reasons, manual meter reading success rates (ratio of actual readings to total readings) in Kampala are on the decline. In this case, customer water use is estimated based on historic consumption trends. While this is a reasonable approach, multiple cycles of meter reading without an actual reading greatly increase the prospect of inaccurate estimates (AWWA 2009; Thornton et al. 2008).

Data capturing audit was carried out to compare the input data used for billing and the readings on the meter reading sheets submitted by meter readers. The readings that were wrongly captured in the billing database were established and their corresponding total volume computed (x m^3). If the water sales for the assessment period were y m^3, the percentage data handling errors were computed as (x/y X 100). A sample of 7,438 customer accounts (or about 6% of total accounts) was analyzed.

Billing errors from poorly estimated volumes that resulted into billing adjustments were generated from the customer billing database and their volume summed up (v m^3). If the water sales for the assessment period were y m^3, the percentage data handling errors were computed as (v/y X 100). Billing errors for the 3 months (October-December 2008) were used for computation of billing errors component of apparent losses.

Billing errors arising from the billing software programming and algorithms were considered negligible and were not assessed.

5.2.3   Assessment of unauthorized water use

Identifying unauthorized consumption in a water distribution system is a challenging task. A proactive approach through investigations of suspicious trends of billing data consumption (zero consumptions, negative consumptions etc) and employing illegal use informers was utilized. These individual site inspections fall under the category of bottom-up auditing of unauthorized consumption. Advertisements were placed in the local newspapers requesting anyone with information on illegal use of water to report to the utility and once confirmed, a cash reward in the range of UShs 50,000 to 200,000 or (US \$25 – 100) was offered depending on the size of the illegal user discovered. In addition historical records on illegal users and their consumption patterns (average monthly per capita consumption of different user groups obtained from customer billing database from 2005 to 2008) were used to quantify the total volume of unauthorized use (q m^3).

The different components of illegal use were broken down as follows:

- Domestic illegal use, q_d = number of properties x 20 m^3 per month
- Commercial illegal use, q_c = number of properties x 500 m^3 per month
- Government Institutions, q_g = number of properties x 500 m^3 per month
- Public Standpipes, q_p = number of properties x 50 m^3 per month
- Total volume of unauthorized use, $q = q_d + q_c + q_g + q_p$

The total volume is then expressed as the percentage of water sales y m^3 during the assessment period. This was the proactive unauthorized use of water component, computed as (q/y x 100).The difference between the apparent losses volume and the sum of the volumes of the above components is the unknown unauthorized use. This methodology is shown in Figure 5.1 and could be used at utility, zonal or DMA level.

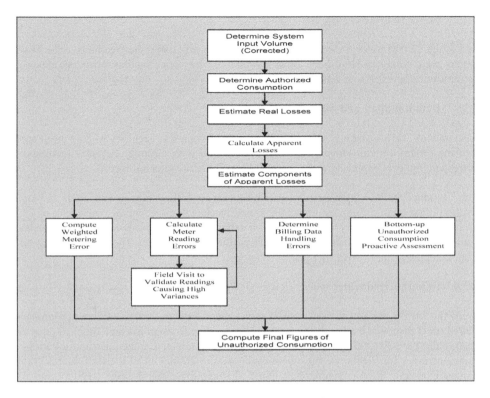

Figure 5.1 Methodology for assessing apparent water loss components

5.3 Application of the Methodology to KWDS

In order to assess the apparent water loss components in the KWDS, metering accuracy, meter reading errors, data handling and billing errors and illegal consumption were analyzed. The results for each of these components of AL are elaborated in the following sections.

5.3.1 Metering accuracy

Based on the methodology of determining weighted average meter accuracy in Chapter 4, the global weighted meter accuracy was estimated at 78% ± 2% or weighted meters error of - 22% ± 2%. This means meters are under-registering consumption by about 22% resulting into significant revenue loss to the utility. The weighted metering error for users with roof tanks was more significant and at 25% while that of users on direct supply was 7%.

The metering efficiency ranged from 80 – 85% (meters with billing index less than 1,000 m^3) to 72% (meters with billing index less than 3,000 m^3). Most domestic meters (>5,000 m^3) were unable to register low flows (< 100 L/h). The proportion of water volume passing through the meters at small flow rates (< 100 L/h) is about 40% for residences with roof tanks (82% of all customers) . For customers on direct supply like public stand pipes or yard taps (18% of all customers) the proportion volumes delivered at small flow rates (< 100 L/h) is much lower (<3%).

5.3.2   Meter reading errors

Out of the 12,000 water meters audited to confirm accuracy of meter readings, only 73 water meters were found to have been incorrectly read. The meter reading error computed as a result was 1.4% ± 0.1% of water sold.

5.3.3   Data handling and billing errors

A sample of 7,438 consumer accounts was audited to confirm accuracy of data input capture into the billing system. Only 8 accounts were found with wrongly captured readings. The corresponding data capturing error computed was 2.5% of water sold.

The customer billing database was queried to assess billing adjustment made arising out wrong billing consumption estimates for 3 months (October-December 2008). Out of 22,855 accounts found to have been billed based on estimates, 643 accounts had billing adjustments made as a result of customer complaints raised for wrong billings. The corresponding billing error computed was 1% of water sold.

5.3.4   Unauthorized water use

From the illegal use records of the last two years (2007 & 2008), the average number of illegal users discovered annually was 939. The increase from an average of 732 illegal cases (2003-2008) to 939 (2007-2008) could be attributed to proactive illegal use investigations and cash incentives for illegal use informers introduced in 2008. Out of these cases, 83.3% were domestic users, 12.5% commercial users, 2.1% public stand pipes (PSPs) and 2.1% government institutions. Using the average monthly consumptions for each consumer category and the annual water sales, the illegal use component was computed as 4% of water sales.

From the known figure of apparent losses computed from the water balance and the sum of all the above computed components, the unknown unauthorized consumption was found to be 6% of water sales.

5.3.5   Apparent losses component breakdown

The total apparent loss component for KWDS was found to be about 37% ± 2% of water sales or revenue water. The breakdown including uncertainties in data acquisition and processing is as follows:

- Metering inaccuracies   –   22±2% of water sales
- Meter reading errors   –   1.4±1% of water sales
- Data handling and billing errors   –   3.5±0.5% of water sales
- Unauthorized use of water   –   10±2% of water sales

The uncertainties were calculated using the ISO Guide to the expression of uncertainty in measurements (ISO 2008) and statistical techniques for algebraically operated data uncertainty (Montgomery and Runger 2007).

5.4   Apparent Losses in Developing and Developed Countries

AL in WDSs are generally not very well understood. For example, the World Bank estimates AL in developing and developed countries to be 40% and 20% of NRW respectively

(Kingdom et al. 2006) while the ADB estimates AL in Asian cities to be 50-65% of NRW (ADB 2007). In order to demystify this myth and try to understand AL in WDSs, some recent case studies in literature have been identified and examined.

5.4.1   Apparent Losses in developing countries

The AL in some developing countries have been reported in various identified studies (Batista and Mendonca Jr 2009; Bidgoli 2009; Dimaano and Jamora 2010; Garzon-Contreras et al. 2005; Garzon-Contreras and Palacio-Sierra 2007; Garzon-Contreras et al. 2009; Lievers and Barendregt 2009; Makara 2009; Medeiros et al. 2010; Mutikanga et al. 2009b; Mutikanga et al. 2011; Schouten and Halim 2010; Seago et al. 2004; Sharma and Nhemafuki 2009; Shin et al. 2005; Wegelin et al. 2011) and are summarized in Table 5.1.

Table 5.1 Apparent losses in some developing countries

Work	Country	City/Utility	NRW (% of SIV)	AL (% of NRW)	UU (% of AL)	MI (% of AL)
Schouten & Halim (2010)	Indonesia	Jarkata/Palyja	48	39	53	47
Sharma & Nhemafuki (2009)	Nepal	Bhaktapur	54	56	17	83
		Dhulikhel	16	25	80	20
Dimaano & Jamora (2010)	Phillipines	Manila/Maynilad	65	20	69	31
Shin *et al* (2005)	South Korea	Busan	26	14	0	100
Makara (2009)	Papua New Guinea	Eda Ranu/Port Moresby	64	16	52	47
Garzon-Contreras & Palacio-Sierra (2007) Garzon-Contreras *et al* (2009)	Colombia	Medellin/EPM	33	32	75	25
		Cali	44	48	56	44
		Bogota	37	43	55	45
Garzon-Contreras *et al* (2005)	Colombia	40 water utilities	41	34	ND	ND
Medeiros *et al* (2010); Batista & Mendonca Jr. (2009)	Brazil	Sao Paulo SABESP	ND	33	26.6% **	20.24%*
Mutikanga *et al* (2009b, 2011).	Uganda	Kampala/NWSC	40	52	43	57 (22*)
Lievers & Berendregt (2009)	Ghana	Accra-Tema	55	55	18%***	ND
Seago *et al* (2004); Wegelin *et al* (2011)	South Africa	All munipalities	23	5 to 50	30	30
Bidgoli (2009)	Iran	Tehran/NWWEC	36	44	26	67

UU = Unauthorized Use; MI = Metering inaccuracies; * = metering error; ** = %age of inspected connections; *** = %age of disconnected customer connections inspected; ND = no data.

From Table 5.1, the following deductions can be made:

- NRW expressed as a percentage of system input volume (SIV) varies from 16% to 65% with an average of 41%. In terms of volume, the highest NRW is 1,497 ML/d in Maynilad, Manila and the smallest is 61 ML/year in Dhulikhel, Nepal.
- AL ranges from 14 to 56% of NRW with an average of 34%.
- The most significant component of AL is metering inaccuracies (MI) with an average of 54% of AL but ranges from 20% to 100% where illegal use of water has been reported to be zero.
- Unauthorized Use (UU) averages 46% of AL with a range of 0 to 80%. Maynilad in Manila has the highest figure in terms of volume lost (202 ML/d) (Dimaano and Jamora 2010). This is more than twice the amount of estimated illegal use for England and Wales.

- In Kampala city, unauthorized consumption occurred in many different ways and was categorized in a chronological order depending on their magnitude as follows: (i) illegal reconnections after disconnection for non-payment of bills (40%); (ii) consumption meter bypass (35%); (iii) illegal connections (16%); and (iv) meter tampering and reversing (9%) (Mutikanga et al. 2011).

5.4.2 Apparent losses in the developed countries

The AL in a few developed countries are summarized in Table 5.2 while additional information on AL trends in England and Wales is shown in Figure 5.2. Generally, NRW and apparent losses are low in developed countries compared to developing countries. Although, apparent losses are low in terms of percentages, they could be significant in terms of volumes and money. For example, in England and Wales, AL are estimated at 118 million m^3/year (Illegal use = 32 million m^3/year and MUR = 86 million m^3/year) (OFWAT 2010); in Lisbon apparent losses are estimated to be 6 million m^3/year (MUR = 4.2 million m^3/year and unauthorized and unbilled consumption 1.8 million m^3/year) (Donnelly 2007; Donnelly et al. 2009); and in Philadelphia, AL is estimated at 21 million m^3/year (unauthorized consumption =6 million m^3/year and metering inaccuracies = 0.4 million m^3/year and systematic data handling error = 14.6 million m^3/year). Philadelphia's fiscal year 2006 water audit indicates that the city's real losses of 21, 620 MG are almost four times its AL based on volume. However, in financial terms, apparent losses were estimated at $ 20 million compared to $ 4 million for real losses. This is because AL are valued at the retail price charged to customers, whereas real losses are valued at the variable production cost (AWWA 2009). Although most developed countries have reduced their leakage to almost economic levels, they are being urged to further reduce these losses to sustainable levels that include not only economic aspects but environmental and social aspects as well. It is therefore prudent for water utilities in developed countries to strive and minimize AL as well due to the huge financial improvement potential at stake.

Table 5.2 Apparent losses in some developed countries

Work	Country	City/Utility	NRW (% of SIV)	AL (% of NRW)	UU (% of AL)	MI (% of AL)
Donnelly (2007) Donnelly *et al* (2009)	Portugal	EPAL/Lisbon	20	35	30	70
Flores & Diaz (2009); Sanchez (2007)	Spain	Madrid/Canal de Isabel II	21	57	ND	92 (14.32*)
Fantozzi (2009)	Italy	Padova	ND	ND	ND	8.84%*
		Pinerolo	ND	ND	ND	28%*
OFWAT (2010)	England & Wales	22 Companies	21	11	27	73 (2.9-7.3*)
AWWA (2009)	USA	Philadelphia	33	20	28	2

*metering under-registration (MUR)

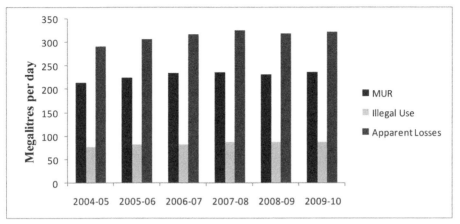

Figure 5.2 Apparent losses trends in England and Wales
(Source: OFWAT, 2010)

In England and Wales, a lot of effort to control water losses has been focused on reducing leakage only. Although leakage reduced by 9% from 2004/05 to 2009/10, AL increased by 11% in the same period (Figure 5.2) (OFWAT 2010). In London, Thames Water, Britain's largest supplier has reported 18,000 cases of unauthorised use of water out of which 1,573 were voluntary confessions during the two months amnesty period to avoid penalty charges (Lea 2011) and total illegal use is estimated at 7.7 ML/day (OFWAT 2010). Unless efforts are made to reduce the apparent losses, the situation is likely to worsen as household meter penetration increases from the current level of 37% to over 80 % by 2030 as projected by some companies (OFWAT 2006; Walker 2009). Meter under-registration can go up to 28% even in developed countries as in Pinerolo, Italy (Table 5.2) (Fantozzi 2009). AL could also exceed real losses as in the case of Canal de Isabel II utility in Madrid (Spain), with 1.2 million installed customer meters making up 92% of apparent water losses (Flores and Diaz 2009; Sanchez 2007). In the El Dorado Irrigation District (EID) in California, customer meter under-registration in the range of 3.1 to 15.7% resulting into estimated AL of $ 300,000 annually have been reported (AWWA 2009).

From Tables 5.1 and 5.2, the most dominant component in developing countries seem to be metering inaccuracies (52%) followed by unauthorized use (46%). This pattern seems to be true for developed countries as well. AL due to metering errors are estimated at 92% in Spain (Sanchez 2007) and 73% in England and Wales (OFWAT 2010). Conversely, in the city of Philadelphia in the USA, customer meter inaccuracy is the least component of AL at 2% and systematic data handling errors are the most dominant at 70% (AWWA 2009). These patterns should therefore not be generalized and could vary from utility to utility depending on local conditions and apparent loss control policies in place.

5.5   Guidelines for Estimating Apparent Losses with Data Limitations

From the Kampala city's case study, reports from other NWSC-Uganda towns and the authors experience in operations of WDSs in developing countries, the framework in Table 5.3 is proposed for estimating the apparent loss components for water utilities in the developing countries with similar context information and utility profile as Ugandan cities and towns.

Table 5.3 Proposed default values for estimating apparent losses in developing countries

Unauthorized water use		Meter billing index and error	With storage tanks	Direct supply	Meter reading, data handling and billing errors	
City (> 100 000 service connections)	10%	Poor (> 7 000 m³)	-28%	-10%	Poor[a]	10%
Municipality (50 000-100 000 service connections)	3%	Average (3 500-7 000 m³)	-20%	-8%	Average[b]	6%
Medium towns (5 000-50 000 service connections)	2%	Good (< 3 500 m³)	-15%	-5%	Good[c]	2%
Small towns (< 5 000 service connections)	0.5%					

[a]No management controls in place, employees are poorly remunerated and inefficient billing system
[b]Management controls in place, fairly remunerated employees and good billing system
[c]No Well function utility with a good customer billing system
All default values are expressed as percentages of water sales or revenue water (m³).

Seago et al. (2004) proposed a similar framework for estimating apparent loss components for South African water utilities based on performance levels (very high, high, low, very low, poor, average, good) to describe data transfer errors and illegal use of water, water quality and meter age as the main factors influencing metering errors and default values expressed as percentages of current annual real losses. Their framework is very difficult to use in practice by water utilities in developing countries due to high uncertainties in estimating real losses as opposed to revenue water.

5.6 Apparent Losses caused By Water Meter Inaccuracies at Low Flow Rates

Most water meters, even new ones, tend to record less water than what actually passes through the meter causing revenue losses to the water utility. This is due to mechanical and electronic limitations. According to Arregui et al. (2006b), the energy transfer from the flowing water to the meter driving mechanism is small at low flow rates and any increase in friction increases the starting flow rate of a meter and could even halt the meter from moving. Many studies have indicated that the amount of meter under-registration is more significant at very low flow rates (Arregui et al. 2006b; Bowen et al. 1991; Noss et al. 1987). AL caused by new meter inaccuracies at ultralow flows have been reported for different meter sizes and models for the USA water industry (Richards et al. 2010). The influence of meter age and low flows induced by private storage tanks on AL caused by in-service meters have been studied in Palermo, Italy (Criminisi et al. 2009). Water meter performance at very low flow rates in the developing countries is not very well understood as meter accuracy is dependent on the condition and operation of the WDS. This study investigates the AL caused by different in-service water meter inaccuracies at very low flow rates in KWDS.

The methodology for meter sampling and testing was described in Chapter 4. However, it should be noted that flow rates being investigated here are below the standard minimum flow rates (Q_{min}) and have not been defined by ISO. The ISO standards do not require any degree of meter accuracy below the minimum flow rate. From the utility perspective of minimizing revenue losses due to metering errors, it is prudent to investigate in-service water meter performance at flow rates below Q_{min}.

The meters investigated were small domestic meters of size 15 mm with nominal flow rates

(Q_n) of 1.5 m^3/h. All meter models were of Class C apart from model 3 which was a Class B meter. Meter models 1 and 2 were of the volumetric (oscillating-piston) types from different manufacturers, model 3 was a multi-jet meter type and model 4 was the single-jet meter type. For these meter models, the minimum flow rate is 15 L/h for Class C meters and 30 L/h for Class B meters (ISO-4064-1 1993) and the maximum permissible error at these flow rates is ±5% (ISO-4064-3 1993). The laboratory results of the different meter models at different billing index groups are presented in Figures 5.3 to 5.6.

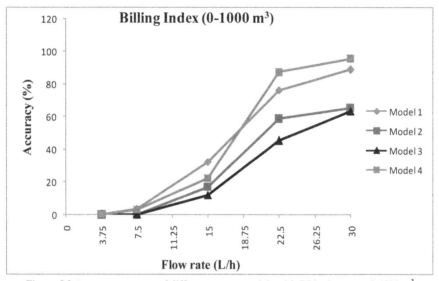

Figure 5.3 Average accuracy of different meter models with BI in the range 0-1000 m^3

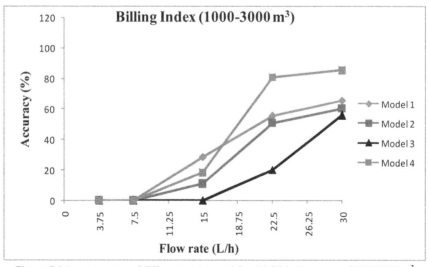

Figure 5.4 Avg. accuracy of different meter models with BI in the range of 1000-3000 m^3

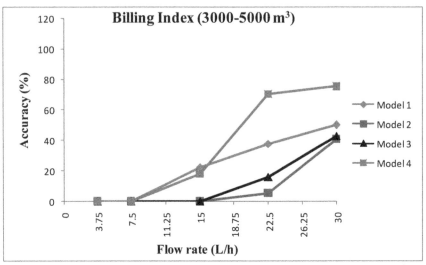

Figure 5.5 Avg. accuracy of different meter models with BI in the range 3000-5000 m³

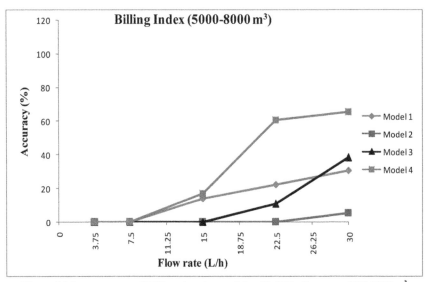

Figure 5.6 Avg. accuracy of different meter models with BI in the range 5000-8000 m³

The following observations have been made from the results:

1. The evolution of metering inaccuracies is different for the different meter types. In KWDS, volumetric meter types (models 1 & 2) tend to degrade more rapidly than velocity-type meters (model 3 & 4). The meter with the least meter accuracy degradation rate (%/m³) was the multi-jet Class B meter type (Table 5.4).

2. The influence of different meter manufacturers for the same metering technology is evident from volumetric meter types (models 1 and 2). The meter performance dominance of model 1 over model 2 across the entire range of low flows and billing index indicates that utility revenues could be compromised depending on which manufacturer you purchase meters from.

3. At billing index (BI) of 5000-8000 m³, model 2 registered the worst performance. It hardly registers any flows below 22.5 L/h and its accuracy at minimum flow rate (30 L/h) is 5.1%.

4. Although model 3 is a Class B meter, its performance with usage (age) is better than Class C meters of volumetric type. At BI of 5000-8000 m³ (Figure 5.6) its accuracy is 38.3% compared to 30.5% (model 1) and 5.1% (model 2). For the KWDS it makes more economical sense to use Class B meters instead of Class C meters whose accuracy at low flow rates degrade rapidly. This is critical if a high proportion of water consumption takes place at very low flows. As indicated in the previous chapter about 20% of domestic consumption in Kampala occurs at flow rates below 30 L/h and significant revenue losses could occur depending on which type of meter is in use.

5. For new meters (BI of 0-1000 m³), the volumetric meter type (model 1) produced the best performance at flow rates of up to 15 L/h. The single-jet meter type (model 4) tended to have the best accuracy at flow rates between 22.5 L/h and 30 L/h.

6. The most robust meter that maintains a fairly high accuracy across the range of low flows is the single jet meter type. This could probably be the best meter for the KWDS as single-jet meter types have been reported to be more resistant to particulates and small debris in water (Arregui et al. 2006b). Although the new nutating disc meter (volumetric type) was found to have produced the best meter performance at low flows out of the six meter types examined in the USA study (Richards et al. 2010), its suitability for use in developing countries is doubtful due to poor water quality often introduced during repairing of leaks and intermittent operations. Volumetric meter types are more susceptible to wear from particulates in water (Arregui et al. 2006b). In their study, Richards et al. (2010) found that, for new meters of size 40 mm (1¹⁄₂-inch) and 50 mm (2-inch) in diameter, single-jet meters tended to have the greatest accuracy at low flows.

Table 5.4 Meter accuracy degradation rates

Meter Model	Accuracy at 0-1000 m³	Accuracy at 5000-8000 m³	Accuracy degradation (%)	Useful life of a meter (m³)	Accuracy degradation rate (%/m³)
1	88.8	30.5	58.3	8000	0.0073
2	65.4	5.1	60.3	8000	0.0075
3	63.4	38.3	25.1	8000	0.0031
4	95.6	65.4	30.2	8000	0.0038

5.6.1 Quantifying apparent loss due to meter inaccuracy at low flow rates

The amount of water lost due to metering inaccuracies at low flows can be significant and varies greatly depending on meter type, billing index, presence of private storage elevated tanks and post-meter leakage flow rates e.g. dripping taps and leaking toilet cisterns. For example, all meter models examined in section (§ 5.5) could hardly register flows at 3.75 L/h and if a household tap is dripping at 3.75 L/h, and there are 10,000 leaking taps in a city, then annual lost revenue of over US $150,000 could be lost (assuming a tariff of 0.5 US $ per m³). However, this example is too simplified as water lost due to metering inaccuracy is a function of the proportion of water consumed at different flow rates (Arregui et al. 2006b; Male et al.

1985). In order to estimate the AL due to meter inaccuracy at low flows, Equation 5.1 and data in Table 5.5 collected from the KWDS was used. The methodology for data collection of domestic consumption profiles was presented in chapter 4.

$$CALI = \sum_{n=1}^{n} \frac{P_W Q_n \varepsilon_n}{(1+r')^{t-1}} k \dots\dots\dots\dots\dots(5.1)$$

where, *CALI* is the cost of AL due to inaccurate meters, n is the total number of meters for each meter model, r' is the real discount rate (6%), P_W is the price of water (0.82 \$/m^3) and assumed to be constant throughout the analysis period, Q_t is the average annual volume of water registered through DN 15 mm domestic meters, t is the number of years the meter is in service, ε_n is the average weighted meter error during the useful life of the meter over the low flow ranges below Qmin, and k is a discount factor for the time the meter is registering other flows (10%). For this study, it was assumed that the distribution of all meter models in the network was the same and equal to 33,000 meters per model. This methodology has been applied by other researchers such as Richards et al. (2010) for estimating water loss due to meter inaccuracy at ultra low flow ranges. The limitations of the method arise from many uncertainties involved in measuring low flows and estimating water use profiles.

The estimated revenue losses due to metering inaccuracies for KWDS are presented in Table 5.6. The most effective meter at measuring low flow rates was the single-jet meter (model 4) with the least revenue loss of US \$561,564 over the 5-year period. These estimates are conservative as sewerage charges were excluded for properties connected to the sewerage system. Table 5.7 shows how much revenue per meter could be recouped by replacing the less accurate meter models with meter model 4.

Table 5.5 Data used for estimating water loss at different low flow ranges

Flow range (L/h)	Proportion of total consumption	Timed flow through meters (%)	Model 1 average accuracy	Model 2 average accuracy	Model 3 average accuracy	Model 4 average accuracy
0-3.75	0.005	0.2	0	0	0	0
3.75-7.50	0.016	1.4	1.75	0	0	1.5
7.50-15.00	0.065	8.1	13.9	6.9	6	9.4
15.00-22.50	0.048	2.1	36.1	26.0	14.5	46.8
22.50-30.00	0.070	78.2	53.3	40.5	36.5	77.6

Table 5.6 Estimated revenue losses at different low flow rates

Meter Model	Year 1	Year 2	Year 3	Year 4	Year 5	Total Losses
1	-\$824,028	-\$777,385	-\$733,382	-\$691,870	-\$652,707	-\$3,679,372
2	-\$934,463	-\$881,569	-\$831,669	-\$784,593	-\$740,182	-\$4,172,476
3	-\$986,980	-\$931,113	-\$878,409	-\$828,687	-\$781,781	-\$4,406,970
4	-\$709,075	-\$668,939	-\$631,074	-\$595,353	-\$561,654	-\$3,166,095

Table 5.7 Estimated revenue recovery potential per meter

Meter Model	Year 1	Year 2	Year 3	Year 4	Year 5
1	-\$3.48	-\$3.29	-\$3.10	-\$2.92	-\$5.76
2	-\$6.83	-\$6.44	-\$6.08	-\$5.73	-\$5.41
3	-\$8.42	-\$7.94	-\$7.49	-\$7.07	-\$6.67

Although meter replacement may not be based on a single criterion of potential revenue recovery at low flow rates, the type of meter should never the less be considered in the decision-making process. From the high meter failure rates of meter models 1 & 2 (Chapter 4) and the high revenue losses of meter model 3, it would be wise to start replacing these meter models progressively with meter model 4 and for any new connections made, meter model 4 should be installed.

5.7 Reducing the Level of Apparent Losses

In order to develop strategies for reducing the level of AL, the factors that influence the different levels of AL must be identified and be well understood. Hereunder are some of the factors that contribute to high levels of AL:

5.7.1 Factors influencing the level of apparent losses

Metered ratio (MR): Using the analogy of leakage with service connections, the more the number of metered customers (or MR defined as ratio of metered customers to total customers) (Cole and Cole 1980; Wallace 1987), the more likely that metering inaccuracies will increase. For example, the UK with 37% metered household properties, meter under-registration (MUR) is estimated at 4.1% for households and 4.9% for non-household properties whereas in Spain, Madrid with universal metering, Canal de Isabel II reports MUR of about 14% and MUR accounts for 92% of AL. The meter management policy is also very critical for proper meter performance over their life time.

Apparent loss management policies: The meter management policy and illegal use control policy in place has a big influence on the level of AL. The frequency of sampling, testing, repair and replacement of customer meters will determine the level of metering inaccuracies. Unauthorized consumption will never be known unless there is a proactive utility strategy to investigate suspicious accounts. Depending on numbers investigated, figures from 18% to 26.6% of the investigated accounts have been reported in Accra, Ghana (Lievers and Barendregt 2009) and Sao Paulo, Brazil (Medeiros et al. 2010) respectively.

Socio-economic aspects: The high number of illegal use cases (or UU) in Colombian cities is attributed to poverty and unplanned settlements (slums)(Medeiros et al. 2010). To the contrary, in Kampala city, Uganda, the records indicate that 12.5% of confirmed illegal cases were commercial users (Mutikanga et al. 2011) who will always do anything to "defeat the system" including bribing utility employees in order to minimize their operating costs. Poor people striving to survive on less than a dollar per day cannot afford to bribe utility employees for illegal connections or meter bypasses. Although the findings in Kampala city indicate that most unauthorized consumption is due to illegal reconnections (40% of the cases) of disconnected customers due to non-payment of bills, actually only 2.1% of the cases were drawing water from public stand-posts that often serve slum dwellers under the poor people category. However, the high number of illegal use cases were found to be also attributed to low pay of utility field plumbers who actually carryout the illegal connections and reconnections to earn extra income. The laying off of these workers as a result of urban water reforms and privatization in 1999 exacerbated the problem (Mutikanga et al. 2009b).

Cultural aspects: The cultural uprightness of citizens also plays a role. For example, in Kampala city, it is not only the poor who use water illegally but also the very rich including five star hotels trying to minimize operating costs. Some individuals have habits of stealing for the sake of it – a habit known as kleptomania. Studies on electricity theft in a sample of 102 countries do confirm these behavioral traits (Smith 2004). In some societies, people will always find ways of evading bills not only for water but for example tax. Some people still perceive water as a social good and not an economic good and will always find ways of using it illegally. For example, why should people use water illegally in England and Wales where most customers' billing is based on flat rate charges assigned by property rates (Walker 2009)? Illegal use of water is a complex social-technical problem that needs social-cultural approaches.

Governance issues: According to McIntosh (2003), poor governance is at the root of the NRW problem. Water theft and electricity theft are synonymous, and it is likely that customers who steal water are likely to steal electricity as well. The only difference is that it is probably easier to see overhead electricity illegal lines than buried illegal water pipes. Higher levels of electricity theft have been reported in countries with high levels of corruption, low government effectiveness, political instability and ineffective accountability (Smith 2004). In Uganda, the laws are too weak and outdated to act as a deterrent to water theft. The fines levied in courts of law are often too low compared to the value of water stolen. In addition, compiling evidence for illegal use cases is not always trivial and can be expensive. On the other hand, the authorities supposed to condone illegal use of water are often the culprits e.g. in Kampala city, the police usually abuse fire hydrants by collecting and selling water to earn extra income. This is particularly common in the developing countries where water service coverage is low, supply is irregular, and salaries are low.

5.7.2 Apparent loss reduction strategies

The reduction of AL is often attractive as it translates into increased revenues and enhanced utility financial viability. Most utilities prefer to address AL to recoup quick revenues needed for controlling real losses and infrastructure rehabilitation. However, apparent loss strategies go beyond conventional engineering approaches to managerial solutions that include social and behavioral sciences. Whereas, AL cannot be totally eradicated, the following interventions have proved to be very helpful in minimizing the losses:

Amnesty: In order to ferret out illegal connections, many utilities have reported successful results because of providing amnesty periods (Lea 2011; Mutikanga et al. 2011). Amnesty periods are usually one to three months when illegal users come forward and confess or face heavy penalties that include bills that go backwards for two to six years or even prosecution in courts of law. Amnesty is usually backed up by public campaigns in the media in order to be effective. McIntosh (2003) suggests that those found with illegal connections after the amnesty period should be convicted in courts of law.

Incentives: Providing utility employees and the public with incentives such as cash rewards for reporting illegal use of water has been used successfully in many cities e.g. Phnom Penh in Cambodia (Biswas and Tortajada 2009) and Kampala in Uganda (Mutikanga et al. 2011). Utility employees need genuine incentives to do their jobs and replace the incentives they have made for themselves through illegal connections, meter bypasses and erroneous meter readings (McIntosh 2003).

Automated meter reading (AMR): In order to reduce AL due to meter reading and data transfer errors, some utilities have reported success stories with AMR technology, for example, Philadelphia Water Department in the USA with over 40,000 properties read remotely via radio transmissions (AWWA 2009). The AMR system includes tamper-detection capabilities to thwart unauthorized consumption. The pilot AMR in Kampala was reported promising with respect to improved understanding of demand patterns, and triggering off real-time intelligent alarms incase of abnormal flows (Mutikanga et al. 2010).

Revenue protection: Establishing revenue protection structures within the utility organizational mainstream have been reported to minimize illegal use in both the water and energy sectors (AWWA 2009; Mutikanga et al. 2009a; Smith 2004). The focus of the revenue protection unit (RPU) is to investigate and monitor suspected accounts more proactively (zero consumption, negative billing, low consumptions etc.). The RPU also enforces effective policies consistently to thwart unauthorized consumption. In the city of Philadelphia, USA, it is reported that after seven years of operations by the RPU, cumulative revenue recoveries of over $17 million were realized (AWWA 2009). Activities included internal restructuring and reorganizing its metering and meter reading groups and controlling illegal use from fire hydrants by installing center compression locks.

Integrated water meter management (IWMM): The IWMM approach aims at addressing the need for better water meter management. Operationally, IWMM includes proper meter selection, sampling and quality control tests of newly purchased meters, adequate meter sizing and resizing, proper meter installation, optimal sampling, testing and replacement (Arregui et al. 2006b; Van Zyl 2011). Most water utilities in the developed countries implement IWMM and the result is low levels of metering inaccuracies. For example, the Philadelphia Water Department in the USA with 2% of AL due to metering inaccuracies (AWWA 2009) and Public Utilities Board (PUB) in Singapore with NRW of 4.4% of water supplied (ADB 2010).

Low flow controllers (LFCs): Recent studies have indicated the potential of minimizing AL due to metering inaccuracies at very low flow rates using LFCs or unmeasured flow reducers (UFR) (Davidesko 2007; Fantozzi 2009; Rizzo et al. 2007a; Yaniv 2009). In a pilot case study in Palermo (Italy), LFCs have been reported to reduce customer meter under-registration from 28.06% to 18.91% recouping additional revenue by 9.15% (Fantozzi et al. 2011). The same study reports that meter under-registration has been reduced to zero by a combined strategy of economic meter replacement (new turbine class C meters of DN 15 mm) and LFCs.

Benchmarking: Water utilities can improve performance by learning good practices from peers through benchmarking. Remarkable transformations have been reported in some Asian cities where different approaches have been used to reduce water losses and particularly AL (ADB 2010). In the city-state of Singapore, the law prohibits illegal connections to the water supply system and this is strictly enforced. PUB Singapore has reduced NRW from 9.5% of water supplied in 1990 to 4.4% in 2008. The overall governance in Singapore is exemplary in terms of transparency and accountability. Laws and regulations are enforced and strictly implemented. There are no illegal connections and corruption is not an issue at PUB (Tortajada 2006). In Phnom Penh, the water utility (PWSA) reduced NRW from 50% in 1999 to 6.6% in 2008. In Manila, Manila Water Company, was able to reduce NRW from 63% in 1997 to 11% in 2011, a remarkable 52% reduction in one and half decades (Luczon and Ramos 2012). One of the factors that led to these successes is their innovative territory management concept and the urban poor, "water for the community" that increased the utility interaction with the community up to the level of informal street leaders who could report and control illegal connections among others. This has proven to be an effective way of minimizing the high rates of illegal connections in depressed communities and informal settlements. This was done with the understanding that AL is not just an engineering problem but a socio-cultural problem that requires changes in community behavior and attitudes toward water use. The remarkable case studies in the eight Asian cities (ADB 2010) indicate that it is possible to reduce apparent losses, inter, alia, NRW and that there is no a unique "one-size-fits-all" solution but a cocktail of solutions depending on local conditions.

5.8 Conclusions and Recommendations

The conclusions based on the study are drawn and recommendations for utility policy to minimize AL and on areas of further research are made in the following sections.

5.8.1 Conclusions

In this Chapter, a methodology for assessing apparent losses in urban WDSs has been presented. Guidelines for assessing apparent losses in water utilities of the developing countries with insufficient resources and data-poor networks have also been established. AL caused by metering inaccuracies at low flow rates was also examined and guidelines for quantifying the losses established. Strategies for minimizing apparent losses have also been presented. The main conclusions drawn from this study are:

1. Assessing components of apparent water losses should be very useful for water utilities in order to develop appropriate strategies for performance improvement.
2. The study confirms that metering inaccuracies and illegal use of water are the major components of AL in KWDS. The results indicate that about 36.9% of water sales in Kampala is due to AL. Although this trend was observed in other countries, under no circumstances should it be generalized as it depends on local conditions e.g. in the city of Philadelphia where systematic data handling errors were the most significant component of AL.
3. The proposed methodology for assessment of apparent loss components is generic in nature and based upon a practical logical sequence. It is expected that water utility managers will find it very helpful while carrying out annual water balances for their water supply systems.
4. AL due to metering inaccuracies at low flows can be significant depending on the meter type and users' water consumption profiles. These losses increase with the ageing of the meter indicated by the cumulative water registered through the meter or billing index. For the KWDS, meter under-registration at low flow ranges from an average of 22.4% to 100% where ultralow flows are not measured at all. Revenue loss as a result is more than US $700,000 annually, although there is potential for recovery by replacing inefficient meters with the most efficient single-jet meter (model 4).
5. The increased revenue and pressure on water utilities for proper water accountability makes the low-flow accuracy of residential water meters one of the key decision variables in selecting a water meter among other criterion.
6. Factors that influence levels of AL are diverse and include metering, data management, illegal use control policies in place, technical and socio-economic and cultural aspects.
7. Minimizing AL is attractive as it enhances the utility financial viability. Strategies to minimize apparent losses range from advanced AMR technology to simple managerial tools of community mobilization and public communication to minimize unauthorized use of water.
8. Performance benchmarking demands uniform methodologies and precise definitions. Water utilities and regulatory institutions should promote use of standardized water loss assessment methodologies and benchmarking to improve WDS operating efficiency.
9. Each water utility should attempt to carry out its own apparent loss assessment and it's only by doing so the factors contributing to AL will be identified, quantified and appropriate reduction strategies developed.

5.8.2 Recommendations

The following three major recommendations are proposed for utilities to consider in minimizing apparent losses:

Apparent Loss Reduction Policies: Policies must be established for proactive interventions to minimize apparent losses. They should include: (i) empowered RPUs that investigate illegal use and enforce the law in combating illegal use, (ii) meter sampling, testing and optimal replacements, (iii) auditing meter readings, billing data entry and adapting AMR technology to minimize systematic data handling errors.

Special programs: Amnesty for illegal users should be explored, cash incentives to reward informers of illegal cases and exposing culprits in the media. Prosecution of illegal users should not be limited to the public alone but include utility employees (plumbers and top managers) who extort money from customers to conceal illegal use and report erroneous meter readings. The Manila Water's "Territory Management Concept" should be adapted and explored by other utilities to address the illegal use of water challenges.

New customer connections timely updates: Systems should be put in place to minimize delays and errors in updating customer service connections on GIS maps and into the customer billing database.

Governance issues: Promoting sound corporate governance is critical and should be integrated with technical and managerial solutions of tackling apparent losses. These include institutional reforms to enhance operating efficiency and fighting entrenched corruption cultures.

Further Research: It is worthwhile developing a systematic framework and guidelines for prioritizing action to reduce apparent losses using techniques such as MCDA. Online monitoring and detection of abnormal events is another promising research area that could help address the problem of illegal use of water (Allen et al. 2011; Armon et al. 2011; Mounce et al. 2010).

5.9 References

ADB. (2007). "Curbing Asia's Non Revenue Water." Asian Development Bank Water Brief (http://www.adb.org/water),retrieved on 15th Novemeber 2010.

ADB. (2010). "Every Drop Counts: Learning from Good Practices in Eight Asian Cities." Asian Development Bank, Manila.

Allen, M., Preis, A., Iqbal, M., Srirangarajan, S., Lim, H. B., Girod, L., and Whittle, A. J. (2011). "Real-time in-network distribution system monitoring to improve operational efficiency " *Journal AWWA*, 103(7), 63-75.

Armon, A., Gutner, S., Rosenberg, A., and Scolnicov, H. (2011). "Algorithmic monitoring for a modern water utility: a case study in Jerusalem." *Water Science and Technology*, 63(2), 233-239.

Arregui, F., Cabrera, E., Cobacho, R., and Garcia-Serra, J. (2006a). "Reducing Apparent Losses Caused by Meters Inacuracies." *Water Practice and Technology*, 1(4), doi:10.2166/WPT.2006093.

Arregui, F., Jr., C. E., and Cobacho, R. (2006b). *Integrated Water Meter Management* IWA Publishing, London.

AWWA. (2003). "Committee report: Applying worldwide BMPs in water loss control." *Journal AWWA*, 95(8), 65-79.

AWWA. (2009). "Water Audits and Loss Control Programs: AWWA Manual M36." American Water Works Association, Denver, USA.

Batista, C. F., and Mendonca Jr, J. C. (2009). "Lowering Under Metering of a Meter Park Practical Tools for Resizing." *Proceedings of the 5th IWA Water Loss Reduction Specialist Conference*, Cape Town, South Africa, 169-175.

Bidgoli, A. M. (2009). "Water losses reduction programme in Iran." *Proceedings of international workshop on drinking water loss reduction: Developing capacity for applying solutions*, UN Campus, Bonn, Germany, 80-87.

Biswas, A. K., and Tortajada, C. (2009). "Water Supply of Phnom Penh: A Most Remarkable Transformation." Lee Kuan Yew School of Public Policy, Singapore.

Bowen, P. T., Harp, J. F., Entwistle Jr, J. M., and Shoeleh, M. (1991). *Evaluating Residential Water Meter Performance*, AWWA Research Foundation., Denver, Colorado, USA.

Cole, E. S., and Cole, G. B. (1980). "Unaccounted-for Water versus Metered Ratio." *AWWA 1980 Annual Conference Proceedings*, Atlanta, Georgia, 1047-1052.

Criminisi, A., Fontanazza, C. M., Freni, G., and La Loggia, G. (2009). "Evaluation of the apparent losses caused by water meter under-registration in intermittent water supply." *Water Science and Technology:WST*, 60(9), 2373-2382.

Davidesko, A. (2007). "Unmeasured Flow Reducer- an innovative solution for water meter under-registration-Case study in Jerusalem, Israel." *Proceedings of the 4th IWA Specialist Water Loss Reduction Conference*, Bucharest, Romania, 704-719.

Dimaano, I., and Jamora, R. (2010). "Embarking on the World's Largest NRW Management Project." *Proceedings of the 6th IWA Water Loss Reduction Specialist Conference*, Sao Paulo, Brazil, CD-ROM.

Donnelly, A. (2007). "Combating NRW in a large multi-functional company; A case study of EPAL; Portugal's largest water supplier." *Proceedings of the 4th IWA Water Loss Reduction Specialist Conference*, Bucharest, Romania, 501-510.

Donnelly, A., Medeiros, M. F., and Franco, B. A. (2009). "Managing Distribution Network; A case study of EPAL in Lisbon since 2005." *Proceedings of the 5th IWA Water Loss Reduction Specialist Conference*, Cape Town, South Africa, 531-538.

Fanner, P., Sturm, R., Thornton, J., and Liemberger, R. (2007). *Leakage Management Technologies*, Awwa Research Foundation Denver, Colorado, USA.

Fantozzi, M. (2009). "Reduction of customer meters under-registration by optimal economical replacement based on meter accuracy testing programme and unmeasured flow reducers." *Proceedings of the 5th IWA Water Loss Reduction Specialist Conference*, Cape Town, South Africa, 233-239.

Fantozzi, M., Criminisi, A., Fontanazza, C. M., and Freni, G. (2011). "Investigations into under-registration of customer meters in Palermo (Italy) and effect of introducing low flow controllers." *Proceedings of the 6th IWA Specialist Conference on Efficient Water Use and Management (CD-ROM)*, Dead Sea, Jordan.

Farley, M., and Trow, S. (2003). *Losses in Water Distribution Networks: A Practitioner's Guide to Assessment, Monitoring and Control*, IWA Publishing, London.

Flores, and Diaz. (2009). "Meter assessment in Madrid." *Proceedings of the 5th IWA Specialist Conference on Efficient Water Use and Management*, Sydney, Australia, October 26-28 CD-ROM.

Garzon-Contreras, F., Gomez-Otero, I., and Munoz-Trochez, C. (2005). "Benchmarking Leakage for Water Suppliers in Valle Del Cauca Region of Colombia." *Leakage 2005*, Halifax, Canada.

Garzon-Contreras, F., and Palacio-Sierra, C. (2007). "A Case Study of Leakage Management in Medellin City, Colombia." *Proceedings of the 4th IWA Water Loss Reduction Specialist Conference*, Bucharest, Romania, 434-443.

Garzon-Contreras, F., Uribe-Preciado, A., Yepes-Enriquez, L., and Agredo-Perdomo, A. (2009). "Unauthorized consumption: The key component of Apparent Losses in Colombia's major cities." *Proceedings of the 5th IWA Water Loss Reduction Specialist Conference*, Cape Town, South Africa, 29-35.

ISO-4064-1. (1993). "Measurement of water flow in fully charged closed conduits - meters for cold water. Part 1: Specifications."

ISO-4064-3. (1993). "Measurement of water flow in closed conduits-meters for cold potable water. Part 3: Test methods and equipment."

ISO. (2008). "Uncertainty of Measurement - Part 3: Guide to expression of uncertainty in measurement (GUM:1995)." International Organization for Standardization(ISO)/International Electrotechnical Commission (IEC), Geneva, Switzerland.

Kingdom, B., Liemberger, R., and Marin, P. (2006). "The Challenge of Reducing Non-Revenue Water (NRW) in Developing Countries ", The World Bank, Washington, DC, USA.

Lambert, A. (1994). "Accounting for losses: the bursts and background concept." *Water and Environment Journal*, 8(2), 205-214.

Lambert, A., and Hirner, W. (2000). "Losses from Water Supply Systems: Standard Terminology and Recommended Performance Measures (IWA's Blue Pages)." The International Water Association (IWA) London.

Lambert, A. O. (2002). "International Report: Water losses management and techniques." *Water Science and Technology: Water Supply*, 2(4), 1-20.

Lea, R. (2011). "Thames sets sights on 18,000 who are paying nothing." The Times, London, February 28.

Lievers, C., and Barendregt, A. (2009). "Implementation of Intervention Techniques to Decrease Commercial Losses for Ghana." *5th IWA Water Loss Reduction Specialist Conference*, Cape Town, South Africa, 490-496.

Luczon, L. C., and Ramos, G. (2012). "Sustaining the NRW reduction strategy: The Manila Water Company Territory Management Concept and Monitoring Tools." *Proceedings of the 7th IWA Water Loss Reduction Specialist Conference*, Manila, Philippines, Feb 26-29, 2012.

Makara, L. (2009). "Successful outcomes from water loss reduction, Eda Ranu, Papua New Guinea " *Proceedings of the 5th IWA Water Loss Reduction Specialist Conference*, Cape Town, South Africa, 73-84.

Male, J. W., Noss, R. R., and Moore, I. C. (1985). *Identifying and Reducing Losses in Water Distribution Systems*, Noyes Publications, New Jersey.

May, J. (1994). "Pressure Dependent Leakage." *World Water and Environmental Engineering, October 1994*, 13.

McIntosh, A. C. (2003). *Asian Water Supplies: Reaching the Urban Poor*, Asian Development Bank and IWA Publishing.

McKenzie, R., Lambert, A., JE., K., and Mtshweni, W. (2002). *ECONOLEAK: Economic Model for Leakage Management for Water Suppliers in South Africa, Users Guide*, WRC Report TT 169/02, South Africa.

Medeiros, M. F., Gondo, J. Y., Rego, A. C., and Padilha, P. S. (2010). "Apparent losses in SABESP-Water meter management system - under-measurement and inevitable loss." *Proceedings of the 6th IWA Water Loss Reduction Specialist Conference*, Sao Paulo, Brazil, CD-ROM.

Montgomery, D. C., and Runger, G. C. (2007). *Applied Statistics and Probability for Engineers, Fourth Edition*, John Wiley and Sons, Inc, Arizona State University, USA.

Mounce, S. R., Boxall, J. B., and Machell, J. (2010). "Development and verification of an online artificial intelligence system for detection of bursts and other abnormal flows." *Journal of Water Resources Planning and Management*, 136(3), 309-318.

Mutikanga, H., Nantongo, O., Wozei, E., Sharma, S. K., and Vairavamoorthy, K (2009a). "Assessing water meter accuracy for NRW reduction." *Proceedings of the 5th IWA Specialist Conference on Efficient Use and Management of Water (CD-ROM)*, Sydney, Australia.

Mutikanga, H., Sharma, S. K., and Vairavamoorthy, K. (2010). "Customer demand profiling for apparent loss reduction." *Proceedings of the 6th IWA Water Loss Reduction Specialist Conference (CD-ROM)*, Sao Paulo, Brazil.

Mutikanga, H. E., Sharma, S., and Vairavamoorthy, K. (2009b). "Water Loss Management in Developing Countries: Challenges and Prospects." *Journal AWWA*, 101(12), 57-68.

Mutikanga, H. E., Sharma, S. K., and Vairavamoorthy, K. (2011). "Assessment of Apparent Losses in Urban Water Systems." *Water and Environment Journal*, 25(3), 327-335.

Noss, R. R., Newman, G. J., and Male, J. W. (1987). "Optimal Testing Frequency for Domestic Water Meters." *Journal of Water Resources Planning and Management*, 113(1), 1-14.

OFWAT. (2006). "Security of Supply, leakage and water efficiency 2005-06 report. Retrieved January 2008 from www.ofwat.gov.uk." OFWAT, London.

OFWAT. (2010). "Service and delivery-performance of the water companies in England and Wales 2009-10 report ", OFWAT, UK.

Puust, R., Kapelan, Z., Savic, D. A., and Koppel, T. (2010). "A review of methods for leakage management in pipe networks." *Urban Water Journal*, 7(1), 25-45.

Richards, G. L., Johnson, M. C., and Barfuss, S. L. (2010). "Apparent losses caused by water meter inaccuracies at ultralow flows." *Journal of American Water Works Association*, 105(5), 123-132.

Rizzo, A., Bonello, M., and Galea St. John, S. G. (2007a). "Trials to quantify and reduce in-situ meter under-registration." *Proceedings of the 4th IWA Specialist Water Loss Reduction Conference*, Bucharest, Romania, 695-703.

Rizzo, A., M., V., Galea, S., Micallef, G., Riolo, S., and Pace, R. (2007b). "Apparent Water Loss Control: The Way Forward." *IWA Water 21, August*

Sanchez, E. H. (2007). "Calculation, estimation and uncertainty in the Apparent loss volume in the water supply system of Canal de Isabel II." *Proceedings of the 4th IWA Water Loss Reduction Specialist Conference*, Bucharest, Romania, 684-694.

Schouten, M., and Halim, R. D. (2010). "Resolving strategy paradoxes of water loss reduction: A synthesis in Jakarta." *Resources Conservation and Recycling*, 54, 1322-1330.

Seago, C., Bhagwan, J., and McKenzie, R. (2004). "Benchmarking leakage from water reticulation systems in South Africa." *Water SA*, 30(5), 25-32.

Seago, C. J., and Mckenzie, R. S. (2007). *An Assessment of Non-Revenue Water in South Africa*, WRC, Report No TT 300/07, South Africa.

Sharma, S. K., and Nhemafuki, A. (2009). "Water Loss Management in Bhaktapur and Dhulikhel Cities in Nepal." *Proceedings of 5th IWA Water Loss Reduction Specialist Conference*, Cape Town, South Africa, 546-552.

Shin, E., Park, H., Park, C., and Hyun, I. (2005). "A case study of leakage management in the city of Busan, Korea." *Leakage 2005*, Halifax, Canada.

Smith, T. B. (2004). "Electricity Theft: A comparative analysis." *Energy Policy*, 32(18), 2067-2076.

Tabesh, M., Asadiyani, Y., and Burrows, R. (2009). "An Integrated Model to Evaluate Losses in Water Distribution Systems." *Water Resources Management*, 23(3), 477-492.

Thornton, J. (2002). *Water Loss Control Manual*, New York, McGraw Hill.

Thornton, J., and Rizzo, A. (2002). "Apparent losses, how low can you go." *Leakage Management Conference proceedings*, Lemesos, Cyprus.

Thornton, J., Sturm, R., and Kunkel, G. (2008). *Water Loss Control*, McGraw-Hill, New York.

Tortajada, C. (2006). "Water Management in Singapore." *Water Resources Development*, 22(2), 227-240.

Van Zyl, J. E. (2011). "Introduction to integrated water meter management." Water Research Commission (WRC TT490/11), South Africa.

Walker, A. (2009). "The Independent Review of Charging for Household Water and Sewerage Services. Interim Report." Department for Environment, Food and Rural Affairs, London.

Wallace, L. P. (1987). *Water and Revenue Losses: Unaccounted for Water*, AWWA, Denver, Colorado, USA.

Wegelin, W., McKenzie, R., Herbst, P., and Wensley, A. (2011). "Benchmarking and Tracking of Water Losses in All Municipalities in South Africa." *Proceedings of the 6th IWA Specialist Conference on Efficient Water Use and Management (CD-ROM)*, Dead Sea, Jordan.

Yaniv, S. (2009). "Reduction of Apparent Losses Using the UFR (Unmeasured-Flow Reducer) - Case Studies." *Proceedings of the 5th IWA Specialist Conference on Efficient Water Use and Management (CD-ROM)*, Sydney, Australia.

Chapter 6 - Pressure Management Planning for Leakage Control

Parts of this chapter are based on:

Mutikanga H.E., Vairavamoorthy K., Sharma S.K., and Akita C.S (2011). "Operational Tools for Decision Support in Leakage Control". *Water Practice and Technology*, 6(3), doi:10.2166/wpt.2011.057.

Mutikanga H.E, Akita C.S., Sharma S.K., and Vairavamoorthy K., (2010). "Pressure Management as a Tool for Water Leakage Reduction". *Proceedings of the 3rd International Perspective on Current & Future State of Water Resources & the Environment (EWRI of ASCE) conference,* IIT Madras, Chennai, India – CD ROM.

Summary

Water utilities, particularly in the developing countries are grappling with challenges of high water losses due to leakage. For some poorly managed and ageing urban networks, leakage levels of up to 50% of water supplied have been reported. Leakage not only represents economic loss but wastage of a precious and scarce natural resource. Pressure management and network hydraulic modeling in conjunction with network zoning have proven to be powerful engineering tools for reducing leakage in many developed countries. Despite their apparent success, these tools have not been applied widely in the developing countries partly due to inadequate information on cost-benefit analyses to support management decision making in implementation of pressure management policies. Economic planning studies are required to promote adoption of proactive leakage control policies in developing countries. In this study a pressure management planning decision support tool for leakage control was developed to promote and justify investment decisions of adopting pressure management strategies. A network hydraulic model was applied to validate the effectiveness of the tool and give users confidence in the tool results. Both methods were applied to predict potential benefits and cost savings for a DMA in Kampala city, Uganda. Predictions by the tool and the network hydraulic model indicate that reducing average zonal pressure by 7 m could result into water savings of 254 m^3/day and 302 m^3/day respectively without compromising customer service levels. In financial terms, this is equivalent to annual net benefits of €56,190 and €66,910 respectively. The results obtained indicate that the predicted water and cost savings compare fairly well. Although conservative in its predictions, the tool will be valuable for engineers and decision-makers planning to implement pressure management strategies for leakage control in the developing countries with inadequate resources for the computationally demanding network hydraulic modeling. Further research to maximize leakage reduction using sophisticated optimization techniques such as genetic algorithms is recommended once proactive control policies such as pressure management are in place and pressure on water resources conservation continues to increase.

6.1   Introduction

The previous Chapters (4 and 5) addressed the apparent loss component of water losses. This Chapter deals with the real loss component of water losses with focus on water distribution leakage.

Leakage accounts for a significant amount of water losses in many cities of the world. It varies from 3% of the water put into the distribution systems in well managed systems to over 50% in poorly managed and deteriorated infrastructure systems (Puust et al. 2010). The economic loss, safety, environmental and social cost implications of leakage reduction are all too obvious. The economic losses due to leakage include the cost of raw water, its treatment, and energy pumping costs. Leaky pipes pose a public health risk as intrusion of contaminants is eminent in case of any pressure drops (Karim et al. 2003). Environmental aspects of leakage include wastage of a scarce water resource, depletion of energy resources and increases in carbon foot-prints of the service provider (Cabrera et al. 2010; Colombo and Karney 2005). Social costs include customer service disruption, property damage and road user's disruption due to main breaks and subsequent repairs. Water leakage not only involves economic, environmental and social costs, but may also trigger premature investment to develop new sources to keep pace with increasing demand (Jowitt and Xu 1990). Reducing leakage will therefore delay costly system expansions and result in lower annual operating costs. In Kampala, more than 8 million m^3 of treated water physically leak from the water supply system every year costing the utility more than US $6 million annually (Mutikanga et al. 2009). Clearly, it is unacceptable, that where public utilities are starving for additional revenues to finance expansion of services particularly for the urban poor and where most connected customers receive water irregularly, that water is also heavily wasted.

Historically, most WDSs have excessive pressures due to the inherent weaknesses in the design methodology that aims to achieve minimum pressure requirements at some critical point during the day and the fact that they have had no active pressure control. Most WDSs experience excess pressure during off-peak periods especially overnight. This is evident from the fact that major breaks tend to occur during the late evening and early morning hours when the system pressures are at their highest. Intuitively, the rate of water lost through leakage increases with increased pressure. Therefore, if the excess pressure in a system can be reduced, then so too can leakage. This is the basic philosophy governing pressure management (PM) in WDSs. PM has been defined by AWWA as "*the practice of managing system pressures to the optimum levels of service ensuring sufficient and efficient supply to legitimate uses and consumers, while reducing unnecessary or excess pressures, eliminating transients and faulty level controls, all of which cause the distribution system to leak unnecessarily*" (AWWA 2009). The objectives of PM for leakage control are three-fold: (i) to reduce the frequency of new breaks within a WDS, (ii) to reduce the flow rates of those breaks and background leakage (weeps and seeps at joints) that cannot be avoided, and (iii) to reduce the risk of further leaks by smoothing pressure variations. PM is the only real loss control tool that can reduce background losses, except for infrastructure replacement (Thornton et al. 2008). PM has been identified as a successful short-, medium- and long-term leakage management tool in the review of international leakage management technologies (Fanner et al. 2007). With advancements in electronic and hydraulic controllers, it is now possible to reduce leakage efficiently and cost-effectively to minimum possible levels by flow modulation pressure reducing valves (PRVs) (AbdelMeguid and Ulanicki 2010; Trow and Payne 2009; Ulanicki et al. 2008).

Pressures into discrete Pressure Managed Zones (PMZs) and/or DMAs can be controlled by PRVs as illustrated in Figure 6.1. Pressure management does not only reduce leakage but extends useful life of infrastructure, reduces operation and maintenance costs through reduced frequency of main breaks and energy consumption, improves customer service as a result of reduced water supply interruptions and is a demand management tool (Girard and Stewart 2007; Lambert and Fantozzi 2010).

While pressure management is known to have a measurable impact on leakage reduction, it is still difficult to quantify additional benefits associated with reduced burst frequency, deferment of capital expenditures, increased asset life, improved levels of service, reduced customer complaints and reduction in other environmental and social impacts. The ability to assess cost-benefits is fundamental to planning and justifying pressure management strategies. Recent studies have proposed a basic prediction equation between pressure reduction and frequency of bursts and leaks as: % reduction in new breaks = BFF x % reduction P_{max} (Lambert and Thornton 2011; Thornton et al. 2008), where BFF is a break frequency factor = 1.4 for mains and services combined and P_{max} is the maximum pressure in a PMZ (static or transient). While using this simplified prediction equation, care must be taken to ensure that the prediction does not reduce the break frequency below the values used in the UARL formula (Lambert et al. 1999).

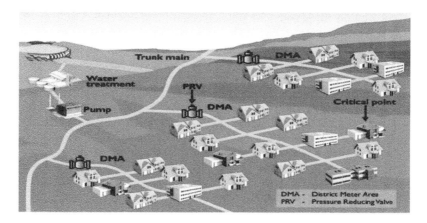

Figure 6.1 Schematic of a typical PMZ
Source: (Burrows 2010)

It is now widely acknowledged that pressure management and network hydraulic modeling in conjunction with DMAs are powerful proactive leakage management tools (Fanner et al. 2007; Thornton et al. 2008; Wu et al. 2011). Many water utilities have reported network pressure reduction, and, inter alia, leakage (Babel et al. 2009; Charalambous 2008; McKenzie et al. 2004; Pilipovic and Taylor 2003). Although these case studies report significant leakage reduction, they did not provide optimal solutions. Research studies have indicated that further leakage reduction could be obtained by applying optimization techniques such as genetic algorithms (GAs) (Awad et al. 2009; Reis et al. 1997; Savic and Walters 1995), mathematical programming (Hindi and Hamam 1991; Jowitt and Xu 1990; Sterling and Bergiela 1984; Ulanicki et al. 2000; Vairavamoorthy and Lumbers 1998), and multi-objective optimization (Nicolini et al. 2011). Pressure management by optimal storage tank levels using a hybrid optimization model (GAs and artificial neural networks) for predicting leakage reduction has also been reported (Nazif et al. 2010).

Whereas optimization techniques are valuable tools for the developed countries where water utilities are under enormous regulatory pressure to reduce leakage further from economical levels to sustainable levels (Mounce et al. 2010), their application in developing countries with limited resources is dubious for the following reasons: (i) they are computationally demanding and expensive, (ii) they are suitable for urban areas where reported leaks and bursts have been eliminated and background leakage control is the main motivation, (iii) they are more effective in zoned networks which are often absent in WDSs of the developing countries, (iv) there are often no PRVs installed in most WDSs of the developing countries to be optimized. Clearly, more appropriate tools that would promote adoption of proactive leakage management policies such as pressure management are required for developing countries.

One of the major barriers to adopting PM strategies is inadequate economical information associated with benefits of PM to support investment decision-making. It is therefore prudent to develop appropriate planning tools and methodologies for predicting potential savings of PM projects (Ulanicki et al. 2000). In their pressure management planning study for leakage control in the city of Mutare in Zimbabwe, Marunga et al. (2006), predicted 25% leakage reduction potential using network hydraulic modeling and MNF techniques. The major drawback of this study was the exclusion of pressure-dependent leakage terms in the network model. This is not realistic as it implies that nodes with low demands will have the same leakage rate even when their nodal pressures are different. Such models cannot be properly calibrated and used for the planning of pressure control schemes (Ulanicki et al. 2000). The inclusion of leakage terms affects the flow and pressure distribution in a network and therefore the resulting optimal PRV settings. This study develops an appropriate pressure management planning decision support tool (DST) (herein referred to as PM-COBT) for predicting the associated cost-benefits of using PRVs to reduce leakage and promote use of pressure management strategies in the water utilities of the developing countries. To evaluate the effectiveness of the DST, a network hydraulic model that is based on sound engineering principles is applied and compared to the tool. For effective prediction of pressure control benefits, the pressure-dependent leakage terms are explicitly included in the tool and hydraulic model. Both methods are applied to a real-developing world case study in Kampala city, Uganda, to predict potential water savings. The water savings are then used as one of the input variables to the economic model (Awad et al. 2008) that estimates the cost-benefits of implementing pressure management strategies.

The rest of the Chapter is organized in the following way. Section 6.2 presents the case study background. Section 6.3 outlines the methodology used to develop the tool and section 6.4 presents the tool. Section 6.5 presents the network hydraulic model of the case study. The application of the DST and network hydraulic model to the case study is presented in section 6.6. The results and discussions including limitations of the tool and some lessons learned are presented in section 6.7. Finally, section 6.8 draws some conclusions based on the study.

6.2   Case Study Background

The details of the Kampala Water Distribution System (KWDS) have been presented in chapter 1. In this section, only the DMAs within the KWDS under study are presented. As there were no DMAs in the KWDS, new DMAs had to be established. In accordance with the goals of the study, two DMAs, namely; Kitintale (DMA1) and Kawuku (DMA2), were selected after examining the network plans, schematics, GIS systems; undertaking field visits and liaising with the operations and distribution staff of the water utility. The DMAs were located on the south eastern part of Kampala City, bordering Lake Victoria, as shown in Figure 6.2. In order to limit the amount of resources required for DMA establishment, these pilot areas were selected in preference to others, due to their ease of isolation, reconfiguration and monitoring. DMA1 had been partially created, as part of the ongoing work to divide the entire network into DMAs, but required only a boundary valve and flow meter, to complete its formation. DMA2 was created from a larger network zone, where no DMA establishment had been attempted. Its size was limited to a small network so that only a single flow meter was required to complete its establishment. In addition to the foregoing prequalification requirements for selection of DMAs, and the need to ensure proper monitoring of flows and pressures, the selected DMAs satisfied the following set of criteria (Farley and Trow 2003):

- Single metered inlet;
- Natural geographic boundaries and less artificial boundaries;
- Hydraulically separable or isolated;
- Leaks and bursts were reportedly common;
- Customer complaints of high pressure;
- Verifiable records of network infrastructure;
- DMA size of 300 – 5000 connections.

The selected DMAs were assumed to be representative of the entire network characteristics with respect to leakage distribution and infrastructure age. Input data for network assets was obtained from the utility GIS and ground level data was collected using mobile GPS units. Flow in DMA1 was measured using the strap-on ultrasonic portable flow meter, installed on the main outlet pipe while in DMA2, the flow was measured using the Sensus Cosmos data logger coupled to a flow meter with a pulse emitter. In both DMAs, pressure was measured at the selected points using the Hydreka Vistaplus pressure logger of type OCTC511LF/30. Flow in DMA2 was also manually recorded from the installed flow meter to facilitate comparison and check consistency of flow data generated by the flow logger. However, in DMA1, flow could not be manually recorded since there was no flow meter installed on the main reservoir outlet pipe.

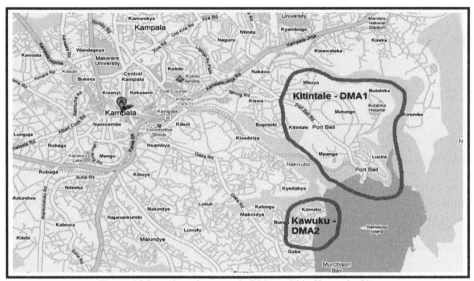

Figure 6.2 Location of case study DMAs within Kampala City

The DMAs system data is presented in Table 6.1. The work presented in this study focuses only on DMA1 as DMA2 did not yield significant benefits to justify investing in pressure management strategies.

Table 6.1 Water supply profile for DMA1 and DMA2

Description	Unit	Kitintale-DMA1	Kawuku-DMA2
Supply Regime		Intermittent	24-Hour
Service Connections	No.	5,443	354
Average Length of Private Connection	m	25	25
Total Pipe Length	km	42.4	2.5
Pipe Sizes (DN)	mm	40-400	40-100
Supply Zone Elevation	m	1136-1222	1143-1174
Average Water Demand	m^3/day	6,167	288
Average Billed Consumption	m^3/day	2,752	225
Non-revenue Water	m^3/day	3,415	63
Average Zonal Pressure	m	61.5	48

6.3 Methodology for the Decision Support Tool (DST)

The methodology used to develop the DST (PM-COBT) is based on:

1. Bursts and Background Estimates (BABE - concepts)
2. Pressure-leakage relationships (FAVAD principles)

   3. Pipe flow-Nodal Head (Q-H) equations

6.3.1   Bursts and background estimates (BABE)

In BABE analyses, components of leakage are considered in three categories (Lambert and Morrison 1996):

- Background (undetectable) leakage – small flow rate, runs continuously
- Reported leaks and bursts – typically high flow rates but short duration
- Unreported leaks and bursts – moderate flow rates, duration depends on intervention policy

The starting point of the DST methodology is the DMA water balance at the time of minimum night flow. In order to accurately measure night flows and apply MNF concepts for leakage assessment, special arrangement was made to ensure 24-hour supply to DMA1 during the study period. Although MNF analysis is not suitable for systems with intermittent supply, usually, most storage tanks are full between 3.00 am and 5.00 am, the period when MNF field tests are carried out and most users are hardly using any water. For the same reason, MNF analysis has been used in other developing countries with similar supply regimes e.g. Brazil (Cheung and Girol, 2009). In order to determine the DMA leakage profile, night flow measurements were carried out and MNF assessed. In DMA1, the highest MNF was 180 m^3/h and the lowest MNF was 105 m^3/h, while the average MNF was 136 m^3/h. The common feature surrounding these flows is that, they all largely occurred during the same hour from 4:00 to 5:00 am, when system pressure was at its maximum. Therefore, the hour at which pressure dependent leakage was at its maximum during the day, was also from 4:00 to 5:00 am. The lowest measured value of MNF, which is an indicator of leakage, was the value closest to the actual night leakage rate. Hence, MNF in DMA1 was taken to be 105 m^3/h. The average night consumption for a households and non-households was derived by analyzing the data logging records of the utility. The average night consumption for households during the MNF period, was 3 L/h, while for a non household it was10 L/h. The ideal way is to physically read the individual meters for the non-households during the hour of MNF as their consumption do vary a lot but this was not possible due to inadequate data logging equipment and the high number of properties involved.

To calculate leakage at MNF time (Q_L), the mass balance Equation 6.1 was applied to the DMA (Farley and Trow 2003).

$$Q_L\,(t_{MNF}) = Q_{DMA}\,(t_{MNF}) - Legitimate\ Night\text{-}Time\ Uses \qquad (6.1)$$

For estimation of legitimate night uses, detailed field investigations are required. In the absence of such detailed studies, McKenzie (2001) proposed use of 6% of total population and average use of 10 L/person/hour at time of MNF. However, these are default values for South African conditions and may not be valid for other countries. The use of active population percentage at time of MNF depends on socio-economic life styles of the population and probably on level of urbanization. The default value of 10 L/person/hour depends on toilet flush capacity. In some countries where water use efficiency is being promoted, toilets have been retrofitted to more efficient sizes of say 6 L/flush. In this study, we used parameters derived from field measurements at the time of MNF for a sample of properties: average of 10 L/property/hr (non-households) and 3 L/household/hr (Table 6.2). The water use is low as most households are in urban poor settlements where houses lack internal plumbing. Use of pit latrines instead of flush toilets is the norm for most households.

Assessment of legitimate night-time use is crucial for accurate leakage reduction predictions. Over-estimation will lead to low leakage levels while under-estimation will lead to high leakage levels, thus exaggerating potential water savings.

6.3.2 Fixed and variable area discharges (FAVAD) principles

Modelling leakage depends on understanding the hydraulics of leaks and how to incorporate the hydraulics into existing models of the WDS. Leakage under pressurised water mains can be represented as orifice flow. Applying the orifice flow equation (Eq. (6.2)) is valid (Wu et al. 2011), but the characteristics of leakage discharge are more complicated than that of orifice flow. The behaviour of leakage flow depends on many factors including pipe materials, type of leak/crack openings, pipe sizes, pressure in the mains, and soil hydraulics (Skipworth et al. 1999; Walski et al. 2006). The hydraulic equation for fully turbulent flow rate (L) through a hole of area (A) subject to static pressure (P) follows the orifice flow square root principle according to Equation 6.2 (Thornton et al. 2008).

$$L = C_d A * (2gP)^{0.5} \qquad (6.2)$$

where L is the flow rate (m^3/s), C_d is the discharge coefficient: a dimensionless factor of less than 1; g is the gravitational constant in m/s^2 ; P is the pressure in metres head.

In practice some types of leaks, C_d and A (and the effective area C_d x A) can be pressure-dependent. This is the premise of the FAVAD paths concept (May 1994). The effect of operating at different pressures is modelled by FAVAD principles. The basic FAVAD equation for analysing and predicting changes in leak flow rate (L_0 to L_1) as average pressure changes from P_0 to P_1 is (Lambert and Fantozzi 2010):

$$L_1/L_o = (P_1/P_0)^{N1} \qquad (6.3)$$

where $N1$ is the pressure exponent. Numerous field and laboratory tests from various countries have shown that $N1$ could vary from 0.5 to 2.3 depending on the type of leak, pipe material and failure type (Greyvenstein and van Zyl 2007; Lambert and Fantozzi 2010). The investigation of the effects of pressure on leak openings of different pipe materials is still a research area (Cassa and Van Zyl 2012; Cassa et al. 2010; Van Zyl and Cassa 2011).

6.3.3 Pressure-dependent and pressure-independent flows

Leakage is now widely acknowledged as pressure-dependent flow (Thornton et al. 2008; Wu et al. 2011). The more the pressure in the mains, the higher the leakage flow rate. In order to assess the impacts of pressure reduction on leakage in a DMA, the total flow into the DMA was separated into two components: (i) bursts and background leakage that is assumed to be pressure-dependent and (ii) consumption that is assumed to be pressure-independent. Bursts and background losses usually occur on service connections, properties and mains while consumption is water used by both households (domestic or residential) and non-households (commercial/industrial and institutions). The level of background leakage is a function of the number of service connections, length of pipes, asset condition, and mains pressure. The background leakage (BL) is obtained from the following empirical equation (Fanner et al. 2007):

$$BL \text{ (L/h)} = UBL * ICF \qquad (6.4)$$

where, *UBL* is the unavoidable background leakage and *ICF* is the infrastructure condition factor. The ICF is the defined as the ratio of the actual background leakage to the unavoidable amount of background leakage. It is an indicator used to describe the condition of the system infrastructure and only relates to background leakage. Background leakage by definition cannot be detected by traditional acoustic techniques and the level of background leakage is high for old systems with high network pressures. A well-managed and maintained WDS has an ICF equal to 1 while a poorly managed system has an ICF close to 4 (Fanner et al. 2007). In this study, the ICF of 1.5 was used in the estimation of background losses in DMA1. The unavoidable background leakage (*UBL*) is obtained from the following empirical equation for WDSs in good condition (Lambert et al. 1999) using *N*1=1.5 for background leakage (May 1994):

$$UBL \text{ (L/h)} = [(20*L_m) + (1.25*N_c) + (0.033*L_c)] * (P_{av}/50)^{1.5} \qquad (6.5)$$

where,

L_m = total length of water mains (km)
N_c = number of service connections (main to curb stop)
L_c = total length of private pipes (curb stop to customer meter) = Nc * Lp
L_p = average service connection pipe length

Although the coefficients in Eq. (6.5) have been derived for the UK water industry and may not be valid for Uganda and other developing countries with often poorly managed WDSs, they have been used by other researchers to provide estimates for leakage in WDSs e.g. in South Africa (McKenzie 1999) and in Zimbabwe (Marunga et al. 2006). According to McKenzie (1999), the default values will provide a realistic estimate of the overall leakage in most cases and the results are not particularly sensitive to individual parameters. It is important to remember that, the aim of the DST is to provide a quick estimate of leakage for a particular DMA given minimal information.

From data in Table 6.1 and equations, 6.1, 6.4 and 6.5, pressure-independent flow for DMA1 was estimated and is presented in Table 6.2. Although it was assumed that all background leakage is pressure-dependent, in practice some component of background leakage is pressure-independent. This is true for consumption as well. So the split of total flow into pressure-dependent and pressure-independent is rather subjective but is necessary to calibrate models used for pressure management planning studies.

Table 6.2 Estimated pressure-dependent and pressure-independent flows at MNF

Description	Number	Rate (L/h)	Estimated Flow (m³/h)
Minimum Night Flow			105.00
Household night use	4,474	3	13.42
Non-household night use	969	10	9.69
Legitimate night use (pressure-independent)			23.11
Boundary meter			15.05
Pressure-dependent leakage			66.84
Background leakage			24.85
Reported and unreported bursts			41.99

The pressure-dependent flow for the hour of MNF is estimated as the difference between total metered flow into the DMA and pressure-independent flow. The pressure-dependent leakage for the 23 hours is then computed using FAVAD principles (Equation 6.6). The hourly leakage rate (Q_L,t) throughout the day is calculated by multiplying the Night-Day-Factor (NDF) with the leakage rate at MNF based on pressure-dependent leakage (Fanner et al. 2007).

$$Q_L(t) = Q_{L,DMA}(t_{MNF}) \times [P(t)/P(t_{MNF})]^{N1} \qquad (6.6)$$

where, $Q_L(t)$ is the leakage rate at the hour t (t $\neq$ t$_{MNF}$), t_{MNF} is the MNF hour, Q_L, (t_{MNF}) is the leakage rate at the MNF hour, $P(t)$ is the average hourly nodal pressure at the hour t (t $\neq$ t$_{MNF}$), $P(t_{MNF})$ is the average hourly nodal pressure at the MNF hour, $N1$ is the pressure exponent. Finally, for each of the remaining 23 hours, the pressure-independent flows are calculated as the difference between total flow into the DMA and the pressure-dependent leakage flow rate.

6.3.4 Flow-head loss (Q-H) equations

When water flows in a pipe network, it loses energy due to internal friction and turbulence. The loss of energy is commonly referred to as head loss. The head loss in a pipe is classified into: (i) frictional head loss and (ii) minor head loss due to minor appurtenances (Bhave and Gupta 2006). The flow-head loss relationship can be expressed as follows:

$$H_L = K*Q^2 \qquad (6.7)$$

where H_L is the headloss (m), K is the head loss coefficient (m^{-5}.h^2) and Q is the flow rate (m^3/h).
The frictional factor (K) for the network can be estimated from the measured hourly inflows at the inlet and pressures at the inlet, the critical point (CP) and Average Zonal Point (AZP). The head loss (H_L) is calculated as the difference in pressure between the inlet point and the CP as well between the inlet and the AZP. The K-factors for each hour of the day for CP and AZP are then calculated using Equation 6.7. These hourly K-factors are assumed to be representative for the entire network for a specific hour of the day.

6.3.5 Analysis of different PRV settings

With known K-factors, nodal demands and nodal outflows (pressure-dependent leakage), it is now possible to assess the impact of different pressure reducing scenarios based on different PRV settings. The objective is to reduce excessive pressure at the inlet point of the DMA while ensuring that the minimum required pressure at the various nodal points especially the CP is not violated. This problem could be straight forward if the relationship between pressure-dependent leakage and nodal heads and pipe flows were linear. Unfortunately, they are not and the problem is a nonlinear programming (NLP) problem which is rather difficult to solve.

The different PRV setting options are analyzed by solving a NLP problem using sequential linear programming (SLP) techniques. SLP is an iterative procedure that involves linearization of the objective function and constraints until a termination criterion is met. The optimal valve control non-linear problem for leakage minimization in WDSs using SLP has long been solved by previous researchers (Hindi and Hamam 1991; Jowitt and Xu 1990) and is not discussed here. The DST methodology has been applied by other researchers working with PM and leakage and further details on the methodology can be found in McKenzie (2001) and Gomez et al. (2011). The DST can be accessed from NWSC-Uganda or UNESCO-IHE Institute for Water Education, Delft on request.

6.4   Decision Support Tool (DST)

A spreadsheet-based decision support planning tool (PM-COBT) using MS Excel® as a platform and coded using visual basic was developed (Appendix A) to predict the potential benefits of pressure management in a given DMA. The main user interface screen shot of the computational worksheet of the tool is shown in Figure 6.3.

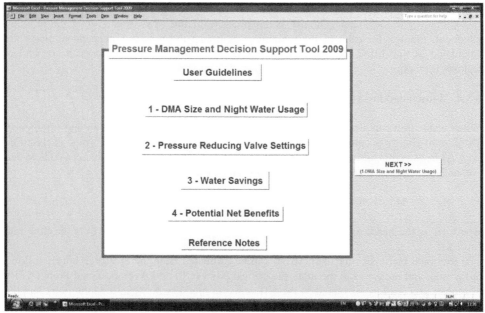

Figure 6.3 Screenshot of the main interface of decision support tool (PM-COBT)

The tool puts emphasis on costs and benefits required to support investment decision-making by answering the following inter-related questions that guide the analysis:

- What does a PM project yield in terms of water savings and other effects?
- What is the value of these project savings?
- How much does the project cost?
- How does the value compare to the cost?

These questions underlie the spreadsheet components of the planning tool. Predicting the costs and benefits of pressure management projects is not all that trivial especially for the initial project where data is not available. With time, the DST predictions will improve through rigorous data collection, testing and refining.

The tool has options of selecting the PM regimes using three options: (i) the fixed outlet control, (ii) the time modulation control valve, and (iii) the flow modulation control valve.

6.4.1   Decision Support Tool data requirements

To be able to use the tool, the following basic infrastructure and system data is required for each DMA to be analyzed. Most of the data are easily available (e.g. number of connections) and other data are not readily available (e.g. leakage coefficients and legitimate night use). In absence of actual fieldwork data, the suggested default values (Lambert 2002), should be used to quickly derive leakage estimates for a particular DMA.

General DMA information

- DMA population
- Length of mains, km
- Number of connections
- Number of properties
- Average zone night pressure (AZNP), m
- Minimum night flow (measured), m^3/h

Night Flow default loss parameters

- Background losses from mains, L/km/h
- Background losses from service connections, L/connection/h
- Background losses from properties, L/property/h
- Active population during the time of MNF, %
- Flush capacity of toilet cisterns, L
- Number of small non-domestic users
- Average use for small non-domestic users, L/h
- Use by large non-domestic users, m^3/h
- Background losses pressure exponent
- Bursts/leaks pressure exponent

Measured flow and pressure and PRV settings

- DMA zonal inflow measured over 24 hours, m^3/h
- Inlet point pressure measured over 24 hours, m
- AZP point pressure measured over 24 hours, m
- Critical point pressure measured over 24 hours, m
- Inlet pressure at MNF, m
- Pressure reducing valve (PRV) settings

Potential net benefits

- Capital cost of PRV, C, €
- Expected life time, M, years
- Interest rate, I, %
- Pipe burst frequency before pressure reduction, BF_o, bursts/year
- Expected customer contacts reduction percentage, N, %
- Cost of water at DMA inlet, AWT, €/m^3
- Cost of repairing a burst, CB, €/burst
- Cost of pumping water, CPW, €/m^3
- Cost of contacting customers, nc, €/year

This information is nowadays easy to collect with aid of pressure and flow data loggers. More information on tool data requirements is provided in the user guideline module of the tool.

In a study supported by the European Commission, TILDE (Tool for Integrated Leakage Detection), various tools (PEACH, PRESMAC, PMARS and PresCalc) for assessing the costs and benefits of leakage reduction via pressure management are briefly presented (Sjovold and Mobbs 2005). Apart from PRESMAC that was developed for the South African Water Research Commission (McKenzie 2001) and is freely accessible, details of the other tools are not known. The PRESMAC tool uses a similar methodology as PM-COBT but was not validated by a network hydraulic model to evaluate its effectiveness.

6.5 Network Hydraulic Modeling (NHM)

The network hydraulic model of the pilot study area (DMA1) was built using input data for network assets obtained from the utility geographic information system (GIS), computer-aided design (CAD) drawings and ground level data collected using mobile GPS units. The NHM was built using EPANET 2.0 (Rossman 2000) hydraulic software that is freely available and guidelines for developing, calibrating and using hydraulic models (Speight et al. 2010; Walski et al. 2003). The case study network model (DMA1) is shown in Figure 6.4. The network model for DMA1 contained a reservoir, a tank, 112 pipes and 94 junctions. The model was applied for Extended Period Simulation (EPS) under steady-state conditions. To enable EPS, hourly demand multipliers or peak factors, derived from the diurnal flow profile at the DMA inlet, were used for the hydraulic analysis. A 24-hour diurnal cycle was applied for modeling.

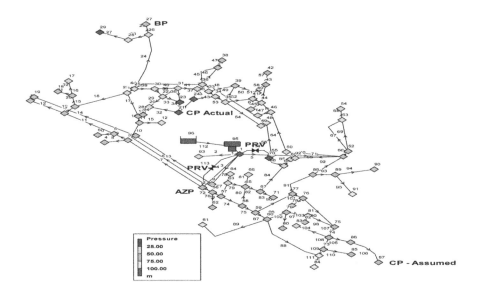

Figure 6.4 Network Model layout of DMA1

6.5.1 Quantifying leakage based on the top-down and bottom-up approaches

In order to quantify leakage in the DMA1, field measurements (flows and pressures) were carried out as earlier explained in section (§6.2). Leakage estimation was based on two methodologies: the UK "bottom-up" MNF analysis (Farley and Trow 2003) and the IWA/AWWA "top-down" water balance methodology (Alegre et al. 2006; AWWA 2009). The night-day factor (NDF) for DMA1 was calculated as 22.9 based on $N1$ value of 1.15. From Equation 6.6 and Q_L, $(t_{MNF}) = 66.84$ m³/h (Table 6.2), the daily leakage is estimated as

1,531 m³/d. This is equivalent to 24.8% when expressed as a percentage of total flow (SIV) into the DMA. This "bottom-up" approach based on MNF analysis compares fairly well with the IWA/AWWA "top-down" water balance approach where leakage was estimated to be 50% of NRW or 1,707 m³/d (Table 6.1). This is equivalent to 28% of DMA system input volume. Since all methods of estimating leakage fairly agree, leakage derived from the "top-down" water balance (19.99 L/s) was used for model calibration and subsequent potential leakage reduction predictions.

6.5.2 Quantifying leakage using the EPANET emitter coefficient

In order to perform hydraulic simulations, leakage was incorporated in the model using the emitter devices of the EPANET 2 hydraulic network solver as outlined by various researchers (Ang and Jowitt 2006; Araujo et al. 2003; Burrows et al. 2003; Rossman 2007; Tabesh et al. 2009; Wu et al. 2010). The emitter allows leakage to be modeled as a pressure-dependent outflow from any node in a hydraulic model as shown in Equation 6.8.

$$Q_i = K_i(P_i)^{N1} \qquad (6.8)$$

where Q_i is the leakage flow at node i, P_i is the pressure at node i and K_i is the emitter coefficient for the node i, estimated as a function of pipe and soil characteristics. $N1$ is the pressure exponent. During leakage calibration, by trial and error, the emitter coefficient (K_i=0.00208) was fixed for all nodes since the pipes from which leakage occurred were not known. The problem with using EPANET emitters is that assumptions have to be made for the empirical emitter parameters. In addition, the method does not take into account system characteristics such as pipe lengths, pipe conditions and the number of service connections. However, the method provides a good starting point for modeling leakage and is considered appropriate for developing countries with data deficient networks and inadequate human resources capacity for sophisticated modeling (Trifunovic et al. 2009). Germanopoulous (1985), using results from field experiments, proposed a non-linear pressure-dependent leakage formula with a pressure exponent of 1.18 (see Chapter 2). Although the method accounts for the effects of pipe length on leakage, it is computationally demanding and was not used in this study.

6.5.3 Nodal demand allocation and calibration

Having estimated leakage rates, the next step is model calibration. The aim of calibration is to minimize the differences between the observed performance and model predictions. The model calibration procedure used in this study was the iterative trial-and-error approach (Walski 1983). Although manual calibration is tedious and has limitations on the number of parameters that can be handled effectively (Abe and Cheung 2009; Wu et al. 2011), it was preferred to advanced calibration tools based on optimization techniques due to its simplicity and the required level of accuracy was deemed to be less stringent for the purpose of the study. Generally, the use of advanced calibration tools is still limited in practice as indicated by recent studies (Savic et al. 2010; Speight and Khanal 2009), probably because they are computationally expensive.

During the calibration process, pipe roughness coefficients and nodal demands were adjusted as calibration parameters while pressure and flow data was collected for model calibration. Although a large amount of "good" observation data is needed for estimating calibration parameters with sufficient confidence (Walski 2000), in this study data collection was limited to three points: inlet point, AZP and CP due to time and financial constraints. In such circumstances, trade-offs between sample design costs, level of model accuracy and real-world constraints is necessary (Kapelan et al. 2003; Speight and Khanal 2009). The calibration procedure applied in this study is shown in Figure 6.5.

Demand allocation: The allocation of base demand is critical for accurate modeling. The base demands were obtained from the utility customer billing database and manually allocated by aggregating consumptions to the nearest nodes. This was enabled by integrating GIS block map information with unique customer account references that are also used in the customer billing database. It was assumed that nodal consumption followed the same pattern as the diurnal flow profile measured at DMA inlet. A system-wide water balance approach with aim of separating real and apparent losses in order to establish appropriate water loss control strategies was applied during calibration of nodal demands in line with the IWA/AWWA water balance methodology. This calibration approach has been applied by various researchers (Almandoz et al. 2005; Cheung and Girol 2009; Nicolini et al. 2011). Total hourly demands in the DMA are tracked while maintaining the overall system demand. The model was calibrated under steady-state and EPS conditions using flow and pressure data.

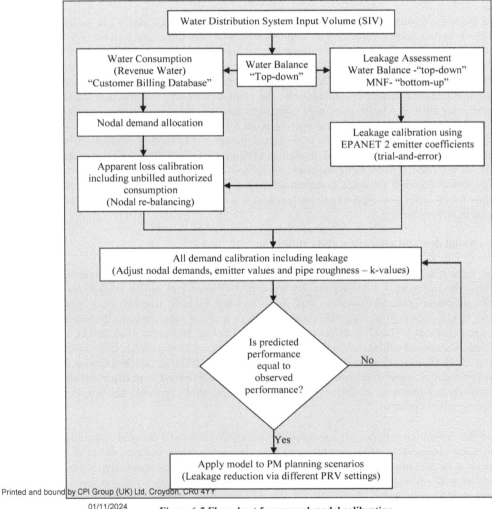

Figure 6.5 Flow chart for manual model calibration

Leakage calibration: With respect to leakage, the models were calibrated by setting all the nodal base demands to zero, followed by adjustment of the emitter coefficient and pressure exponent (Equation 6.13), till the correct value of leakage estimated from the utility top-down annual water balance closely matched the simulated leakage within the desired accuracy limits. It was assumed that leakage in the network was pressure-dependent and occurs throughout the network. Leakage flows are then added to real demand to obtain new global consumption values. Nodal demands and pipe roughness coefficients were further adjusted till the simulated flows and pressures closely matched the measured flows at the DMA inlet and pressures at the AZP and CP.

Apparent losses calibration: The nodal base demands were increased uniformly until the value of authorized consumption and apparent losses as indicated in Table 6.3 was achieved. In the analysis, it was assumed that the apparent losses and unbilled authorized consumption followed the same pattern as the billed nodal consumptions (i.e. DMA SIV diurnal pattern).

Table 6.3 Measured system demand and estimated leakage for model calibration

Average water demand (L/s) Q_{SIV}	Billed consumption (L/s) Q_{BAC}	Authorized consumption and Apparent losses (L/s) $Q_{BAC} + Q_{UAC} + Q_{AL}$	Real losses (Leakage) (L/s) Q_L
71.38	31.85	51.39	19.99

For each successive time step, nodal consumption is re-balanced to satisfy the water balance in Equation 6.9 and diurnal patterns adjusted until the measured values could be simulated within the desired precision.

$$Q_{SIV} = Q_{BAC} + Q_{UAC} + Q_{AL} + Q_L \qquad (6.9)$$

Where, Q_{SIV} is the total flow into the DMA, Q_{BAC} is the billed authorized consumption, Q_{UAC} is the unbilled authorized consumption, Q_{AL} is the apparent losses volume and Q_L is the leakage volume. The water balance and diurnal water use patterns enabled proper calibration for EPS.

6.5.4 Model Validation

The model was verified for EPS and Figure 6.6 shows the simulated and observed pressure (goodness-of-fit) while Figure 6.7 depicts the observed and simulated inflow and leakage.

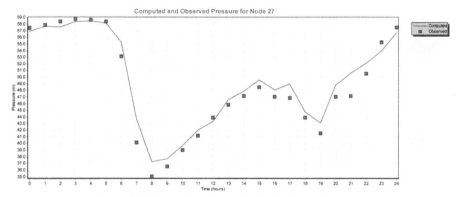

Figure 6.6 Comparison of computed and observed pressure

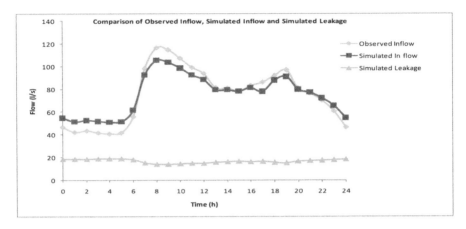

Figure 6.7 Comparison of simulated and observed flow into DMA1

The calibration methodology resulted in a maximum error of 3.8 m for pressure and 11 L/s for flows. The calibration criteria used to accept the errors were: (i) less than 5 m for pressure and (ii) less than 10% of the MNF measured at the inlet point (29 l/s) i.e. < 2.9 L/s. The mismatch between the integral areas under the observed and simulated inflows particularly at the hour of MNF in Figure 6.7 could be due to variations in the actual night use by users, errors in the measurement equipment, demand patterns based on a limited data set, uncertainty in the estimated calibration parameters, and localized calibration to a small part of the network.

Table 6.4 Calibration statistics for flow and pressure for DMA1

Location Node	1	2	87	27
No. of Observations	24	24	24	24
Observed Mean	76.08	61.49	52.97	48.32
Computed Mean	76	60.78	53.17	49.2
Error Between Means (%)	-0.11	-1.15	0.38	1.82
Correlation Coefficient	0.998	0.899	0.852	0.988

Although there was a mismatch between observed and simulated flows, the correlation coefficients (Table 6.4) were very close to one indicating acceptable performance and good model representation of the system behaviour and suitable for the intended purpose of the study.

6.6 Application to case study

The DST and NHM have been applied to DMA1 of the KWDS using field data and information from the NWSC procurement department to compare the estimated costs of the PM project against cost-savings of the projected benefits. Where data was not available, gaps were filled using data from literature. The potential for leakage reduction by PM was examined based on three PRV options: Fixed outlet, Time modulation and Flow modulation. The advantages and disadvantages of these PRVs were discussed in Chapter 2.

The net benefits derived from a pressure reduction scheme proposed for implementation in DMA1, were estimated from the difference between related costs before and after introduction of the scheme, using the following cost model (Awad et al. 2008):

$$NPM = CLR + CBR + CCR + CDER - CPRV - CMM - CED \qquad (6.10)$$

Where, NPM = net benefit from pressure reduction (€/year), CLR = benefit from leakage reduction (€/year), CBR = benefit from reduced pipe burst frequency (€/year), CCR = benefit from customer complaints reduction (€/year), $CDER$ = benefit from direct energy reduction (€/year), $CPRV$ = annual cost of installation, construction and commissioning pressure reducing valves (€/year), CMM = annual cost of maintenance and monitoring (€/year), and CED = initial cost of engineering and design (€/year).

6.7   Results and Discussion

Only results of DMA1 are presented here as DMA2 (Table 6.1) did not yield significant benefits to justify the investments related with introduction of pressure management. The measurement of flows and pressures, at the selected points in the pilot DMAs, indicated that the average flow in DMA1 was 257 m³/h. The flow pattern and pressure profiles, obtained from the field measurements for DMA1, are shown in Figure 6.8.

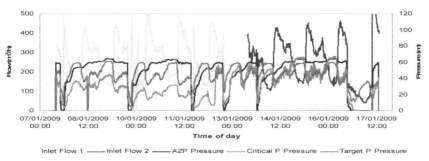

Figure 6.8 Flow pattern and pressure profile for DMA1

6.7.1   Comparison of leakage estimation by different methods

Table 6.5 shows the estimated leakage using the different methods. From Table 6.5, all methods compare fairly well but the model gives a more conservative figure of leakage and is likely to result in more benefits of water savings in practice than those predicted by MNF analysis and water balance techniques. Leakage in this particular DMA averages 25% of system input volume but could even be as high as 50% in the older parts of the network e.g. Kansanga-Kiwafu (average pressures greater than 90 m) that were difficult and costly to isolate into DMAs within the limited time of the study.

Table 6.5 Comparison of estimated leakage by different approaches

	EPANET	"Bottom-up" MNF	"Top-down" IWA/AWWA Water Balance
Leakage (m³/d)	1425	1531	1707
% of System Input Volume (6,167 m³/d)	23	25	28

6.7.2   Comparison of water savings predicted by the NHM under different PM options

To assess the impact of pressure reduction on leakage levels, three types of PRVs were analyzed, namely; fixed outlet control, time modulated control and flow modulated control PRVs. The valves were introduced in the network models to control or reduce inlet pressure, as shown in Figure 6.4. The results are presented in Table 6.6. The results are average figures derived directly from 24-hour simulation runs after pressure reduction. The primary criterion was to ensure the availability of flow at all nodes, throughout the DMA at all times, including the maximum consumption period, even if the pressure at that moment violated the required minimum pressure of 10 m (DWD 2000).

Table 6.6 Comparison of estimated water and cost-savings for different PRVs

PRV Options	Average Pressure (m)		Average Leakage (L/s)	Water Savings		Net Benefits (€/yr)
	AZP	CP		L/s	%	
No Control	63.42	9.31	19.99			
Fixed Outlet	56.00	4.50	16.50	3.49	17.5	66,910
Time Modulation	51.25	1.83	15.72	4.27	21.4	81,100
Flow Modulation	54.28	2.95	15.52	4.47	22.4	82,230

From Table 6.6, the following can be deduced:

1. Significant cost savings were obtained in all cases by introducing PM. The most cost-effective valve of pressure control and leakage reduction is the flow modulated outlet PRV with net benefits of about 82,230 €/year. Whereas flow and time modulation PRVs provide more benefits over the fixed-outlet PRV, the extra costs and operational issues do not always justify the added benefits (Trow and Payne 2009). The fixed-outlet PRVs are considered more appropriate for water utilities in the developing countries that are just starting to work with PM systems. They are relatively cheap in terms of investment cost and easy to operate and maintain. Further reductions could be realized in future by adopting "intelligent" PM. Trials in the UK suggest an average 20% saving can be achieved when intelligent PM is applied to an existing pressure managed zone (Trow 2010).

2. Lowering average pressure in the network by 7 to 12 m reduces leakage by about 20% of its original value.

3. All PRV options did violate minimum service pressure requirements of 10 m. The violations were observed in 5% of the supply area. In order to maximize leakage reduction the DMA could be reconfigured, such that the 5% of the area is served by other neighboring DMAs. Actually in practice, most areas in Kampala receive water at very low pressures of about 4.0 m and some parts only get water at night or nothing at all.

The cost-benefit results obtained should not be generalized because of the uncertainty and subjectivity of data used and assumptions made. For instance, variation in repair and maintenance costs will have a significant influence on cost savings of the pressure management scheme. Lastly, socio-economic cost-benefits like traffic disruptions and others were not quantified.

6.7.3   Comparison of water savings predicted by the DST and NHM

The decision support tool (DST) and network hydraulic modelling (NHM) have been applied to DMA1 in Kampala. The results presented herein are based on conservative predictions of pressure reduction using Fixed-Outlet PRVs.

The analysis results presented in Tables 6.7, 6.8 and 6.9 indicate that lowering pressure by 7 m yields more than 250 m^3/day of water savings and annual net financial benefits of over €56,000. The DST predictions are more conservative as it predicts about 48 m^3/d and €10,000 (or 16%) less water and cost savings benefits compared to the NHM. Although the model may not be very precise in predicting financial benefits it is still useful as the predicted savings are generally within 10 to 20% of those actually achieved in practice (McKenzie 2001).

Table 6.7 DST water savings for PRV settings (P_1= 63.5 , P_2= 56 m)

Hour	DMA Inflow (m^3/h)		Water savings (m^3/h)
	Before PM	After PM	
0 - 1	167.05	156.87	10.18
1 - 2	150.72	140.75	9.97
2 - 3	155.97	146.36	9.61
3 - 4	149.55	140.85	8.70
4 - 5	146.63	138.10	8.53
5 - 6	149.45	140.09	9.36
6 - 7	200.21	189.89	10.32
7 - 8	352.68	340.28	12.40
8 - 9	419.58	405.46	14.12
9 - 10	412.09	398.00	14.09
10 - 11	385.45	372.03	13.42
11 - 12	357.25	344.74	12.51
12 - 13	337.41	325.25	12.16
13 - 14	292.00	280.16	11.84
14 - 15	291.42	279.91	11.51
15 - 16	281.60	277.09	4.51
17 - 18	298.32	288.40	9.92
18 - 19	309.80	300.65	9.15
19 - 20	330.61	318.57	12.04
20 - 21	347.91	335.29	12.62
20 - 21	290.16	279.93	10.23
21 - 22	275.47	267.00	8.47
22 - 23	252.82	243.94	8.88
23 - 24	219.08	209.18	9.90
Daily Totals	6573.23	6318.79	254.44

Table 6.8 NHM water savings for PRV settings (P₁= 63.5 , P₂= 56 m)

Hour	DMA Leakage (L/s)		Water savings (L/s)
	Before PM	After PM	
0 - 1	24.64	18.29	6.35
1 - 2	24.86	18.45	6.41
2 - 3	25.47	18.44	7.03
3 - 4	25.91	18.76	7.15
4 - 5	25.94	18.77	7.17
5 - 6	24.86	18.65	6.21
6 - 7	22.52	17.94	4.58
7 - 8	16.54	15.35	1.19
8 - 9	14.40	13.93	0.47
9 - 10	14.34	13.94	0.40
10 - 11	14.76	14.28	0.48
11 - 12	15.78	14.70	1.08
12 - 13	16.02	14.91	1.11
13 - 14	17.47	15.58	1.89
14 - 15	18.39	16.06	2.33
15 - 16	19.44	16.63	2.81
17 - 18	18.79	16.23	2.56
18 - 19	18.95	16.38	2.57
19 - 20	17.46	15.42	2.04
20 - 21	16.77	14.99	1.78
20 - 21	19.07	16.41	2.66
21 - 22	21.63	16.96	4.67
22 - 23	22.28	17.24	5.04
23 - 24	23.35	17.62	5.73
Daily Mean	19.99	16.50	3.49

Table 6.9 Summary of comparison of DST and NHM predictions of PM benefits

	Average Pressure (m)	Total Inflow (m³/d)	Savings (m³/d)	Savings (%)	Net Benefits (€/yr)
No PM	63	6167			
DST	56	5913	254	4	56,190
NHM	56	5865	302	5	66,910

The analyses of different pressure profiles at the critical point are shown in Figure 6.9. It is evident from Figure 6.9 that the critical point pressure predicted by the DST is higher than that predicted by network hydraulic modeling confirming the conservativeness of the tool in its predictions. In the absence of measured data, the network hydraulic modeling results could be used as input data for the tool, thus complimenting each other.

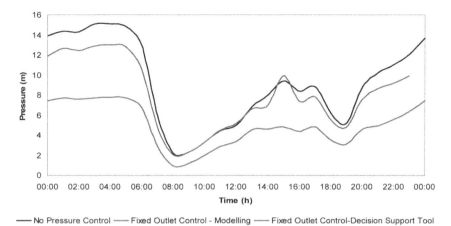

— No Pressure Control —— Fixed Outlet Control - Modelling —— Fixed Outlet Control-Decision Support Tool

Figure 6.9 Comparison of CP pressure predicted by NHM and the DST

6.7.4 Limitations of the Decision Support Tool (DST)

The decision support tool has the following limitations:

- It may under estimate the potential net water savings and divert attention and expectations away from pressure management strategies. This shortcoming is an inherent property in the tool's methodology that can be overcome by progressively adopting network hydraulic modeling techniques;

- The head loss–flow relationship used by the tool to determine network friction factors, considers all influences on head loss to be lumped in one parameter. This may lead to under estimation of head loss and subsequent prediction of high values of critical point pressure;

- The tool uses coefficients and default values developed from a series of field testing in the developed countries (e.g. FAVAD *N1* values, BABE typical flow rates of leaks and bursts at some standard pressure) – all of which may cumulatively lead to erroneous estimates of net benefits. It is important to make necessary modifications to suit local conditions;

- The tool does not provide information such as flow and pressure in the rest of the DMA. Calibration is localized at only three points (inlet, AZP, and CP) and this could lead to erroneous predictions.

6.7.5 Key lessons learned during the study

The key lessons learned from the two DMAs of varying sizes and different hydraulic characteristics can be used as guidelines to assess whether there is potential for pressure management:

I.   **Pressure reduction potential:** The difference between the existing maximum and minimum pressure in a DMA during a day should be greater than 10 m in order to achieve a good rate of return on investment. PRVs are highly cost-effective in reducing excessive pressures in DMAs with widely varying pressure. It is the ratio of average pressures that influence reliable predictions.

II.   **Leakage reduction potential:** the ratio of maximum to minimum flows into the DMA should be lower than 10 in order to get a positive effect of pressure management on leakage reduction. Ratio of MNF to average flow must be greater than 0.20. A high ratio is an indicator of high leakage. However, in some cases this ratio may be misleading as a high ratio could be associated with a high factory demand and may only indicate a high night use. The overall water use for the system must therefore be taken into consideration.

III.   **Real loss component of water delivered:** Leakage must be more than 20% of system input volume or more than 200 L/connection/day. In the Kampala DMAs, where leakage management is reactive, thus indicating a backlog of unreported and background leakage, real losses are over 200 L/con/d. In the UK for instance where the regulator sets stringent targets for leakage and where intelligent PM is in use and leakage levels are approaching economic levels, real losses average 113 L/con/d at average pressures of about 50 m (Thornton et al. 2008).

IV.   **Size of DMA:** optimal number of properties should be between 2,500 and 5,000 for a good return on investment. Although DMA2 had high real losses (322 L/con/d), it did not yield substantial benefits to justify the cost of investment due to the small number of properties (354 No.). The smaller the size of the DMA, the more likely that the total recoverable leakage volume will be too low to justify the investment. However, when the DMA has more than 5,000 properties, it becomes difficult to discriminate small leaks from minimum consumption hour flow data, making the DMA less effective (AWWA 2009). Pressure reduction is best carried out in small DMAs, supplied by mains that can be reduced to accept a PRV not in excess of 200 mm in diameter. PRVs become excessively expensive above the 200 mm size (Mistry 2009).

V.   **Reduction of failure frequency potential:** the average age of pipelines must exceed 15 years with main and service connection breaks exceeding 50 breaks/100 km/year and 50 breaks/1000 service connections/year respectively. WDSs having less than 40 breaks/100 km/year are considered to be in acceptable state (Pelletier et al. 2003). Age alone is not a reliable indicator of leakage and must be complimented with pipe failure data. From the DMA records, significant break frequency was reported on pipes of more than 15 years old.

VI.   **Challenges of leakage modeling:** modeling leakage in a WDS is still a difficult task and assumptions made are not very valid. Leaks are not distributed along all pipes and are not proportional to real consumptions, nor are they evenly spread over a system. In addition, some leaks such as toilet cistern leaks are not pressure-dependent. Another limitation is the high costs for collecting model calibration data.

VII.   **Water Supply:** Continuous or regular water supply is preferred. The efficiency of pressure management decreases with increasing number of supply interruptions. In addition, most tools and methods have been developed for 24-hour water supply regimes.

6.8   CONCLUSIONS

In this Chapter an appropriate decision support tool for evaluating potential net benefits of implementing pressure management strategies in WDSs of the developing countries is presented. The study clearly illustrates the benefits of pressure management using the pilot study area. Over 250 m³/day of water savings and €56,000 net financial benefits have been predicted for the Kampala DMA1. The DST predictions rely heavily on accuracy of data before and after pressure management implementation. Most of the cost model parameters used were estimated and will need to be updated in future. It is anticipated that the developed DST will act as a stimulus to promote use of pressure management strategies as part of the broader leakage management policies in the water utilities of the developing countries to recuperate water losses. Advances in computing power, accessibility of affordable and user friendly software, and capacity building are opening up the field of modelling to water utilities in developing countries. For operational synergies however, it is advisable to compliment the DST with network hydraulic modelling as reliable data and resources become available with time for more accurate predictions of pressure management benefits. In order to maximize leakage reduction and benefits of pressure management, the following further research areas are recommended:

- Optimal PRV selection and location;

- Intelligent pressure management using optimization techniques to maximize pressure reductions without compromising customer service levels;

- Testing and refining the prediction methods for quantifying the economic benefits in order to understand fully the real impacts of pressure management;

- Establishing leakage coefficients and default values such as normal household use, background losses from mains, service connections, properties, pressure exponents, etc. that are relevant to developing countries with often poorly managed and maintained WDSs. This will improve the quality of leakage estimates and subsequent planning studies.

6.9   References

AbdelMeguid, H., and Ulanicki, B. (2010). "Pressure and Leakage Management in Water Distribution Systems via Flow Modulation PRVs." *Proceedings of the 11th Water Distribution System Analysis Conference (WDSA 2010)*, Tuscon, Arizona.

Abe, N., and Cheung, P. B. (2009). "Epanet Calibrator - An integrated computational tool to calibrate hydraulic models." Integrating Water Systems, Boxall and Maksimovic, eds., Taylor and Francis Group, London, 129-133.

Alegre, H., Baptista, J. M., Cabrera, E. J., Cubillo, F., Hirner, W., Merkel, W., and Parena, R. (2006). *Performance Indicators for Water Supply Services, IWA Manual of Best Practice*, IWA Publishing.

Almandoz, J., Cabrera, E., Arregui, F., Cabrera Jr, E., and Cobacho, R. (2005). "Leakage Assessment through Water Distribution Network Simulation." *Journal of Water Resources Planning and Management*, 131, 458-466.

Ang, W. K., and Jowitt, P. W. (2006). "Solution for Water Distribution Systems under Pressure-Deficient Conditions." *Journal of Water Resources Planning and Management*, 132(3), 175-182.

Araujo, L. S., Coelho, S. T., and Ramos, H. M. (2003). "Estimation of distributed pressure-dependent leakage and consumer demand in water supply networks." Advances in Water Supply Management, Maksimovic, Butler, and Memon, eds., Swets & Zeitlinger, Lisse, 119-127.

Awad, H., Kapelan, Z., and Savic, D. (2008). "Analysis of Pressure Management Economics in Water Distribution Systems." *Proceedings of the 10th Annual Water Distribution System Analysis Conference WDSA2008, August 17-20*, Kruger National Park, South Africa, 520-531.

Awad, H., Kapelan, Z., and Savic, D. A. (2009). "Optimal setting of time-modulated pressure reducing valves in water distribution networks using genetic algorithms." Integrating Water Systems, Boxall and Maksimovic, eds., Taylor and Francis Group, London, UK, 31-37.

Awad, H., Kapelan, Z., and Savic, D. A. (2010). "Optimal setting of time-modulated pressure reducing valves in water distribution networks using genetic algorithms." Integrating Water Systems, Boxall and Maksimovic, eds., Taylor and Francis Group, London, 31-37.

AWWA. (2009). "Water Audits and Loss Control Programs: AWWA Manual M36." American Water Works Association, Denver, USA.

Babel, M. S., Islam, M. S., and Gupta, A. D. (2009). "Leakage Management in a low-pressure water distribution network of Bangkok." *Water Science and Technology:Water Supply*, 9(2), 141-147.

Bhave, P. R., and Gupta, R. (2006). *Analysis of Water Distribution Networks*, Alpha Science International Ltd., Oxford, UK.

Burrows, A. (2010). "Bringing Intelligence to Water Control." *Proceedings of the 4th Annual Global Leakage Summit (CD-ROM)*, London, UK.

Burrows, R., Mulreid, G., and Hayuti, M. (2003). "Introduction of a fully dynamic representation of leakage into network modelling studies using EPANET." Proceedings of the International Conference on Advances in Water Supply Management, C. Maksimovic, D. Butler, and F. A. Memon, eds., Swets & Zeitlinger, Lisses, 109-118.

Cabrera, E., Pardo, M. A., Cobacho, R., and Cabrera Jr, E. (2010). "Energy Audit of Water Networks." *Journal of Water Resources Planning and Management*, 136(6), 669-677.

Cassa, A. M., and Van Zyl, J. E. (2012). "Predicting the pressure-area slopes and leakage of cracks in pipes." *Proceedings of the 7th IWA Water Loss Reduction Specialist Conference, February 26-29*, Manila, Philippines.

Cassa, A. M., Van Zyl, J. E., and Laubscher, R. F. (2010). "A numerical investigation into the effects of pressure on holes and cracks in water supply pipes." *Urban Water Journal*, 7(2), 109-120.

Charalambous, B. (2008). "Use of district metered areas coupled with pressure optimization to reduce leakage." *Water Science and Technology: Water Supply*, 8(1), 57-62.

Cheung, P. B., and Girol, G. V. (2009). "Night flow analysis and modeling for leakage estimation in a water distribution system." Integrating Water Systems, Boxall and Maksimovic, eds., Taylor and Francis Group, London.

Colombo, A. F., and Karney, B. W. (2005). "Impacts of Leaks on Energy Consumption in Pumped Systems with Storage." *Journal of Water Resources Planning and Management*, 131(2), 146-155.

DWD. (2000). "Water Supply Design Manual." Directorate of Water Development (DWD), Kampala.

Fanner, P., Sturm, R., Thornton, J., and Liemberger, R. (2007). *Leakage Management Technologies*, Awwa Research Foundation Denver, Colorado, USA.

Farley, M., and Trow, S. (2003). *Losses in Water Distribution Networks: A Practitioner's Guide to Assessment, Monitoring and Control*, IWA Publishing, London.

Germanopoulos, G. (1985). "A Technical Note on the inclusion of Pressure Dependent Demand and Leakage terms in Water Supply Network Models." *Civil Engineering and Environmental Systems*, 2(3), 171-179.

Girard, M., and Stewart, R. A. (2007). "Implementation of Pressure and Leakage Management Strategies on the Gold Coast, Australia: Case Study." *Journal of Water Resources Planning and Management*, 133, 210.

Gomes, R., Marques, A. S., and Sousa, J. (2011). "Estimation of the benefits yielded by pressure management in water distribution systems." *Urban Water Journal*, 8(2), 65-77.

Greyvenstein, B., and van Zyl, J. E. (2007). "An experimental investigation into the pressure-leakage relationship of some failed water pipes." *Journal of water supply: Research and Technology-AQUA*, 56(2), 117-124.

Hindi, K. S., and Hamam, Y. M. (1991). "Pressure control for leakage minimization in water supply networks." *International Journal of systems science*, 22(9), 1573-1585.

Jowitt, P. W., and Xu, C. (1990). "Optimal Valve Control in Water-Distribution Networks." *Journal of Water Resources Planning and Management*, 116(4), 455-472.

Kapelan, Z., Savic, D. A., and Walters, G. A. (2003). "Multiobjective sampling design for water distribution model calibration." *Journal of Water Resources Planning and Management*, 129(6), 466-479.

Karim, M. R., Abbaszadegan, M., and LeChevallier, M. (2003). "Potential for Pathogen Intrusion During Pressure Transients." *Journal American Water Works Association*, 95(5), 134-146.

Lambert, A., and Morrison, J. A. E. (1996). "Recent developments in application of "Bursts and Background Estimates" concepts for leakage management." *Water and Environment Journal*, 10(2), 100-104.

Lambert, A., and Thornton, J. (2011). "The relationships between pressure and bursts - a "state-of-the-art" update." IWA Magazine Water 21, April, 37-38.

Lambert, A. O., Brown, T. G., Takizawa, M., and Weimer, D. (1999). "A review of performance indicators for real losses from water supply systems." *Aqua- Journal of Water Services Research and Technology*, 48(6), 227-237.

Lambert, A. O., and Fantozzi, M. (2010). "Recent Developments in Pressure Management." *IWA Specialized Water Loss Conference*, Sao Paulo, Brazil, CD-Rom.

Marunga, A., Hoko, Z., and Kaseke, E. (2006). "Pressure management as a leakage reduction and water demand management tool: The case of the City of Mutare, Zimbabwe." *Physics and Chemistry of the Earth*, 31(15-16), 763-770.

May, J. (1994). "Pressure Dependent Leakage." *World Water and Environmental Engineering, October 1994*, 13.

McKenzie, R. (1999). *Development of a Standardised Approach to Evaluate Burst and Background Losses in Water Distribution Systems in South Africa*, WRC, Report TT 109/99, South Africa.

McKenzie, R. (2001). *PRESMAC: Pressure Management Program*, WRC, Report TT 152/01, South Africa.

McKenzie, R. S., Mostert, H., and de Jager, T. (2004). "Leakage reduction through pressure management in Khayelitsha: two years down the line." *Water SA*, 30(5), 13-17.

Mistry, P. H. (2009). "To boldly go to a new way of operating water supply networks: South East Queensland Experience." *Proceedings of the 5th IWA Specialist Conference on Efficient Use and Management of Water (CD-ROM)*, Sydney, Australia.

Mounce, S. R., Boxall, J. B., and Machell, J. (2010). "Development and verification of an online artificial intelligence system for detection of bursts and other abnormal flows." *Journal of Water Resources Planning and Management*, 136(3), 309-318.

Mutikanga, H. E., Sharma, S., and Vairavamoorthy, K. (2009). "Water Loss Management in Developing Countries: Challenges and Prospects." *Journal AWWA*, 101(12), 57-68.

Nazif, S., Karamouz, M., Tabesh, M., and Moridi, A. (2010). "Pressure Management model for Urban Water Distribution Networks." *Water Resources Management*, 24, 437-458.

Nicolini, M., Giacomello, C., and Deb, K. (2011). "Calibration and Optimal Leakage Management for a Real Water Distribution Network." *Journal of Water Resources Planning and Management*, 137(1), 134-142.

Pelletier, G., Mailhot, A., and Villeneuve, J. P. (2003). "Modeling water pipe breaks-three case studies." *Journal of Water Resources Planning and Management*, 129(2), 115-123.

Pilipovic, Z., and Taylor, R. (2003). "Pressure management in Waitakere City, New Zealand-a case study." *Water Science and Technology: Water Supply*, 3(1/2), 135-141.

Puust, R., Kapelan, Z., Savic, D. A., and Koppel, T. (2010). "A review of methods for leakage management in pipe networks." *Urban Water Journal*, 7(1), 25-45.

Reis, L. F. R., Porto, R. M., and Chaudhry, F. H. (1997). "Optimal Location of Control Valves in Pipe Networks by Genetic Algorithm." *Journal of Water Resources Planning and Management*, 123(6), 317-326.

Rossman, L. A. (2000). *EPANET 2 users manual*, USEPA, Cincinnati.

Rossman, L. A. (2007). "Disscussion of "Solution for Water Distribution Systems under Pressure-Deficient Conditions" by Wah Khim Ang and Paul W. Jowitt." *J. Water Resour. Plann. and Manage.*, 133(6), 566-567.

Savic, D. A., Kapelan, Z., and Jonkergouw, P. (2010). "Quo vadis water distribution model calibration." *Urban Water Journal*, 6(1), 3-22.

Savic, D. A., and Walters, G. A. (1995). "An Evolution Program for Optimal Pressure Regulation in Water Distribution Networks." *Engineering Optimization*, 24(3), 197-219.

Sharma, S. K., and Vairavamoorthy, K. (2009). "Urban water demand management: prospects and challenges for the developing countries." *Water and Environment Journal*, 23, 210-218.

Sjovold, F., and Mobbs, P. (2005). "TILDE: D21 Econometric Tools (Project No. 22111403)." Trondheim, Norway.

Skipworth, P. J., Saul, A. J., and Machell, J. (1999). "The effect of regional factors on leakage levels and the role of performance indicators." *Water and Environment Journal*, 13(3), 184-188.

Speight, V., and Khanal, N. (2009). "Model calibration and current usage in practice." *Urban Water Journal*, 6(1), 23-28.

Speight, V., Khanal, N., Savic, D., Kapelan, Z., JonKergouw, P., and Agbodo, M. (2010). "Guidelines for Developing, Calibrating, and Using Hydraulic Models." Water Research Foundation, Denver, Colorado.

Sterling, M. J. H., and Bergiela, A. (1984). "Leakage reduction by optimised control valves in water networks." *Transactions of the Institute of Measurement and Control*, 6(6), 293-298.

Tabesh, M., Asadiyani, Y., and Burrows, R. (2009). "An Integrated Model to Evaluate Losses in Water Distribution Systems." *Water Resources Management*, 23(3), 477-492.

Tanyimboh, T., Tahar, B., and Templeman, A. (2003). "Pressure-driven modelling for water distribution systems " *Water Science & Technology: Water Supply*, 3(1-2), 255-261.

Thornton, J., Sturm, R., and Kunkel, G. (2008). *Water Loss Control*, McGraw-Hill, New York.

Trifunovic, N., Sharma, S., and Pathirana, A. (2009). "Modelling Leakage in Distribution System using EPANET." *Proceedings of the 5th IWA Water Loss Reduction Specialist Conference*, Cape Town, South Africa, 482-489.

Trow, S. (2010). "A summary of experience in the use of intelligent pressure management systems." *Proceedings of the 6th IWA Water Loss Reduction Specialist Conference (CD-ROM)*, Sao Paulo, Brazil.

Trow, S. W., and Payne, A. (2009). "Intelligent Pressure Management - A New Development for Monitoring and Control of Water Distribution Systems." *Proceedings of the 5th IWA Water Loss Reduction Specialist Conference*, Cape Town, South Africa, 302-314.

Ulanicki, B., Bounds, P. L. M., Rance, J. P., and Reynolds, L. (2000). "Open and Closed Loop Pressure Control for Leakage Reduction." *Urban Water*, 2, 105-114.

Vairavamoorthy, K., and Lumbers, J. (1998). "Leakage reduction in water distribution systems: Optimal valve control." *Journal of Hydraulic Engineering*, 124(11), 1146-1154.

Van Zyl, J. E., and Cassa, A. M. (2011). "Linking the power and FAVAD equations for modelling the effect of pressure on leakage." *Proc. of the 11th Int. Conference on Computing and Control of the Water Industry (CCWI 2011) - Urban water management challenges and Opportunities, September 5-7*, Exeter, UK.

Walski, T., Bezts, W., Posluszny, E. T., Weir, M., and Whitman, B. E. (2006). "Modeling Leakage Reduction through Pressure Control." *Journal American Water Works Association*, 98(4), 147-155.

Walski, T. M. (1983). "Technique for Calibrating Network Models." *Journal of Water Resources Planning and Management*, 109(4), 360-372.

Walski, T. M. (2000). "Model Calibration Data: The Good, the Bad, and the Useless." *Journal AWWA*, 92(1), 94-99.

Walski, T. M., Chase, D. V., Savic, D. A., Grayman, W., Beckwith, S., and Koelle, E. (2003). *Advanced Water Distribution Modeling and Management*, HAESTAD PRESS, Waterbury, CT, USA.

Wu, Z. Y., Farley, M., Turtle, D., Kapelan, Z., Boxall, J., Mounce, S., Dahasahasra, S., Mulay, M., and Kleiner, Y. (2011). *Water Loss Reduction*, Bentley Institute Press, Exton, Pennsylvania, USA.

Wu, Z. Y., Sage, P., and Turtle, D. (2010). "Pressure-dependent leak detection model and its application to a district water system." *Journal of Water Resources Planning and Management*, 136(1), 116-128.

Chapter 7 - Multi-criteria Decision Analysis for Strategic Water Loss Management Planning

Parts of this chapter are based on:

Mutikanga, H.E, Sharma, S.K., and Vairavamoorthy, K (2011). "Multi-criteria Decision Analysis: A strategic planning tool for water loss management". *Water Resources Management*, 25(14), 3947-3969.

Mutikanga, H.E, and Sharma, S.K. (2012). "Strategic Planning for Water Loss Reduction with Imprecise Data". *Proceedings of the 7ʰ IWA Water Loss Reduction Specialist Conference*, February 26-29, Manila, Philippines, CD-ROM Edition.

Summary

Water utilities particularly in the developing countries continue to operate with considerable inefficiencies in terms of water and revenue losses. With increasing water demand and scarcity, utilities require effective strategies for optimum use of available water resources. Diverse water loss reduction options exist. Deciding on which option to choose amidst conflicting multiple criteria and different interests of stakeholders is a challenging task. In this chapter, an integrated multi-criteria decision-aiding framework methodology for strategic planning of water loss management is presented. The PROMETHEE II outranking method of the MCDA family was applied within the framework in prioritizing water loss reduction options for Kampala city. A strategic plan that combines selective mains and service lines replacement and pressure management as priorities is the best compromise based on preferences of the decision makers and seven evaluation criteria characterized by financial-economic, environmental, public health, technical and social aspects. The results show that the most preferred options are those that enhance water supply reliability, public health and water conservation measures. This study demonstrates how decision theory coupled with operational research techniques could be applied in practice to solve complex water management and planning problems. The developed multi-criteria decision-aid framework methodology will not only be a valuable tool in helping data-deficient water utilities to evaluate and prioritize water loss reduction strategies but will also facilitate the understanding of how to structure and evaluate decision problems.

7.1 Introduction

Water losses represent a major problem for water utilities, customers and the environment. In Chapter 2, alternative water loss reduction options were highlighted. However, selecting which option to apply and when, remains a challenging decision problem for most water utility managers. In this chapter, we develop an integrated multi-criteria decision analysis framework methodology for evaluating and prioritizing water loss reduction strategy options.

Since water loss is inevitable, every water utility should have a strategic plan for water loss management. Strategic planning has proven to be a valuable tool for sustainable urban water management (Malmqvist et al. 2006). However, water utilities in developing countries often lack the necessary capabilities to carryout strategic planning (Mugabi et al. 2007; Schouten and Halim 2010). Strategic planning is about setting a long-term direction based on sound predictions, analysis of options and key decisions about the future of an organisation. The NWSC-Uganda corporate plan (2009-2012) broadly categorises water loss reduction among other sub-goals under the main goal of revenue maximization (NWSC 2009). This traditional way of evaluating water loss reduction strategies based on a single criterion of maximising revenue is unrealistic as water services management takes place in a multiple-criteria environment. Clearly, a better structured approach for strategic planning is required to safeguard against asset stripping by a utility driven mainly to maximize profits.

Water loss reduction options are characterized by different multiple criteria and deciding on which strategy option to apply and when to apply it is no simple task amidst conflicting interests of different actor groups involved in the water services sector. Kain et al. (2006) suggested use of multi-criteria decision-aid techniques in such a situation.

MCDA is a tool developed in the field of decision theory for resolving operational research problems with a finite number of decision options based on a set of evaluation criteria (Figueira et al. 2005; Lu et al. 2007; Simonovic 2009). The discrete MCDA techniques take into account a range of quantitative and qualitative criteria beyond the financial criterion during the optimal decision-making process (NAMS 2004). They have the potential to provide well-structured, rational, consistent, transparent and objective solutions to complex decision problems in water resources management and planning (Afshar et al. 2011; Akbari et al. 2011; Cabrera Jr et al. 2011; Hajkowicz and Collins 2007; Silva et al. 2010).

The application of MCDA techniques to solve drinking water problems has been reported by various researchers (Al-Barqawi and Zayed 2008; Baur et al. 2003; Bouchard et al. 2010; Kodikara et al. 2009; Morais and Almeida 2007). Of particular interest, is the leakage management study in Brazil (Morais and Almeida 2007). Although Morais and Almeida (2007) addressed the leakage strategy in WDSs, they did not tackle water loss in totality as the apparent water loss component that is often significant in developing countries (Mutikanga et al. 2009) was not considered. In addition, they did not carry out a water balance to identify whether the problem was leakage or apparent losses. They instead indicate a water loss of 60% for the case study. Use of percentage as a water loss indicator can be misleading as it is heavily influenced by consumption and has nothing to do with water loss control (Fanner et al. 2007).

The application of decision theory to real-world problems in practice is by far not trivial and could be one of the hardest parts of the decision-making process. According to Hajkowicz and Higgins (2008), the main challenge is not in developing more sophisticated MCDA methodologies but to develop more systematic frameworks to support the initial structuring of the decision problem in terms of selecting criteria and decision options. Clearly, there remains a gap between developed decision theories and applications. The knowledge gap is even bigger especially in developing countries where these methods and tools are hardly applied and therefore not very well understood.

This study attempts to bridge this knowledge gap between decision theory and applications by developing a simple but rather intuitive MCDA framework methodology for identifying, evaluating and prioritizing water loss reduction options. The methodology is evaluated by application to a real-developing world WDS in Kampala city, Uganda. The results indicate that the integrated multi-criteria aiding framework methodology is capable of solving complex WLM and planning problems under uncertain environments with imperfect data.

In this chapter, the decision making process is reviewed in section 7.2, the MCDA basic concepts and methods are presented in section 7.2 and the Preference Ranking and Organization Method for Enrichment Evaluation (PROMETHEE) methodology is discussed in section 7.4. Section 7.5, presents the integrated MCDA framework methodology for strategic WLM planning (SWLMP). The application of the MCDA framework methodology to evaluate and prioritize water loss reduction strategy options in the KWDS is presented in section 7.6. The results are discussed in section 7.7. Lastly, conclusions are drawn in section 7.8.

7.2   The Decision Making Process

Decision making has been defined differently by various researchers. Harris (1998) formally defines decision making as *"the study of identifying and choosing alternatives based on the values and preferences of the decision makers"*. According to Lui et al. (2007), decision making is *"the cognitive process leading to the selection of a course of action among alternatives"*. It can be deduced from these two definitions that decision making involves decision-makers (DMs), options and selecting a final solution in a clear and transparent manner.

Many engineering problems such as water losses in water distribution networks are rather complex in nature, where most decisions involve multiple objectives (financial, economic, environmental, public health, technical and socio-economic) and various actors (DMs, regulators, customers, environmentalists, politicians, experts and other stakeholders). In such situations, a well-structured, disciplined and transparent decision-making process which incorporates the use of credible evaluation methods is required to deliver sound decisions. According to Backer et al. (2002), using such a disciplined approach can help to avoid misunderstandings that question the validity of the analyses and ultimately slow progress.

7.2.1 Steps in decision making

The following nine-steps have been recommended from literature on this subject to help DMs understand and easily follow a decision-making process (Lu et al. 2007; Simonovic 2009):

1. Define the problem.
2. Determine the requirements.
3. Establish objectives and goals.
4. Generate options.
5. Determine criteria.
6. Select a decision-making method or tool.
7. Evaluate options against criteria.
8. Validate solutions against problem statements.
9. Implement the problem.

The final stage is to apply the obtained solution to the decision problem. From the process, decision-making is all about choice that decision-makers have to make. There is no "one-size- fits-all" decision-making process and decision problems will be site specific based on the local context information.

7.3 Multi-criteria Decision Analysis

Decisions in real-world contexts are usually made in the presence of multiple, conflicting, and uncertain environments. Decision-making in such environments is complex and requires appropriate decision support tools. Multi-criteria decision analysis (MCDA) is a decision making technique used in solving decision problems with the following characteristics (Lu et al. 2007; Simonovic 2009):

- multiple and conflicting criteria
- incommensurable criteria (different units)
- overall goal of ranking a finite number of decision options based on a family of evaluation criteria.

Guitouni and Martel (1998) described the MCDA methodology as a non-linear recursive process made up of four steps: (i) structuring the decision problem, (ii) articulating and modeling the preferences, (iii) aggregating the alternative evaluations (preferences) and (iv) making recommendations. MCDA is not a prescriptive answer but a transparent and informative decision process which helps to uncover how peoples' intuitive decision procedures can be informed by a structured rational analytic process (Ananda and Herath 2009).

7.3.1 Definition and terminologies of basic terms of MCDA methods

The following definitions and terminologies are used in MCDA methods:

1. *Goals:* goals are broad statements of intent and desirable programmatic values.
2. *Criteria:* are measures of performance by which options are judged.
3. *Objectives*: reflections of the desire of decision-makers.
4. *Options*: constitutes the object of the decision. Other, terms such as alternatives, actions, scenarios, plans, and programs exist. The term alternatives can only be used if the choices available are mutually exclusive, meaning that only one of them can be implemented at a time.
5. *Actor:* a person involved in the decision process, playing any role, be it a DM, an expert, or a stakeholder. Actors are all people involved in the process other than the analyst.
6. *Analyst:* the person responsible for the decision-aid process and usually referred to as the facilitator or researcher.

7.3.2 Multi-criteria problems

A multi-criteria decision problem can be expressed as (Brans and Mareschal 2005):

$$\max\{g_1(a), g_2(a),..., g_j(a),..., g_k(a) \,|\, a \in A\}, \tag{7.1}$$

where A is a finite set of possible alternatives $\{a_1, a_2,...,a_i,..., a_n\}$ and $\{g_1(.), g_2(.),..., g_j(.),...g_k(.)\}$ a set of evaluation criteria.

The basic data of a multi-criteria problem (Eq. (7.1)) consists of an evaluation matrix (EM) shown in Table 7.1. The evaluation matrix may contain a mix of ordinal and cardinal data. An MCDA model (Table 7.1) requires at least two non dominated decision options and discriminating criteria (Hajkowicz and Higgins 2008). Combining preference information (thresholds and weights) with performance data in the evaluation matrix table provides the results.

Table 7.1 Multi-criteria evaluation matrix

a	$g_1(.)$	$g_2(.)$	...	$g_j(.)$	...	$g_k(.)$
a_1	$g_1(a_1)$	$g_2(a_1)$	...	$g_j(a_1)$	...	$g_K(a_1)$
a_2	$g_1(a_2)$	$g_2(a_2)$	...	$g_j(a_2)$	...	$f_K(a_2)$
...	...	...	...	...	...	...
a_i	$g_1(a_i)$	$g_2(a_i)$	...	$g_j(a_i)$	...	$g_K(a_i)$
...	...	...	...	...	...	...
a_n	$g_1(a_n)$	$g_2(a_n)$	...	$g_j(a_n)$	...	$g_K(a_n)$

The natural dominance relation associated to a multi-criteria problem of type (Eq. (7.1)) is defined as follows:

For each $(a,b) \in A$:

$$\begin{cases} \forall_j: g_j(a) \ge g_j(b) \\ \exists k: g_k(a) > g_k(b) \end{cases} \leftrightarrow aPb,$$

$$\forall_j: g_j(a) = g_j(b) \leftrightarrow aIb, \qquad (7.2)$$

$$\begin{cases} \exists s: g_s(a) > g_s(b) \\ \exists r: g_r(a) < g_r(b) \end{cases} \leftrightarrow aRb,$$

where P, I, and R respectively stand for preference, indifference and incomparability. This means that an option is better than another if it is at least as good as the other on all criteria. If an option is better on a criterion s and the other one better on criterion r, it is impossible to decide the best option without additional information. Both options are said to be incomparable. The maximization of the DMs satisfaction is analogous to the optimization of an objective function over a set of feasible solutions as in classical operational research (Guitouni and Martel 1998).

Usually, no one option is best for all criteria and this often leads to incomparability in most pairwise comparisons. It is impossible to decide without additional information. All MCDA methods usually start with the same evaluation matrix (Table 7.1) but differ on additional required information. This information includes, but not limited to the following:

- Trade-offs between criteria;
- A value function aggregating all the criteria in a single function in order to obtain a mono-criterion problem for which an optimal solution exists;
- Weights giving the relative importance of the criteria;
- Preferences associated to each pairwise comparison within each criterion;
- Thresholds fixing preference limits.

Some requisites for building an appropriate multi-criteria method can be found in Brans and Mareschal (2005).

7.3.3 Multi-criteria Decision Analysis Methods

There are various MCDA methods and most of them belong to the discrete MCDA category where a decision problem is defined by a finite number of decision options. Lai et al. (2008) in their review paper on MCDA categorized these methods as follows:

1. Elementary methods - are the simplest form of MCDA and are hardly applied in water resources management due to inadequacy of their simple preference models (e.g. weighted sum, conjunctive and disjunctive methods etc).
2. Single synthesizing criterion approach – they reduce all criteria to a single criterion for comparison. They include multi-attribute utility theory (MAUT) methods, technique for order preference by similarity of ideal solution (TOPSIS), SMART, Fuzzy (weighted sum and maxmin), AHP etc. These methods belong to the American school of thinking (Roy and Vanderpooten 1996).
3. Outranking methods – use pairwise relations to compare actions, identifying preferences for one over the other and preference aggregation. They include novel approach to imprecise assessment and decision environments (NAIDE), ELECTRE, PROMETHEE, ORESTE, REGIME etc. These methods belong to the European school of thinking (Roy and Vanderpooten 1996).

4. Goal or reference point method – identifies decision options that are closest to the ideal and furthest from the anti-deal. They include goal and compromise programming.
5. Fuzzy set theory – the fuzzy set approach uses imprecise and uncertain information that provides a rigorous and flexible approach to complex resource management problems. Fuzzy set is used as a tool that can be applied to any MCDA methodologies rather than as a specific MCDA methodology itself.

7.3.4 Strengths and Weaknesses of MCDA Methods

Critics of MCDA say that the method is prone to manipulation, is very technocratic and provides a false sense of accuracy while proponents claim that MCDA provides a systematic, transparent approach that increases objectivity and generates results that can be reproduced (Guitouni and Martel 1998; Janssen 2001; Macharis et al. 2004). The main elements of criticism are as follows:

1. *Aggregation algorithms:* different methods yield different solutions when applied to the same problem. The choice among the different methods is often not straight forward and could easily influence the outcome of the decision-making process.
2. *Compensatory methods:* complete aggregation methods of the additive type (e.g. AHP) allow for trade-offs between good performance on one criterion and poor performance on some other criterion. Often important information is lost by such aggregation (e.g. in PROMETHEE II complete ranking). For example, poor performance on water quality could be compensated with good performance on investment cost. The underlying value judgments of the aggregation procedure are therefore debatable and probably not acceptable from the public health and regulatory point of view. A multi-criteria problem is mathematically ill-defined since an action *a* may be better than an action *b* according to one criterion and worse according to another. This is because complete axiomatization of multi-criteria decision theory is very difficult (Munda et al. 1994).
3. *Elicitation process:* the way subjective information (weights and preference thresholds) is elicited is not trivial and is likely to influence the results.
4. *Incomparable options*: as the purpose of all MCDA is to reduce the number of incomparability, MCDA problems are often reduced to single-criterion problems for which an optimal solution exists completely changing the structure of the decision problem which is not realistic. In addition alternatives are often reduced to a single abstract value during data aggregation resulting in loss of useful information. To a lay person it may be easy to understand the cost of an alternative in monetary values rather than an abstract value indicating that option A is better or worse than option B by a value of say 0.45.
5. *Scaling effects*: some MCDA methods derive conclusions based on scales in which evaluations are expressed which is unacceptable. For example if two strategy options (A and B) with the same weight (0.5) have different costs (A=10,000, B=18,000) and their impact on water quality improvement is (A=0.2, B=0.8), their overall performance would be (A=5000.1 and B= 9000.4). If costing were scaled back to a 0-1 scale, then the relative importance of the two criteria would be better represented.
6. *Problem structuring*: results could be manipulated by omission or addition of some relevant criteria or options. MCDA methods have been reported to suffer from rank reversals by introduction of new options (De Keyser and Peeters 1996; Dyer 1990).
7. *Additional required information*: depending on how much additional information is required by the different MCDA methods, "black box" effects are likely to occur thus

compromising the ability of the decision-maker to clearly follow the decision process and evaluate the results.

8. *Uncertainty*: the results are often provided to two decimal places which give a false sense of accuracy considering the uncertainties in the input data used and their error propagation in the model. Uncertainty is also inherent in the decision-making process in that it is difficult to quantify and represent performance of most options by a single value.

However, MCDA, as mentioned earlier, is a decision-aid tool and does not take away the DMs role of decision-making. It is important that while using MCDA methods, its pros and cons are made clear to all participants.

7.3.5   How to select an appropriate MCDA method

There are various MCDA methods and selecting an appropriate method can be a multi-criteria problem itself (Abrishamchi et al. 2005). There is no single MCDA method that can claim to be a superior method for all decision making problems. Different researchers have provided different views on this issue. Guitouni and Martel (1998) argue that different methods will yield different recommendations while Hajkowicz and Higgins (2008) argue that the ranking of decision options is unlikely to change markedly by using a different MCDA method provided ordinal and cardinal data are handled appropriately. However, the guidelines provided by Guitouni and Martel (1998), may still be helpful in selecting an appropriate MCDA method. A recent review of MCDA for water resource planning and management has shown that MCDA is mostly used for water policy evaluation, strategic planning and infrastructure selection (Hajkowicz and Collins 2007). The same review indicates that the most commonly applied methods were fuzzy set analysis, compromise programming (CP), AHP, ELECTRE and PROMETHEE respectively. In this study, the PROMETHEE method was selected based on guidelines provided by Guitouni and Martel (1998).

7.4   The PROMETHEE Preference Modelling Information

The family of the PROMETHEE (Preference Ranking Organization Method for Enrichment Evaluations) outranking methods were first developed by Brans in 1982 and extended by Brans and Vincke (1985); Brans and Mareschal (1994) and recently updated in Brans and Mareschal (2005).

7.4.1   Principles of the PROMETHEE Method

The evaluation matrix (Table 7.1) is the starting point of the PROMETHEE method. The options are evaluated based on information provided in the evaluation matrix Table. The outranking approach uses an *A-F-E* model; in this case; *A* is the set of alternatives, *F* a consistent family of criteria and *E* the evaluation matrix (Guitouni and Martel 1998). Outranking methods are based on the so-called partial comparability axiom. According to this axiom, preferences can be modeled by means of four binary relations: indifference, strict preference, large preference, and incomparability. They allow the following assumptions made in implicit in-value based approaches (e.g. MAUT and AHP) to be relaxed: (i) compensation between criteria and (ii) existence of a true ordering of alternatives (Ananda and Herath 2009).

7.4.2   The weights

The weights are non-negative numbers, independent from the measurement units of the criteria. The higher the weight, the more important is the criterion. Usually weights are normalized such that their total sum is equal to one.

7.4.3   The preference function

The preference structure of PROMETHEE is based on pairwise comparisons which is usually the deviation between evaluations of two options based on a particular criterion. The larger the deviation, the more the preference. The preference degree may be considered as real numbers varying between 0 and 1. This means that for each criterion, the DM has in mind a function

$$P_j(a, b) = F_j[g_j(a) - g_j(b)] \quad \forall a, b \in A, \quad (7.3)$$

and for which:

$$0 \leq P_j(a, b) \leq 1. \quad (7.4)$$

In case a criterion is to be maximized, this function shows the preference of a over b for observed deviations between their evaluations on criterion $g_j(.)$.

The pair $\{g_j(.), P_j(a,b)\}$ is known as the generalized criterion associated to criterion $g_j(.)$. This generalized criterion should be defined for each criterion. In order to facilitate the selection of a specific preference function, six basic types have been proposed by the authors of PROMETHEE (Brans and Mareschal 1994; Brans and Mareschal 2005; Brans and Vincke 1985) and are presented in Appendix B. The PROMETHEE procedure can be applied once the evaluation matrix table $\{g(.)\}$ is established, the weights w_j and the generalized criteria $\{g(.), P_j(a,b)\}$ are defined for $i = 1,2,...,n; j = 1,2,...,k$.

7.4.4   The individual stakeholder group analysis

PROMETHEE permits the computation of the following quantities for each stakeholder r ($r = 1,..., R$) and options a and b:

$$\begin{cases} \pi_r(a, b) = \displaystyle\sum_{j=1}^{k} P_j(a, b)w_{r,j}, \\ \emptyset_r^+(a) = \displaystyle\sum_{x \in A} \pi_r(x, a), \\ \emptyset_r^-(a) = \displaystyle\sum_{x \in A} \pi_r(a, x), \end{cases} \quad (7.5)$$

For each option a, belonging to the set A of options, π (a,b) is the overall preference index of a over b, taking into account all the criteria, $\emptyset_r^+(a)$ and $\emptyset_r^-(a)$. These measure respectively the strength and weaknesses of a compared to other options. The net outranking flow ($\emptyset_r(a)$) of option a for stakeholder k can be calculated as follows:

$$\emptyset_r(a) = \emptyset_r^+ - \emptyset_r^-(a) \quad (7.6)$$

For each stakeholder, the three main PROMETHEE tools can be used to analyse the evaluation problem:

- the PROMETHEE I partial ranking;
- the PROMETHEE II complete ranking;
- the GAIA (Geometric Analysis for Interactive Aid) plane

The PROMETHEE I partial ranking is prudent and does not decide which option is best. The partial ranking is obtained from the positive and negative outranking flows. When the information of all outranking flows is consistent, there is clear preference of one option over the other. There is indifference when both positive and negative flows are equal. When information is inconsistent, incomparability occurs. This often happens when option *a* is good on a set of criteria on which *b* is weak and reversely *b* is good on some other criteria on which *a* is weak. In this case, the DM takes full responsibility for the decision.

When PROMETHEE II is considered, all the options are comparable. PROMETHEE II provides a complete ranking of options from best to worst. The net outranking flow is the basis for complete ranking of options. However, information is likely to be lost, resulting into disputable solutions. In real-world applications, Brans and Mareschal (2005), recommend use of both PROMETHEE I and PROMETHEE II in the decision analysis. The GAIA plane displays graphically the relative position of the options in terms of contributions to various criteria. The reliability of information provided by the GAIA plane is indicated by delta (δ). According to Brans and Mareschal (2005), in most cases they have treated, delta was larger than 60% and in many cases larger than 80%. In this case, the information is quite rich and helps to understand the structure of the multi-criteria problem.

7.4.5 The PROMETHEE GDSS procedure

The PROMETHEE Group Decision Support System (GDSS) has been developed to provide decision aid to a group of decision makers (DM_1), (DM_2), ..., (DM_r),...,(DM_R) (Brans and Mareschal 2005; Macharis et al. 1998). In this phase, the points of view of the different actors are pooled and analyzed using the same tools as in the individual stakeholder analysis. Any conflicts among various actor's points of view are visualized in the GAIA plane. This provides insight into the trade-offs that will need to be made among the various actors' interests.

The global net flow $\emptyset_G$ is calculated as a weighted average of the individual flows:

$$\emptyset_G(a_i) = \sum_{r=1}^{R}\sum_{j=1}^{k}\emptyset_{r,j}(a_i)w_j\omega_r, \quad i = 1,2,...,n, \quad (7.7)$$

where ω_r represents the relative importance of stakeholder *r*.

7.4.6 The decision sights software

Decision Sights (D-Sight) is the new software implementation of the PROMETHEE and the GAIA methods. Decision Sights is a spin-off company of the Universite Libre de Bruxelles in Belgium. The older versions of the PROMETHEE method software were PROMCALC and Decision Lab 2000 (Brans and Mareschal 1994; Geldermann and Zhang 2001).

D-Sight is a Windows application that uses a typical spreadsheet interface to manage the data of a multi-criteria problem. For a single DM case, three basic input data for the software are required: decision matrix (Table 7.1), weights of criteria and performance functions of criteria. In addition to this information from each of the DMs, for Group Decision Making (GDM), the software also requires the weights assigned for the voice of each DM. The software displays the output results in four different ways, each complementing each other:

- PROMETHEE I Diamond
- PROMETHEE II rankings;
- Profiles of options;
- GAIA plane (for single DM) or scenario plane for GDM.

Additional tools such as the "walking weights" and the "decision stick" are available and may be used for further analysis of the decision problem including sensitivity analysis.

The PROMETHEE method is generally easy to comprehend compared to other outranking methods and is widely applied in practice. A recent literature review identifies 217 scholarly papers based on the PROMETHEE method and about 90% were categorized as application papers solving real-life problems in water resources, environmental management, logistics and transport, forestry, chemistry, finance etc. (Behzadian et al. 2010). These methods are highly appreciated by end-users because they are easy to use, intuitive, auditable, and transparent with several graphical and interactive tools. Like all methods, PROMETHEE has some drawbacks that users must be aware of including lack of guidelines on how to determine criteria weights and it suffers from rank reversal on addition or removal of options particularly PROMETHEE II (De Keyser and Peeters 1996; Macharis et al. 2004; Mareschal et al. 2008). To avoid "black box" effects in PROMETHEE, Macharis et al. (2004) suggests use of a more manageable figure of seven evaluation criteria and seven alternatives. However, this could be another weakness of the PROMETHEE by limiting complex MCDA problems to a few options and criteria. This weakness could be improved by hybrid methods that combine two or more methods for operational synergies such as PROMETHEE-AHP (Macharis et al. 2004), Fuzzy-AHP (Wang et al. 2011), and fuzzy multi-objective decision-making methods (Lu et al. 2007).

7.5   The MCDA Framework Methodology for SWLMP

In this section we describe an integrated MCDA group decision-aiding framework for SWLMP in water utilities. The proposed decision-aiding framework is shown in Figure 7.1. Integrative framework is the problem structuring method which provides guidance for stakeholder engagement, criteria selection and alternative development (Lai et al. 2008). The method aids water utility DMs in selecting and prioritizing water loss reduction strategy options. The framework explicitly includes sustainability dimensions of WLM such as economic, technical, environmental and social. The prioritizing is enabled by PROMETHE II outranking method of the MCDA family. The framework method comprises of seven phases: (1) problem structuring phase; (2) design phase; (3) choice phase; (4) group decision phase; (5) testing phase; (6) implementation phase; and (7) monitoring phase. The developed seven-step framework methodology is briefly described in the next sub-sections.

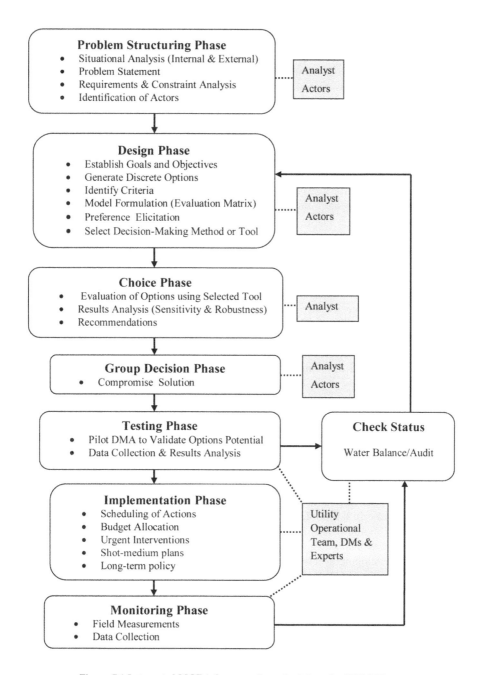

Figure 7.1 Integrated MCDA framework methodology for SWLMP

7.5.1   Problem structuring phase

The aim at this stage is to come up with a clear and concise problem statement. To be able to do so requires identifying root causes, understanding the organizational context information, managerial assumptions and any related initial and desired conditions. Identification of actors and their participation in the decision process starts at this stage. Actors often referred to as stakeholders are people who have an interest, financial or otherwise, in the consequences of any decision taken (Macharis et al. 2010). In this study, actors refer to people other than the analyst involved in the decision process (DMs, experts and stakeholders). The analyst (in this case researcher) is responsible for the decision-aid process as a facilitator ensuring that the responsibility for the decision and its outcomes is of the DMs. The ultimate goal of this step is to come up with a clear problem statement agreed upon by all actors.

7.5.2 Design phase

In this phase, the following decision tasks are accomplished:

Establishment of objectives and goals: Good decisions need clear objectives and goals. Goals are broad statements of intent and desirable long-term plans. The goals and objectives are derived from the utility's vision and mission statements. The goals and objectives should include sustainability dimensions such as economic, environmental and social aspects. In practice objectives are often conflicting and may be realized over a short, medium and long-term period.

Identification of options or alternatives: At this stage options that may contribute to the achievement of the objectives are identified. Water loss reduction options are selected after carrying out a water balance/audit. The water balance reveals the nature and magnitude of the decision problem and provides guidance on which strategy options to adopt. The strategy options can then be selected from a rich menu developed by the International Water Association (IWA) and the American Water Works Association (AWWA) based on many years of research (Alegre et al. 2006; AWWA 2009; Fanner et al. 2007). Each option should be clearly described as to how it will aid in solving the decision problem and how each differs from others.

Identification of Evaluation Criteria (EC): In this step, criteria for evaluating the performance of each option are defined. The choice of the evaluation criteria is based on the established objectives and the purpose of options considered. Criteria should be able to discriminate among options, complete, non-redundant, operational, measurable and minimal for sound decision making (Georgopoulou et al. 1998). Criteria should not be too many or too few. This means the selected criteria should be meaningful, relevant and completely cover all aspects of the objectives. Although large infrastructure projects may require over a hundred criteria, a more manageable typical range is six to twenty (UKDCLG 2009). Generally, actors should be engaged in developing criteria but in most cases analysts do develop the criteria to avoid influences from powerful DMs on other actors and when actors are not technically competent to generate criteria. The developed criteria should be able to measure the sustainability dimensions and are usually categorized in a similar manner as options into economic, environmental, technical and social aspects – usually in the AHP hierarchical structure.

Model Formulation: This step basically involves building the evaluation matrix. Each entry in the evaluation matrix represents the evaluation of an option according to its performance based on a criterion. Scoring of options is a technical activity, performed by a competent team especially when real outcomes of the decision options are often uncertain (Flug et al. 2000). The team may be composed of the analyst and experts on the subject matter. It requires experience and at times literature research especially in planning studies where outcomes are hard to predict with certainty. The evaluation matrix could include quantitative (cardinal) data, qualitative (ordinal) data or both depending on the context information as most MCDA methods can work with both data. It has been urged that the presence of qualitative information in real-world problems involving environment, socio-economic and physical planning aspects, is a rule rather than an exception (Munda et al. 1994). Working with qualitative data usually requires scaling techniques of manipulating linguistic terms into interval scales. Linguistic terms reflect uncertainty, inaccuracy and fuzziness of DMs, and fuzzy sets coupled with MCDA could be explored to handle complex water resources planning problems – fuzzy multi-criteria methods (Lu et al. 2007).

Preference Elicitation: In this step, weights and preference functions for the evaluation criteria and objectives reflecting their relative importance are defined by the actors. The actors must assign preference functions to each criterion using appropriate tools such as the generalized criteria functions in PROMETHEE (Brans and Mareschal 2005).

Weighting of criteria is subjective and has direct influence on the results of prioritizing strategy options. It is therefore critical that criteria weights are determined rationally and truthfully. Generally, there are two methods for weighting of criteria: the equal weights and the rank-order method. The equal weights method ($w_i = 1/n$, $i = 1, 2,\ldots, n$) is the most widely applied due to its simplicity. It requires minimal knowledge of the DM's priorities and minimal input from the DM. The problem with this method is that it assumes equal importance for all criteria which is unrealistic and often not the case in practice. The rank-order method ($w_1 \geq w_2 \geq \ldots \geq w_n \geq 0$ and sum of all criteria weights must equal to one).

Wang et al. (2009), classifies the rank-order method into three categories: subjective weighting method, objective weighting method and combination weighting method. The subjective methods determine criteria weights based on the preferences of the DMs. They explain the elicitation process more clearly and are the most used for MCDA in water resources management. They include SMART, AHP, SIMOS and the Delphi method. The objective weights are obtained by mathematical methods based on the analysis of initial data. The objective weight procedure is not very clear and includes methods such as least mean square (LMS), minmax deviation, entropy, TOPSIS and multi-objective optimization. The combination or optimal weighting methods are a hybrid of methods that include multiplication and additive synthesis.

In this study, we apply the SIMOS procedure for weighting criteria (Figueira and Roy 2002). The main innovation of this procedure is relating a "playing card" to each criterion. The procedure can be summarized into four main steps as follows:

1. Each DM is given n colored cards (or n criteria). Each card has the criterion name inscribed on it and objective of the criterion. A number of white cards (blank cards) are also provided.
2. The DM is then asked to rank the cards from the least important to the most important. If certain criteria are perceived to be of equal importance (same weighting), the cards are grouped together (same rank position).
3. The DMs are asked to insert the white cards between two successively ranked colored cards (or group of cards) in order to express their strong preference between criteria. The number of white cards is proportional to the difference between the importance of the considered criteria.
4. The DM is finally asked to answers the question "how many times more important the first ranked criterion (or group of criteria) is, relative to the last ranked criterion (or group of criteria)?"

The ranking of criteria are then transformed using an appropriate algorithm that attributes a numerical value to the weights of each criterion (Figueira and Roy 2002). The main advantage of the SIMOS weighting method lies in its simplicity but yet intuitive procedure. The fact that the DM handles the cards in order to rank them, inserting white ones, allows indirect understanding of the aim of the procedure. In addition, DMs express the relative importance of criteria using ordinal preferences, allowing determining indirectly numerical values for weights. The respondents often find it easier to express their weightings on an ordinal scale rather than on a numerical scale. This procedure has been applied by other researchers in Greece (Georgopoulou et al. 1998), Austria (Madlener et al. 2007) and Australia (Kodikara 2008) and was found to be very well accepted by DMs. The procedure and its algorithms are presented in appendix B4.

Selecting the Decision-making Method:

There are many MCDA methods for solving a decision problem. Munda et al., (1994) categorizes them as: (i) utility based models, (ii) outranking methods, (iii) the lexicographic model, and (iv) ideal point approaches and aspiration level models. The guidelines on how to chose an appropriate MCDA method can be found in Guitouni and Martel (1998). Generally, selecting an appropriate MCDA method depends on the decision problem at hand, expertise, experience and the preference of the analyst or DM. Preferably, the simpler method should be chosen for the decision analysis.

7.5.3 The choice phase

At this stage, the evaluation matrix is complete with performance data of each option against each criterion. The options are then evaluated against objectives using the defined criteria, weightings of criteria, preference functions and thresholds by running the selected MCDA model. The results may be the ranking of options or a selected subset of the most promising options depending on the decision problem. The evaluation is usually of technical nature and must be executed by the analyst or any other experienced expert. Because the assignment of criteria weights is subjective and subject to errors during elicitation and in the aggregation process, it is important to verify the stability of the results by performing a sensitivity analysis to verify whether the result changes when the weights are modified. The results obtained in this phase must be recommended by the analyst to the actors for a final group decision.

7.5.4 Group decision phase

This stage is concerned with conflict resolution and reaching consensus. The analyst presents the results to the group with aid of visualization tools for clarity. Individual actor results as well as group decision results are presented including sensitivity analysis. In case of disagreements among the actors, the group could be asked to assign new weights to criteria and the evaluation repeated. If disagreements still arise, then new options and criteria could be selected and the evaluation repeated. Generally, the whole process is iterative and it is up to the analyst to determine the termination criteria in case of continued disagreements.

7.5.5 Testing phase

Once the decision is made, pilot testing follows to validate whether the chosen option truly solves the identified decision problem based on real field measurements. This phase is very important in planning studies particularly when new strategy options have not been applied before in the organization and the MCDA results are based on uncertain performance predictions. In addition, implementation of water loss reduction strategies is very costly and must be based on accurate and reliable data. Depending on the findings, the earlier results of prioritized strategies could change before implementation. Once the solution is validated, it is recommended to DMs for consideration and onward implementation.

7.5.6 Implementation phase

This final stage is to implement the final decision via well-scheduled action plans and available budgets. The implementation of prioritized strategy options could change along the way as more insight into the impact of different options on different components of water losses become available through the water balance. This provides flexibility during implementation. In practice, there is always some degree of overlap, where a mixture of strategies is applied concurrently to maximize the level of water loss reductions and cost optimization. For example pressure management could first be done at a macro level and later at a micro level after network zoning and establishment of DMAs.

7.5.7 Monitoring phase

The final monitoring phase provides a continuous feedback loop on performance of each strategy based on actual field measurements and data collection to update the decision-making process. The whole MCDA framework methodology is actually iterative (Figure 7.1) and should be considered as an exercise, which can always be revised and refined in the light of better data and information to support design of new options.

7.6 Application of the Integrated Framework Methodology

The methodology was applied to the NWSC's Kampala water distribution system in Uganda. The profile of the KWDS has been presented in Chapter 1 (§1.2.1) and is not discussed here.

7.6.1 Problem formulation for the KWDS

Based on the case study information the agreed upon problem statement was documented as *"identify and prioritize strategies to reduce water losses in the KWDS"*.

Requirements

The problem definition dictated the following key requirements for the decision problem:

1. Strategy options should address both real and apparent losses.
2. Cost of implementing a set of selected strategies should not exceed €3.6 million per year.
3. Strategies should lead to a water loss reduction of at least 12 million m^3 per year.
4. Implementation period to achieve water loss reduction target should not exceed 10 years.

7.6.2 Identifying actors

The actors were proposed by the analyst and approved by the General Manger of NWSC's Kampala Water. They were all selected from within the utility apart from one actor. Three actor groups were identified to represent utility DMs, water users and environmentalists. In total eight DMs were selected for the preference elicitation process. In this study, actors are referred to as the eight DMs (DM1 to DM8). The senior manager in-charge of the Kampala Finance and Accounts Department together with three senior managers from the Kampala

Water Supply Department represented utility interests (DM1, DM2, DM3, and DM4). One Branch Manager from the Commercial and Customer Care Department together with the GTZ Technical Adviser to NWSC represented customer's interests (DM5, DM6). The NWSC's Water Quality Manager and the Kampala Water Production Manager represented the environmental interests (DM7 and DM8). The environmental group was a hypothetical group due time constraints to mobilize real environmental officials from relevant government departments.

7.6.3 Establishing goals and objectives

The goals and objectives of the study were derived from the utility's mission of "*providing safe and reliable water services to customers at a fair price and in an environmentally friendly manner*". The goal of this study was to reduce water losses in the KWDS. The goals were viewed in the broader national water sector policy of utility financial viability, environmental protection, public health protection, technically acceptable level of service and socio-economic aspects.

In light of the utility's mission and problem statement, five main objectives were established:

1. Maximize revenues and minimize costs.
2. Maximize water savings.
3. Maximize good quality water.
4. Maximize water supply reliability.
5. Maximize affordability of water.

7.6.4 Generating options

In order to generate appropriate water loss reduction strategy options, a water balance was established by the utility water loss unit using the IWA/AWWA water balance methodology and the proposed methodology for assessing the apparent loss component (Mutikanga et al. 2011). The water balance for KWDS was presented in Chapter 1 (Figure 1.2). Based on the KWDS water balance, the following seven strategy options were proposed by the analyst and accepted by the DMs.

1. Meter replacement (S1).

2. Illegal use control (S2).

3. Improved speed and quality of repairs (S3).

4. Selective mains and service line replacements (S4).

5. Network zoning and establishing District Meter Areas (DMAs) (S5).

6. Pressure management (S6).

7. Active leakage control (S7).

The strategy options address both real and apparent losses. WLM is rather peculiar due to the dynamics and migratory attributes of water losses. When you tackle real losses (leakage) alone apparent losses often increase and vice-versa (Vermersch and Rizzo 2007). It was therefore prudent to adopt holistic strategies that address both real and apparent losses.

7.6.5   Determining evaluation criteria

A brain-storming session with actors was arranged to derive the criteria relevant for performance evaluation of the strategy options with aim of realizing the set objectives and goals. The objectives and criteria decision-making hierarchy is presented in Figure 7.2. The disturbance index definition can be found in Baur et al. (2003). Due to insufficient performance evaluation data and to avoid "black box" effects, not all criteria (Figure 7.2) were used in the decision process although it was important to highlight and discuss all potential evaluation criteria with the stakeholders. The selected seven key evaluation criteria are shown in Table 7.2.

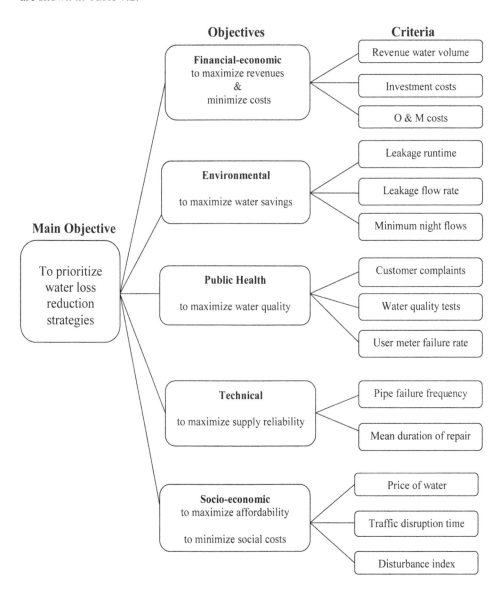

Figure 7.2 Objective-tree hierarchy of the decision problem with multiple criteria

Table 7.2 Evaluation criteria

Code	Criteria	Description
EC1	Revenue generation	The ability of option to improve revenue. The higher the potential, the most preferable the option
EC2	Investment cost	The cost needed to implement the option. The lower the cost, the most preferable the option
EC3	Operation & Maintenance costs	The costs associated with adopting the option. The lower the cost, the most preferable the option
EC4	Water saved	The ability of option to reduce leakage. The higher the potential, the most preferable the option
EC5	Water quality	The ability of option to improve water quality. The higher the potential, the most preferable the option
EC6	Supply reliability	The ability of option to minimize supply interruptions. The fewer the frequency of bursts and leaks, the most preferable the option
EC7	Affordability	The impact of option on water tariff. The lower the impact on tariff, the most preferable the option

7.6.6   Predicting performance

The evaluation matrix for the case study is shown in Table 7.3. Each entry in the evaluation matrix represents the evaluation of an option according to its performance based on a criterion.

Table 7.3 Evaluation matrix

Objective	Criteria	Direction	S1	S2	S3	S4	S5	S6	S7
Finacial-economic									
	EC1 Revenue	Maximise	5	3	3	3	1	2	1
	EC2 Investment Cost	Minimise	2	1	1	5	3	3	2
	EC3 O & M Costs	Minimise	1	4	5	1	2	2	4
Environmental									
	EC4 Water Saved	Maximise	1	1	4	5	2	4	3
Public Health									
	EC5 Water Quality	Maximise	1	1	2	5	2	4	3
Technical (Level of Service)									
	EC6 Supply Reliability	Maximise	1	1	3	5	2	4	3
Socio-economic									
	EC7 Affordability	Maximise	5	4	1	2	2	3	2

Scoring of options was based on experience and literature research as real outcomes of the decision options will be realized in future. In addition there was lack of reliable quantitative data for some options that are currently in use due to institutional challenges coupled with database limitations. To ensure accurate and objective evaluations, the evaluation matrix was completed by a team of the utility personnel in the water loss control unit who are experts on this subject matter. Since the criteria were qualitative in nature, the strategy options were evaluated by first transforming the linguistic terms to interval scale and using criteria measured on a Likert Scale ranging from 1 (poor performance) to 5 (very good performance) as shown in Table 7.4. This being an interval scale, the intervals between statements are meaningful but scale scores have no meaning.

Table 7.4 Transformation of linguistic terms to interval scale

Linguistic Terms	Interval Scale
Very Poor (very low)	1
Poor	2
Fair	3
Good	4
Very Good (very high)	5

7.6.7 Selecting the multi-criteria method and preference modelling

The MCDA method used was the PROMETHEE II and its D-Sight software tool. The preference elicitation process in the case study comprised of an interviewer-assisted questionnaire survey to derive preference functions and weights for the evaluation criteria (EC) and objectives. A survey was conducted on eight DMs and weights were assigned for each criterion and objective to reflect their relative importance to the decision. As the criteria were qualitative, the preference function applied in this study was the type I (usual criterion) of the six generalized criteria as defined by the authors of the PROMETHEE method (Brans and Mareschal 2005). The preference thresholds can be chosen by means of the D-Sight software. In this way, a lot of flexibility is provided to represent the preferences of the decision-makers.

7.6.8 Determining criteria weights

The weights were derived using the "Revised Simos" procedure (Figueira and Roy 2002) and the criteria weight values are presented in Table 7.5. The details of the "Revised Simos" procedure are presented in Appendix B4. For the group decision, the median was considered as the representative value since it agrees with the majority view of the group.

Table 7.5 Evaluation criteria weights assigned by each DM

Criteria	Weight Values									
	DM1	DM2	DM3	DM4	DM5	DM6	DM7	DM8	Mean	Median
C1	19	12	6	18	0	4	15	11	11	12
C2	5	11	7	5	9	0	2	9	6	6
C3	10	8	7	10	4	0	9	9	7	9
C4	26	25	14	33	7	15	12	14	18	15
C5	13	14	16	7	26	15	15	14	15	15
C6	19	20	32	20	35	38	40	24	29	28
C7	7	10	8	7	19	27	8	19	13	9

7.6.9 Evaluating options

The water loss reduction strategy options were evaluated and prioritized with the D-Sight Software tool, which uses the PROMETHEE algorithm. The PROMETHEE II individual decision and group decision rankings are shown in Tables 7.6 and Figure 7.3. The GAIA plane and PROMETHEE I Diamond ranking in Figures 7.4 and 7.5 respectively do provide additional information on the solution of the decision problem. The prioritized strategies for the group decision are graphically depicted in Figure 7.6. The evaluation of options was done by the analyst as it is a task of technical nature that needs expertise.

Table 7.6 PROMETHEE II rankings for individual DMs and group scenario

Rank	Individual Decision Maker								Group Decision
	DM1	DM2	DM3	DM4	DM5	DM6	DM7	DM8	
Rank1	S4	S4	S4	S4	S4	S4	S4	S4	S4
Rank2	S6	S6	S6	S6	S6	S6	S6	S6	S6
Rank3	S3	S3	S3	S3	S7	S7	S3	S1	S3
Rank4	S1	S7	S7	S1	S3	S3	S7	S3	S7
Rank5	S7	S1	S5	S7	S5	S1	S1	S7	S1
Rank6	S5	S2	S1	S5	S1	S5	S5	S2	S5
Rank7	S2	S5	S2	S2	S2	S2	S2	S5	S2

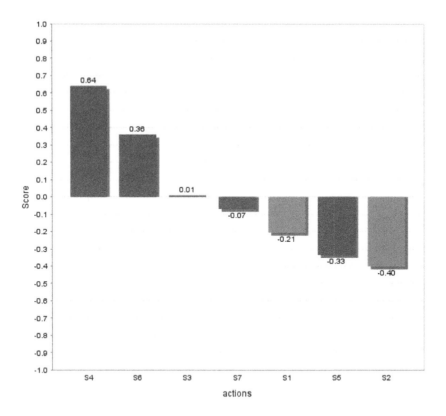

Figure 7.3 Global profile ranking of options for the group decision

Figure 7.4 GAIA plane for group decision

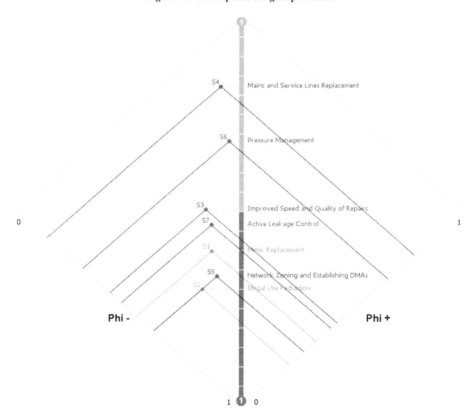

Figure 7.5 The PROMETHEE I diamond for group decision

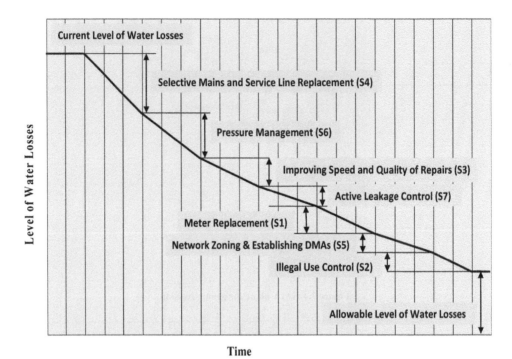

Figure 7.6 Prioritized water loss reduction options for Kampala city (not to scale)

7.6.10 Sensitivity analysis

The sensitivity of the results was analyzed using the capabilities of the D-Sight Software in-built tools. The results are presented in Table 7.7. The stability intervals indicate the range in which the weight of a criterion can be changed without affecting the ranking. For example, the technical criterion with an initial normalized weight of 30.4% may be weighted between 19.7% and 44.7% (stability interval of 25%) without affecting the ranking provided all other factors remain constant. The sensitivity of the ranking can be considered as marginal with respect to the criteria weight values assigned.

Table 7.7 Weight sensitivity analysis of group on strategy ranking

Criteria	Min. Weight	Value	Max. Weight	Stability Interval
Financial-economic	19.5%	28.3%	35.4%	15.9%
Environmental	2.7%	15.8%	33.0%	30.3%
Public Health	6.3%	15.8%	27.1%	20.8%
Technical	19.7%	30.4%	44.7%	25.0%
Social-economic	0.2%	9.8%	15.4%	15.2%

7.6.11 Group decision-making

The PROMETHEE and GAIA plane provide both descriptive and prescriptive tools and were exploited during a group meeting organized to discuss the PROMETHEE II ranking of options, sensitivity of weighting criteria and agree on a compromise solution. During the discussions, most DMs were surprised that the current strategies (S1 and S2) being implemented by the utility were ranked among the last. Although the DMs were surprised by the outcome, consensus was reached that the global prioritized ranking was a group satisfactory solution that fully represented their preferences and met the set objectives and goals. Reaching consensus was rather easy as there was no dispute on the first two best options (S4 and S6) and minor disagreements on the third ranked option (S3) and last option (S2) as shown in Table 7.6. As a way forward, the DMs also unanimously agreed that since scoring of options was based on ordinal data and literature research, a testing phase (Figure 1) was warranted to validate and quantify more precisely the impact of each strategy option before the implementation phase. This will provide information required to re-construct Figure 7.6 on scale. This pilot testing approach has been used to justify implementation of pressure and leakage management strategies on the Gold Coast, Australia (Girard and Stewart 2007).

7.6.12 Compromise solution testing

The water utility has now embarked on the testing phase by incorporating the prioritized water loss reduction strategy options from this study into the on-going Network Management Improvement and Action Research Project (NMIARP) in Kampala. The NMIARP is a very ambitious and challenging project with a result chain shown in Figure 7.7.

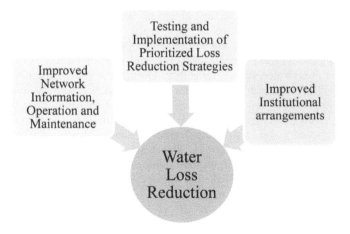

Figure 7.7 NMIARP result chain

This is a demanding phase that will acquire real detailed data for reliable assessment and quantifying performance values with certainty for each strategy option. For example, if 5 km of mains replacement is carried out, what is its water loss reduction potential in terms of volume? Finding such answers during this phase will facilitate re-construction of the water loss reduction framework (Figure 7.6) on scale. Other performance indicators such as costs, burst frequency, revenue generated etc. should be quantified and used to validate and refine the results of the PROMETHEE II ranking.

7.6.13 Implementation phase

The decision of implementation is the task of the DMs. Based on actual quantitative data of the testing phase, an action plan with clear time schedules, cost estimates and responsibilities will be drawn for implementation of the prioritized strategies. Although Figure 7.6 shows a prioritized order of implementing each strategy option, in practice there will be some degree of overlap, where a mixture of strategies is applied concurrently to maximize the level of water loss reductions. For example mains replacement could be done at the same time with network zoning and establishing DMAs. Also pressure management could first be done at a macro level and later at a micro level after network zoning and DMA establishment. Since consensus on the solution was arrived at during the group decision phase through a transparent and interactive decision process, it is anticipated that the implementation phase will be smooth and successful.

7.6.14 Monitoring phase

The final monitoring phase is to provide continuous feedback on performance of each strategy based on actual field measurements and data collection to update the decision-making process. The whole MCDA framework methodology is actually iterative (Figure 7.1) and should be considered as an exercise, which can always be revised and refined in the light of better data and information to support design of new options. It should be noted that water loss control can be very expensive, and water utilities must seek to determine how far they can go based on economic theory (economic levels of water loss reduction).

7.7   Results Discussion

From Figure 7.3, the strategy option S4 (mains and service lines replacement) is the best and strategy option S2 (illegal use control) is the least preferred. Whereas option S2 is the most preferred in terms of investment cost (Table 7.3), it is the least favorable in terms of performance. All DMs ranked option S4 as the best strategy (Table 7.6) and six DMs out of eight, ranked S2 last. This could be explained by the criteria weights in Table 7.5, where the most important criterion was water supply reliability. During the group discussion, the consensus was that reliable water supply must first be available before it is stolen and measured accurately. Furthermore, benefits of pipeline rehabilitation result from a better hydraulic performance of the system, from improved system reliability, improved water quality, reduced pumping energy and a generally improved quality of service (Baur et al. 2003; Cabrera et al. 2010; Herz and Lipkow 2003). The current strategies (S1 and S2) being applied by NWSC-Kampala, are valid based on the traditional cost-benefit analysis (CBA) method rather than on the MCDA method that includes sustainability measures and incorporate other stakeholder views such as customers.

From Table 7.5, the most important criteria according to the DMs preferences are supply reliability, water savings and water quality respectively. According to Christodoulou et al. (2010a), "*the driving forces behind pipe-replacement capital improvement projects have primarily been the mandate to safeguard the health of urban populations, the need to increase the reliability of the pipe networks and the service provided to people, as well as socioeconomic factors in relation to the cost of operations and maintenance of pipe networks*". This explanation coupled with the DMs' preferences on EC clearly justifies why the most preferred options were S4 and S6 (pressure management) respectively. Since PROMETHEE II is a decision support tool, one could probably argue that since option S6 is less costly compared to option S4, it could be the best alternative, should there be budget constraints during implementation.

The third ranked strategy of improved speed and quality of repairs (S3) is more of a reactive strategy after failure has occurred. Although the first two ranked strategies are very good they don't actually repair leaks and main breaks in practice. Strategy S3 is the only one that repairs actual failures and will always be used to supplement other strategies. The duration of real losses can be conceptually divided into awareness, location and repair times (AWWA 2009). The run time of a leak is a major factor contributing to the overall real loss volume and early detection and repair is essential. Although online monitoring and event detection have been reported to be effective tools for early detection of bursts and leaks (Allen et al. 2011; Armon et al. 2011; Mounce et al. 2010) and are likely to trigger quick repair response times, its application in the KWDS is dubious due to so many visible leaks that take long to be repaired. Good leardership at the operational level coupled with online monitoring and detection techniques offers hope for minimizing leakage in Kampala city in future among other options.

The fourth ranked strategy of active leakage control (S7) emerged low probably due to the fact that there is no sense in putting too much effort in detecting invisible underground leaks in a city like Kampala where the utility is overwhelmed by visible leaks everywhere. However, this is a very valuable strategy that enables utilities to address leakage proactively, in order to attain economic and sustainable levels of leakage. The sixth ranked strategy of network zoning and DMA establishment (S4) was considered of less importance probably because it is a complementary strategy rather than a stand alone strategy. DMAs also require costly on-line hydraulic measurements (real-time) using wireless sensors, telemetry (e.g. SCADA) and other communication infrastructure for effective leakage management (Christodoulou et al. 2010b). This probably explains why most DMs ranked the DMA strategy low.

Although not determined scientifically but based on many years of experience, the general guidance provided to utilities in selecting the order of implementing real loss reduction strategies (Trow and Farley 2004), is in agreement with the developed prioritized water loss reduction strategy framework for Kampala city (Figure 7.6). However, Fanner et al. (2007) suggest that mains replacement should always be carried out last as they require high investment costs. The decision on when to replace pipes should be based on local conditions, consideration of sustainability dimensions and preference of all stakeholders. According to Christodoulou et al. (2010a), proactive replacement of small-diameter pipes takes precedence over replacement of large diameter pipes. This seems to make much sense especially for Kampala city where most leaks occur on service lines and quick gains in terms of water savings are likely to be realized by prioritizing replacement of customer service lines and small distribution pipes.

The GAIA-criteria plane

From the GAIA plane (Figure 7.4), five kinds of observations can be made:

1. Structure of the MCDA problem: a high delta value of 93 % is an indicator of the high reliability and richness of information used to help in understanding the structure of the multi-criteria problem.

2. Relative position of options: close options in the plane have similar profiles while distant options strongly differ. In our case study, options S4 and S6 are comparable with respect to the considered criteria while S3 and S1 are totally different.

3. Relative position of criteria: close criteria axis in the plane is similar while opposite criteria are in conflict. In our case for example, technical and public health criteria are close to each other. This means that, on average, an option that will be good for technical (e.g. mains replacement) will also be good for public health and vice versa, considering the preferences of the DMs. It is also clear that the financial-economic criterion is in conflict with the technical and public health criteria and on average, an

option that will be good for say the public health criterion will be bad with respect to the financial-economic criterion, still based on the preferences of the DMs.

4. Decision Stick: the red stick indicates the current best direction for the compromise solution. In this case, the mains and service lines replacement strategy (S4) located furthest away from the origin and in the direction of the decision stick is the best option while illegal use control (S2) is the worst option, confirming the results of the PROMETHEE II rankings.

5. Clusters: There are two distinctive clusters of options: cluster 1 (S4, S6) cluster 2 (S3, S7), cluster 3 (S1, S2) and Cluster 4 (S5). Cluster one and two options address real losses while cluster 2 options are strategies for apparent losses. Cluster 4 indicates an isolated option (S5) with no criterion apparently supporting it as it is a complimentary strategy.

However, the GAIA plane helps only to understand the underlying structure of the decision problem and conclusions should always be confirmed with use of other tools in the D-Sight software.

The PROMETHEE I Diamond

In order to avoid loss of information and to have confidence in the results, PROMETHEE I partial rankings were also examined in the decision analysis. The results are shown in Fig. 7.5. The results are in agreement with PROMETHEE II rankings that strategies S4, S6, S3, S7 and S1 are the best in that order. However, it reveals more information that the sixth best option is not obvious as the network zoning and establishment of DMAs strategy (S5) is incomparable to the illegal use control strategy (S2). This ambiguity is not noticeable in the PROMETHEE II complete rankings.

7.7.1 Challenges and lessons learned

The main challenges and lessons learned during application of the integrated multi-criteria decision-aiding framework methodology for SWLMP were as follows:

1. Handling of ordinal data and subsequent transformation into cardinal data to enable evaluation of the options based on criteria (Table 7.3) was apparently not trivial. Imprecise data is quite common in utilities of the developing countries and careful attention is required to meaningfully use MCDA methods especially with PROMETHEE.
2. The decision process exposed data gaps and has compelled the utility to establish a knowledge base to support future decision-making processes
3. Most participants found the preference elicitation process difficult especially defining the generalized criteria. In addition PROMETHEE does not provide specific guidelines for determining criteria weights, suffers from rank reversal and "black box" effects when confronted with more than seven criteria and seven options.
4. The foundation of the decision-making process is the problem structuring and design phases and their efficient execution is decisive for success.
5. Although the aggregation algorithm for data was questionable at the start and perceived by some DMs as a "black box"; with time the rigorous process moves towards building trust and commitment among the DMs which is vital for successful implementation.

6. MCDA is a very powerful decision-aid tool that facilitates structured group discussions where diverse practical and local knowledge is utilized to evaluate different options and gain information regarding different courses of action.

7. Participants demonstrated high levels of trust in the PROMETHEE MCDA process as the procedure allows for in-depth group discussions clearly pointing out factual differences among alternatives enabling consensus on criteria and final group decision to be attained in a more transparent way. It is also capable of working with limited data.

8. The D-Sight software is simple to use and its GAIA graphical descriptive interfaces were very valuable for the interactive group discussions.

9. Public participation is not institutionalized in the Ugandan water utility which does not suit the MCDA consensus seeking-approach. This denies customers and other stakeholders chance to influence the decision process. For example road users are often inconvenienced whenever there is a pipe failure and their views could probably have added more social criteria (social costs) that could have probably reversed the ranking.

10. Although participation of more stakeholders makes the decision processes more credible, the process can be resource consuming if not well designed and carefully planned.

11. Mobilizing DMs for brainstorming sessions away from their routine work was not easy. For effective participation of DMs in the decision-making process, workshop venues should be far from their places of work and at least a day off leave should be granted to allow meaningful brainstorming sessions.

12. The utility corporate strategic plan (2009-2012) evaluates water loss reduction strategies based on a single criterion of revenue maximization. This conventional way of handling water losses is unrealistic and this study has shown that water loss reduction is actually a multi-criteria problem and should be treated likewise to raise its profile in the utility. Adopting application of MCDA in utility corporate planning as a policy may be a valuable tool to break barriers toward sustainable improvement of WDS efficiency.

7.8 Conclusions

This Chapter presents an integrated MCDA framework methodology for strategic water loss management planning with emphasis on structuring the decision problem including careful selection of criteria and decision options. The proposed framework explicitly considers the triple-bottom line sustainability dimensions of economic, environmental, and social objectives. The framework is capable of handling qualitative information and as such is useful for water loss reduction planning where numerical data is often lacking and imprecise. The framework includes the PROMETHEE outranking method with its D-Sight software tool in solving the decision problem. It also incorporates expert knowledge, DMs preferences and relevant local data to optimize selection of water reduction strategies.

The multi-criteria decision-aiding framework methodology was illustrated on a real-developing world case study clearly highlighting the step-by-step group decision-making process in practice. For Kampala city, the prioritized options for water loss reduction were mains and service lines replacement followed by pressure management and improved speed and quality of repairs. The results show that the most preferred options are those that enhance water supply reliability, public health and water conservation measures. In addition, the results demonstrate that the cheapest option is not necessarily the best when multiple-criteria are considered in an explicit way. The D-Sight software proved to be a valuable tool for handling the complex water loss decision planning problem.

It can be concluded that the developed multi-criteria decision-aiding framework methodology for strategic water loss management planning is a valuable tool envisaged to help water utilities in evaluating and prioritizing water loss reduction strategies in urban water distribution systems particularly in the developing countries where the problem is more prominent.

7.9   References

Abrishamchi, A., Ebrahimian, A., Tajrishi, M., and Marino, M. A. (2005). "Case Study: Application of Multicriteria Decision Making to Urban Water Supply." *Journal of Water Resources Planning and Management*, 131(4), 326-335.

Afshar, A., Marino, M. A., Saadatpour, M., and Afshar, A. (2011). "Fuzzy TOPSIS Multicriteria Decision Analysis Applied to Karun Reservoir System." *Water Resources Management*, 25, 545-563.

Akbari, M., Afshar, A., and Mousavi, S. J. (2011). "Stochastic multiobjective reservoir operation under imprecise objectives: multi-criteria decision-making approach." *Journal of Hydroinformatics*, 13(1), 110-120.

Al-Barqawi, H., and Zayed, T. (2008). "Infrastructure Management: Integrated AHP/ANN Model to Evaluate Municipal Water Mains' Performance." *Journal of Infrastructure Systems (ASCE)*, 14(4), 305-318.

Alegre, H., Baptista, J. M., Cabrera, E. J., Cubillo, F., Hirner, W., Merkel, W., and Parena, R. (2006). *Performance Indicators for Water Supply Services, IWA Manual of Best Practice*, IWA Publishing.

Allen, M., Preis, A., Iqbal, M., Srirangarajan, S., Lim, H. B., Girod, L., and Whittle, A. J. (2011). "Real-time in-network distribution system monitoring to improve operational efficiency " *Journal AWWA*, 103(7), 63-75.

Ananda, J., and Herath, G. (2009). "A critical review of multi-criteria decision making methods with special reference to forest management and planning." *Ecological Economics*, 68, 2535-2548.

Armon, A., Gutner, S., Rosenberg, A., and Scolnicov, H. (2011). "Algorithmic monitoring for a modern water utility: a case study in Jerusalem." *Water Science and Technology*, 63(2), 233-239.

AWWA. (2009). "Water Audits and Loss Control Programs: AWWA Manual M36." American Water Works Association, Denver, USA.

Baker, D., Bridges, D., Hunter, R., Johnson, G., Krupa, J., Murphy, J., and Sorenson, K. (2002). "Guidebook to Decision-Making Methods." WSRC-IM-2002-00002, Department of Energy, USA. http://emi-web.gov//Nissmg/Guidebook_2002.pdf (accessed on 15 February 2011).

Baur, R., Le Gauffre, P., and Saegrov, S. (2003). "Multi-criteria decision support for annual rehabilitation programmes in drinking water networks." *Water Science and Technology: Water Supply*, 3(1-2), 43-50.

Behzadian, M., Kazemzadeh, R. B., Albadvi, A., and Aghdasi, M. (2010). "PROMETHEE: A comprehensive literature review on methodologies and applications." *European Journal of Operational Research*, 200, 198-215.

Bouchard, C., Abi-zeid, I., Beauchamp, N., Lamontagne, L., and Desrosiers, J. (2010). "Multicriteria decision analysis for the selection of a small drinking water system." *Journal of water supply: Research and Technology-AQUA*, 59(4), 230-242.

Brans, J. P., and Mareschal, B. (1994). "The PROMCALC and GAIA decision support system for MCDA." *Decision Support Systems*, 12, 297-310.

Brans, J. P., and Mareschal, B. (2005). "PROMETHEE Methods." Multiple Criteria Decision Analysis: State of the Art Surveys, J. Figueira, S. Greco, and M. Ehrgott, eds., Springer, New York, 163-189.

Brans, J. P., and Vincke, P. (1985). "A Preference Ranking Organisation Method: (The PROMETHEE Method for Multiple Criteria Decision-Making) " *Management Science*, 31(6), 647-656.

Brans, J. P., Vincke, P., and Mareschal, B. (1986). "How to select and how to rank projects: The PROMETHEE method." *European Journal of Operational Research*, 24, 228-238.

Cabrera, E., Pardo, M. A., Cobacho, R., and Cabrera Jr, E. (2010). "Energy Audit of Water Networks." *Journal of Water Resources Planning and Management*, 136(6), 669-677.

Cabrera Jr, E., Cobacho, R., Estruch, V., and Aznar, J. (2011). "Analytical hierarchical process (AHP) as a decision support tool in water resources management." *Journal of Water Supply: Research and Technology-AQUA*, 60(6), 343-351.

Christodoulou, S., Agathokleous, A., Charalambous, B., and Adamou, A. (2010a). "Proactive Risk-Based Integrity Assessment of Water Distribution Networks." *Water Resources Management*, 24, 3715-3730.

Christodoulou, S., Agathokleous, A., Kounoudes, A., and Mills, M. (2010b). "Wireless Sensor Networks for Water Loss Detection." *European Water*, 30, 41-48.

De Keyser, W., and Peeters, P. (1996). "A note on the use of PROMETHEE multicriteria methods." *European Journal of Operational Research*, 89, 457-461.

Dyer, J. S. (1990). "Remarks on the analytic hierarchy process." *Management Science*, 36(3), 249-258.

Fanner, P., Thornton, J., Liemberger, R., and Sturm, R. (2007). *Evaluating Water Loss and Planning Loss Reduction Strategies*, Awwa Research Foundation, AWWA, Denver, USA

Figueira, J., Greco, S., and Ehrgott, M. (2005). *Multicriteria Decision Analysis: State of the Art Surveys*, Springer, New York.

Figueira, J., and Roy, B. (2002). "Determining weights of criteria in the ELECTRE type methods with a revised Simos' procedure." *European Journal of Operational Research*, 139, 317-326.

Flug, M., Seitz, H. L., and Scott, J. F. (2000). "Multicriteria Decision Analysis Applied to Glen Canyon Dam." *Journal of Water Resources Planning and Management*, 126(5), 270-276.

Geldermann, J., and Zhang, K. (2001). ""Software review: "Decision Lab 2000""." *Journal of Multicriteria Decision Analysis*, 10, 317-323.

Georgopoulou, E., Sarafidis, Y., and Diakoulaki, D. (1998). "Design and implementation of a group DSS for sustaining renewable energies exploitation." *European Journal of Operational Research*, 109, 483-500.

Girard, M., and Stewart, R. A. (2007). "Implementation of Pressure and Leakage Management Strategies on the Gold Coast, Australia: Case Study." *Journal of Water Resources Planning and Management*, 133, 210.

Guitouni, A., and Martel, J. M. (1998). "Tentative guidelines to help choosing an appropriate MCDA method." *European Journal of Operational Research*, 109, 501-521.

Hajkowicz, S., and Collins, K. (2007). "A Review of Multiple Criteria Analysis for Water Resource Planning and Management." *Water Resources Management*, 21, 1553-1566.

Hajkowicz, S., and Higgins, A. (2008). "A comparison of multiple criteria analysis techniques for water resources management." *European Journal of Operational Research*, 184, 255-265.

Harris, R. (1998). "Introduction to Decision Making." Virtual Salt. http://www.virtualsalt.com/crebook.htm. (accessed 15 February 2011).

Herz, R. K., and Lipkow, A. T. (2003). "Strategic water network rehabilitation planning." *Water Science and Technology: Water Supply*, 3(1-2), 35-42.

Janssen, R. (2001). "On the use of Multi-criteria Analysis in Environmental Impact Assessment in The Netherlands." *Journal of Multicriteria Decision Analysis*, 10(2), 101-109.

Kain, J.-H., Karrman, E., and Soderberg, H. (2006). "Integration of common knowledge." Strategic Planning of Sustainable Urban Water Management, Malmqvist, Heinicke, Karrman, Stenstrom, and Svensson, eds., IWA Publishing, London.

Kodikara, P. N. (2008). "Multi-Objective Optimal Operation of Urban Water Supply Systems," PhD thesis, Victoria University, Australia.

Kodikara, P. N., Perera, B. J. C., and Kularathna, M. D. U. P. (2009). "Optimal operation of urban water supply systems: A multi-objective approach using PROMETHEE method." Water and Urban Development Paradigms, J. Feyen, K. Shannon, and M. Neville, eds., CRC Press, Leiden.

Lai, E., Lundie, S., and Ashbolt, N. J. (2008). "Review of multi-criteria decision-aid for integrated sustainability assessment of urban water systems." *Urban Water Journal*, 5(4), 315-327.

Lu, J., Zhang, G., Ruan, D., and Wu, F. (2007). *Multi-Objective Group Decision Making: Methods, Software and Applications with Fuzzy Set Techniques*, Imperial College Press, London.

Macharis, C., Brans, J. P., and Mareschal, B. (1998). "The GDSS Promethee Procedure." *Journal of Decision Systems*, 7, 283-307.

Macharis, C., De Witte, A., and Turcksin, L. (2010). "The Multi-Actor Multi-Criteria Analysis (MAMCA) application in the Flemish long-term decision making process on mobility and logistics." *Transport Policy*, 17, 303-311.

Macharis, C., Springael, J., De Brucker, K., and Verbeke, A. (2004). "PROMETHEE and AHP: The design of operational synergies in multicriteria analysis. Strengthening PROMETHEE with ideas of AHP." *European Journal of Operational Research*, 153, 307-317.

Madlener, R., Kowalski, K., and Stagl, S. (2007). "New ways for the integrated appraisal of national energy scenarios: the case of renewable energy use in Austria." *Energy Policy*, 35, 6060-6074.

Malmqvist, P.-A., Heinicke, G., Karrman, E., Stenstrom, T. A., and Svensson, G. (2006). "Urban water in context." Strategic Planning of Sustainable Urban Water Management, Malmqvist, Heinicke, Karrman, Stenstrom, and Svensson, eds., IWA Publishing, London.

Mareschal, B. (1986). "Stochastic multicriteria decision making and uncertainty." *European Journal of Operational Research*, 26, 58-64.

Mareschal, B., De Smet, Y., and Nemery, P. (2008). "Rank Reversal in the PROMETHEE II Method: Some New Results." *Proceedings of the 2008 IEEE IEEM*, Singapore, 959-963.

Morais, D. C., and Almeida, A. T. (2007). "Group Decision Making for Leakage Management Strategy of Water Network." *Resources Conservation and Recycling*, 52, 441-458.

Mounce, S. R., Boxall, J. B., and Machell, J. (2010). "Development and verification of an online artificial intelligence system for detection of bursts and other abnormal flows." *Journal of Water Resources Planning and Management*, 136(3), 309-318.

Mugabi, J., Kayaga, S., and Njiru, C. (2007). "Strategic planning for water utilities in developing countries." *Utilities Policy*, 15, 1-8.

Munda, G., Nijkamp, P., and Rietveld, P. (1994). "Qualitative multicriteria evaluation for environmental management." *Ecological Economics*, 10, 97-112.

Mutikanga, H. E., Sharma, S., and Vairavamoorthy, K. (2009). "Water Loss Management in Developing Countries: Challenges and Prospects." *Journal AWWA*, 101(12), 57-68.

Mutikanga, H. E., Sharma, S. K., and Vairavamoorthy, K. (2011). "Assessment of Apparent Losses in Urban Water Systems." *Water and Environment Journal*, 25(3), 327-335.

NAMS. (2004). *Optimised Decision Making Guidelines*, NZ National Asset Management Steering Group, Auckland, New Zealand.

NWSC. (2009). "Corporate Plan (2009-2012): "maximizing the cash operating margin"." National Water and Sewerage Corporation, Kampala, Uganda.

Roy, B., and Vanderpooten. (1996). "The European School of MCDA: Emergency, Basic Features and Current Works." *Journal of Multicriteria Decision Analysis*, 5, 22-38.

Schouten, M., and Halim, R. D. (2010). "Resolving strategy paradoxes of water loss reduction: A synthesis in Jakarta." *Resources Conservation and Recycling*, 54, 1322-1330.

Silva, V. B. S., Morais, D. C., and Almeida, A. T. (2010). "A multicriteria group decision model to support watershed committees in Brazil." *Water Resources Management*, 24, 4075-4091.

Simonovic, S. P. (2009). *Managing Water Resources: Methods and Tools for a Systems Approach*, UNESCO Publishing, Paris.

Trow, S., and Farley, M. (2004). "Developing a strategy for leakage management in water distribution systems." *Water Science and Technology: Water Supply*, 4(3), 149-168.

UKDCLG. (2009). "Multi-criteria Analysis: a Manual." UK Department of Communities and Local Government (UKDCLG), London. ISBN 978-1-4098-1023-0.

Vermersch, M., and Rizzo, A. (2007). "An Action Planning Model for Control of Non-Revenue Water." *Water Loss 2007 Conference Proceedings*, Bucharest, Romania, 94 - 107.

Wang, J.-J., Jing, Y.-Y., Zhang, C.-F., and Zhao, J.-H. (2009). "Review on multi-criteria decision analysis aid in sustainable energy decision-making " *Renewable and Sustainable Energy Reviews*, 13, 2263-2278.

Wang, Y., Li, Z., Tang, Z., and Zeng, G. (2011). "A GIS-based spatial multi-criteria approach for flood risk management in the Dongting Lake Region, Hunan, Central China." *Water Resources Management*, 25(13), 3465-3484.

Chapter 8 - Conclusions and Recommendations

Summary

Water and revenue losses have become an increasingly important problem for utilities particularly in developing countries. The goal of this study was to develop a suite of appropriate tools and methodologies (toolbox) for water loss management in developing countries. The main output of this study is a water loss management toolbox comprising of: (i) a performance assessment system for evaluating and improving water distribution system efficiency, (ii) an integrated water meter management model to help address the problem of metering inefficiencies and improve utility revenues, (iii) a methodology for assessing apparent water losses in urban water distribution systems, (iv) a pressure management planning decision support tool for leakage control, and (v) an integrated MCDA framework methodology for evaluating and prioritising water loss reduction strategy options. Although the tools and methodologies developed in the research have been tested and validated on the Kampala water distribution system, they are generic and easily adaptable to suit local conditions in other developing countries. It is envisaged that the developed toolbox will be useful to water utility managers and other decision-makers in developing countries trying to reduce water losses in their distribution systems. Whereas good progress has been made in developing tools and methodologies for managing losses in water distribution, considerable work remains. Water distribution losses represent the next frontier of research needs and technology challenges for the drinking water industry especially in the developing countries and this study makes recommendations for further work to sustainably reduce water losses and improve urban water distribution system efficiency.

8.0   Introduction

This Chapter presents the major findings and conclusions from the study. The following sections summarize work done in the previous Chapters. Lastly, recommendations for future research are made.

8.1   Water loss management in developing countries: challenges and prospects

Chapter 1 of this study provides insight into the challenges and prospects of water loss management. It showed that water losses vary widely worldwide. They range from as low as 3-7% of distribution input in the Netherlands (Beuken et al. 2006) to 55% in Latin America (Corton and Berg 2007), 63.8% in Maynilad, Manila in Asia (ADB 2010) and 70% in Liberia Water and Sewer Corporation in Africa (WSP 2009). Clearly, the problem of water losses is more prominent in the developing countries. The World Bank estimates that more than 16 billion m^3 of treated water physically leak from urban WDSs of the developing countries, while over 10 billion m^3 are delivered to customers but not paid for (Kingdom et al. 2006). With the increasing global change pressures (urbanization, climate change, increasing population), there is a high likelihood of a further reduction in the available water resources in the future. This is likely to be compounded by the high rate of infrastructure deterioration which will result in greater loss of treated drinking water. The impact of poorly managed urban WDSs coupled with global changes pressures, could result in extreme scarcity scenarios. According to WHO/UNICEF (2010), 884 million people in the world do not have access to safe drinking water, almost all of them in the developing regions. Clearly, it is unacceptable, that where public utilities are starving for additional revenues to finance expansion of services particularly for the urban poor and where most connected customers receive water irregularly, that water is also heavily wasted.

The challenges and prospects for WLM in developing countries have been presented by various researchers (Mutikanga et al. 2009c; Schouten and Halim 2010; Sharma and Vairavamoorthy 2009). One of the main challenges facing water utilities is lack of appropriate tools and methodologies for WLM including: (i) performance assessment of WDSs (ii) quantifying and assessing AL and (iii) comparing and prioritising water loss reduction strategy options. The main goal of this research was to develop decision support toolbox (tools and methods) for WLM in developing countries. This study mainly focused on the following specific objectives: (i) to develop an appropriate performance assessment system for evaluation and efficiency improvement of urban water distribution systems in the developing countries, specifically focusing on water loss reduction and to validate its effectiveness by application to a real-developing world case study, (ii) to investigate water meter performance in the Kampala water distribution system and develop generic intervention tools for minimizing the associated revenue losses in the developing countries, (iii) to develop a methodology for assessing apparent losses in urban water distribution systems based on field data and investigations in the Kampala water distribution system; and investigate the apparent water losses caused by metering inaccuracies at ultralow flow rates, (iv) to develop a decision support tool for pressure management planning to control leakage in urban water distribution systems of the developing countries by application of economic analysis and network hydraulic modeling techniques, and (v) to develop an integrated multi-criteria decision-aiding framework methodology for strategic water loss management planning in developing countries and evaluate its effectiveness in prioritizing water loss reduction strategy options by application to the Kampala water distribution system.

8.2   Review of tools and methods for managing losses in water distribution systems

In Chapter 2, a critical review of the WLM tools and methodologies was carried out, research gaps identified and appropriate tools and methodologies for reducing water and revenue losses in the developing countries developed. Several tools and methods for WLM have been developed in the last two decades (Alegre et al. 2006; Arregui et al. 2006; AWWA 2009; Berg 2010; Cabrera Jr et al. 2011; Fanner et al. 2007a; Fanner et al. 2007b; Farley and Trow 2003; Thornton et al. 2008; Wu et al. 2011). There are various methods and tools available for WLM that include the water balance, minimum night flow analysis, component-based analysis, leak detection and localization using acoustics, network hydraulic modelling, statistical flow analysis, optimization-based models for leakage hotspots, optimization of system pressures to minimize leakage, optimal pipeline replacement and renewal, transient-based leak detection, integrated water meter management, multi-objective optimization, multi-criteria decision analysis, online monitoring and detection, performance evaluation and benchmarking. However, decision support guidelines on which appropriate tool or method to choose for given local conditions are still lacking. The review indicated that although tools and methodologies have been developed based on empirical data for well-managed urban WDSs in the developed countries, they may not be appropriate or directly applicable for reducing water losses in the developing countries (Mutikanga et al. 2012).

The focus of this research was to close the knowledge gaps identified particularly in the areas of performance evaluation and benchmarking, apparent losses, pressure management and strategic planning for WLM in developing countries. It can be can be concluded that, although not exhaustive, this review could be a valuable reference resource for practitioners and researchers dealing with water loss management in distribution systems and provides a road map for future research.

8.3   Water distribution system performance evaluation and benchmarking

In Chapter 3, a water distribution performance assessment system (PAS) was developed focusing on how well water resources are utilized (Mutikanga et al. 2009b; Mutikanga et al. 2010a; Mutikanga et al. 2010c). The PAS comprises of performance indicators for water loss management, the water balance, guidelines for estimating uncertainty in the water balance input variables and uncertainty propagation into the computed NRW, methods for selecting and developing new PIs, and the DEA-multiple-measure and benchmarking methodology for performance evaluation and improvement of water distribution system efficiency. In addition, apparent loss indices analogous to the infrastructure leakage index (ILI) were proposed to facilitate benchmarking efforts made in reducing apparent losses across utilities. The UAAL benchmark of 7% of revenue water has been proposed for developing countries.

The results in Chapter 3 revealed high uncertainties in the system input volume measurements for KWDS and significant inefficiencies in the Ugandan urban WDSs. The potential WDS efficiency gains were established (water savings estimated at 42.6 ML/day) and performance improvement targets proposed. Some policy implications for the Ugandan urban water sector have been proposed. They include tariff regulation to provide incentives to utilities to recover revenue losses resulting from inefficiencient management of WDSs.

8.4   Water meter management for reduction of revenue losses

Chapter 4 investigated the performance of in-service customer water meters of size 15 mm and developed an integrated water meter management (IWMM) framework to assist water utilities address the problem of water meter management and maximize revenues. The framework includes all aspects of metering from acquisition to replacement with aim of maximizing benefits of a meter during its lifecycle. The influence of factors such as demand profiles, metering technology, storage elevated household tanks and sub-metering on water meter performance in the Kampala water distribution system were examined.

The findings indicate high metering errors (-21.5%) and high meter failure rate (6.6%/year). Over 75% of failures were observed in the volumetric (oscillating-piston) water meter types with the main cause of meter failure being particulates in water. The study also indicates an average reduction in revenue water registration of 18.0% due to sub-metering. To minimize revenue losses due to metering inaccuracies and failure, an IWMM framework was developed. The framework includes the optimal meter replacement period model (I-WAMRM), guidelines for optimal meter selection and sizing, and guidelines for effective sub-metering to minimize meter under-registration (Mutikanga et al. 2009a; Mutikanga et al. 2010b; Mutikanga et al. 2011a; Mutikanga et al. 2011c). The findings of this study are expected to be useful to both utility managers and meter manufacturers who work in the water industry especially in the developing countries to make appropriate metering and sub-metering decisions.

8.5   Assessment of apparent losses in urban water distribution systems

In Chapter 5, a methodology for assessing apparent losses in urban WDSs was developed (Mutikanga et al. 2011b). The methodology was then applied to the KWDS to estimate different apparent loss components. The results indicate that about 37% of water sales in Kampala are due to apparent losses. The major apparent loss components for Kampala were found to be high metering inaccuracies (-22% ± 2%) and illegal use (-10% ± 2%), expressed as a percentage of revenue water. Meter reading errors (-1.4% ± 0.1%) and data handling and billing errors (-3.5% ± 0.5%) were low. Guidelines were also established for assessing apparent losses in resource constrained and data-deficient water utilities of the developing countries to help them quickly estimate apparent losses (Mutikanga et al. 2011b). The influence of ultralow flows on apparent losses due to metering inaccuracies in the KWDS was also examined and found to be significant. Meter under-registration at low flow rates varied from an average of 22.4% to 100% where ultralow flows were not registered at all. Estimated revenue loss as a result was more than US $700,000 annually. The study indicated potential for renuenue recovery by replacing inefficient meters with the most efficient single-jet meter model (> US $6 revenue recovery potential per meter per year). The findings revealed that AL caused by meter inaccuracies at low flow rates are influenced by the meter type and manufacturer and rapidly increase with usage and/or age of the meter. The proposed methodology for assessment of apparent losses in urban WDSs is generic in nature and based upon a practical logical sequence. Water utility managers will find it very helpful in assessing the annual water balance components of water losses.

8.6   Pressure management and network hydraulic modelling for leakage control

In Chapter 6, a decision support tool (PM-COBT) for pressure management planning to control leakage was developed (Mutikanga et al. 2011e). A network hydraulic model was applied to validate the effectiveness of the tool and give users confidence in the tool results. Both methods have been applied to predict potential benefits and cost savings for a real DMA in the KWDS. Predictions by the tool and the network hydraulic model indicate that reducing average zonal pressure by 7 m could result into water savings of 254 m^3/day and 302 m^3/day respectively without compromising customer level of service. In financial terms, this is equivalent to annual net benefits of €56,190 and €66,910 respectively. The results obtained indicate that the predicted water and cost savings compare fairly well. Although conservative in its predictions, the tool would be valuable for engineers and decision-makers planning to adopt pressure management as a strategy for leakage control in the developing countries with inadequate resources for the computationally demanding network hydraulic models. The tool will specifically help utility managers predict the potential benefits of implementing pressure management, thus justifying the investment decisions.

8.7   Multi-criteria decision analysis (MCDA) for water loss management

In Chapter 7, an integrated MCDA framework methodology for strategic water loss management planning was developed (Mutikanga et al. 2011d). The methodology was applied to KWDS to compare and prioritize water loss reduction strategy options. For the case study, seven evaluation criteria under five main non-commensurate and conflicting objectives that include supply reliability, cost efficiency, water conservation, environmental and public health protection were identified and examined. A strategic plan that combines selective mains and service lines replacement and pressure management as priorities was found to be the best compromise based on the preferences of the DMs. This study demonstrated that decision theory coupled with operational research techniques could be applied in practice to solve complex water management and planning problems. The developed MCDA decision-aiding framework will be a valuable tool for strategic planning of water loss reduction in developing countries with often data-poor WDSs.

8.8   Application Guidelines for the Water Loss Management Toolbox

In the previous chapters, the details of the various tools and methods for WLM were presented. In this section we provide application guidelines to aid users navigate to the most relevant tools and methods within the decision support toolbox for WLM, depending on user's needs. The toolbox application guidelines are summarized in Table 8.1. The tools and methods could be used as standalone tools independent of the others in solving the decision problem at hand.

Table 8.1 Toolbox application guidelines

What would you like to do?	Available Tools and Methods	Relevant Chapter
Assess and improve water distribution system efficiency	**Performance Assessment System (PAS)** Water balance model (WLA-PI Tool) Performance Indicators (PIs) Uncertainties and uncertainty propagation DEA-multiple-measure and benchmarking tool	3
Decision on which water meter to buy and when to replace it; Assess in-situ water meter performance	**Integrated Water Meter Management (IWMM) Framework Tool** Optimal meter sampling and testing Demand profiling and optimal sizing techniques Optimal replacement period model (I-WAMRM)	4
Assess components of Apparent water losses	Methodology and guideline frameworks (including data-poor networks and metering inaccuracies at ultralow flows)	5
Predict benefits of pressure management schemes and justify investment decisions	Pressure management cost-benefit decision support tool (PM-COBT)	6
Evaluate and prioritize water loss reduction strategy options	Integrated multicriteria decision-aiding framework methodology for strategic water loss management planning	7

8.9 Recommendations for Future Research

Whereas this study has developed some tools and methodologies and answered some research questions on water loss management, it has also opened debate on some important but yet unanswered questions and studies that could not be covered here due to time limitations that need further research. The following recommendations are proposed for future research work on water loss management:

- The water balance is an effective tool for evaluating WDS performance. However, the usefulness of the results depends heavily on the accuracy of data used. Although the concept of quantifying uncertainties in the water balance is well recognized by researchers, in practice the usefulness of uncertainty analysis is not yet well acknowledged and widespread. Further research on how to communicate in a simple language and foster awareness of the usefulness of uncertainty analysis will be valuable;

- In evaluating WDS performance, only DEA and PI–based benchmarking methodologies were applied. There is need to compare results using other methods such as stochastic frontier analysis (SFA) and corrected ordinary least squares (COLS), and to assess the inclusion of other input and output parameters on the outcome of the efficiency scores;

- The optimal meter replacement model developed during this research is based on the degradation rate for one water meter type model. There is need to gather more data and develop other degradation rate models for other water meter types and meter cohorts. The uncertainties in the degradation model could be minimized further by more data collection and incorporating other variables that influence water meter performance such as water quality. The guidelines for testing in-service meters are also still lacking and are another research area;

- While considerable effort has been invested in tackling real losses in the last two decades, apparent losses have received relatively less attention. Although this study has made a contribution to increase understanding of apparent losses, much work still remains to match apparent loss interventions with real loss (leakage) interventions;

- In estimating apparent losses due to ultralow flow rates, only in-situ small meters of 15 mm size were considered. Further work is required for other customer water meter sizes;

- In the economic model applied for predicting pressure management net benefits for leakage control, many assumptions were made and most data used was from literature. Further testing and refining of the prediction models is required based on actual field data in order to fully understand the real benefits of pressure management (e.g. impact of pressure reduction on frequency of bursts and leaks, deferment of capital costs, energy savings, reduced repair costs etc.);

- The benefits of traditional DMAs are increasingly being challenged and they may no longer be relevant in future. Further research to investigate more open network scenarios, development and optimal placement of multi-parameter sensors (flow, pressure, water quality) for efficient leakage management and other water utility objectives is needed;

- In ranking and prioritizing of water loss reduction strategies for KWDS, the research applied the PROMETHEE outranking method of the MCDA family. It would therefore be interesting to find out if other MCDA methods such as AHP would produce the same results;

- This research did not address areas of economical levels of water losses and active leakage control. With advancement in technology for leak detection and flow measuring equipment, pressure sensors, telemetry, computation power and communication facilities, opportunities for further research to minimise water losses to economic and sustainable levels exist based on application of optimization and real-time control techniques. Along with collection of online hydraulic measurements, a field of work is emerging to improve data analysis and event detection and reduce on number of spurious alerts. Detection of slowly progressive leaks and bursts is also still an intresting research area;

- Finally, even though many appropriate tools and methodologies for water loss management have been developed, they are often not applied by most water utilities. The question therefore is whether the problem is lack of tools and methodologies or good utility leadership to foster application? Research in this direction to identify institutional challenges and barriers would help in bridging the gap between academic research and applications. There is need to develop a web-based decision support software integrating the different component functions of the WLM toolbox. Web-based applications are emerging as a communication strategy and could be one of the bridge-gap avenues towards sustainable water loss reduction in water utilities.

8.10 References

ADB. (2010). "Every Drop Counts: Learning from Good Practices in Eight Asian Cities." Asian Development Bank, Manila.

Alegre, H., Baptista, J. M., Cabrera, E. J., Cubillo, F., Hirner, W., Merkel, W., and Parena, R. (2006). *Performance Indicators for Water Supply Services, IWA Manual of Best Practice*, IWA Publishing.

Arregui, F., Jr., C. E., and Cobacho, R. (2006). *Integrated Water Meter Management* IWA Publishing, London.

AWWA. (2009). "Water Audits and Loss Control Programs: AWWA Manual M36." American Water Works Association, Denver, USA.

Berg, S. (2010). *Water Utility Benchmarking: measurements, methodologies and performance incentives* IWA Publishing, London.

Beuken, R. H. S., Lavooij, C. S. W., Bosch, A., and Schaap, P. G. (2006). "Low leakage in the Netherlands Confirmed." *Proceedings of the 8th Annual Water Distribution Systems Analysis Symposium (ASCE)*, Cincinnati, USA, 1-8.

Cabrera Jr, E., Dane, P., Haskins, S., and Theuretzbacher-Fritz. (2011). *Benchmarking Water Services: Guiding water utilities to excellence*, IWA Publishing, London.

Corton, M. L., and Berg, S. V. (2007). "Benchmarking Central American Water Utilities." Public Utility Research Centre, University of Florida, Gainesville, Florida.

Deb, A. K., Hasit, Y. J., and Grablutz, F. M. (1995). *Distribution System Performance Evaluation*, AWWA Research Foundation and AWWA, Denver, USA.

Fanner, P., Sturm, R., Thornton, J., and Liemberger, R. (2007a). *Leakage Management Technologies*, Awwa Research Foundation Denver, Colorado, USA.

Fanner, P., Thornton, J., Liemberger, R., and Sturm, R. (2007b). *Evaluating Water Loss and Planning Loss Reduction Strategies*, Awwa Research Foundation, AWWA, Denver, USA

Farley, M., and Trow, S. (2003). *Losses in Water Distribution Networks: A Practitioner's Guide to Assessment, Monitoring and Control*, IWA Publishing, London.

Kingdom, B., Liemberger, R., and Marin, P. (2006). "The Challenge of Reducing Non-Revenue Water (NRW) in Developing Countries ", The World Bank, Washington, DC, USA.

McIntosh, A. C. (2003). *Asian Water Supplies: Reaching the Urban Poor*, Asian Development Bank and IWA Publishing.

Mutikanga, H., Nantongo, O., Wozei, E., Sharma, S. K., and Vairavamoorthy, K. (2009a). "Assessing water meter accuracy for NRW reduction." *Proceedings of the 5th IWA Specialist Conference on Efficient Use and Management of Water (CD-ROM)*, Sydney, Australia.

Mutikanga, H., Sharma, S. K., and Vairavamoorthy, K. (2009b). "Performance Indicators as a Tool for Water Loss Management in Developing Countries." *Proceedings of the 5th IWA Water Loss Reduction Specialist Conference*, Cape Town, South Africa, 22-28.

Mutikanga, H., Sharma, S. K., and Vairavamoorthy, K. (2010a). "A Comprehensive Approach for Estimating Non-Revenue Water in Urban Water Supply Systems " *Proceedings of the IWA World Water Congress and Exhibition* Montreal, Canada, September 19-24, CD-ROM.

Mutikanga, H., Sharma, S. K., and Vairavamoorthy, K. (2010b). "Customer demand profiling for apparent loss reduction." *Proceedings of the 6th IWA Water Loss Reduction Specialist Conference (CD-ROM)*, Sao Paulo, Brazil.

Mutikanga, H., Sharma, S. K., Vairavamoorthy, K., and Cabrera Jr, E. (2010c). "Using performance indicators as a water loss management tool in developing countries." *Journal of Water Supply: Research and Technology-AQUA*, 59(8), 471-481.

Mutikanga, H., Vairavamoorthy, K., Kizito, F., and Sharma, S. K. (2011a). "Decision Support Tool for Optimal Water Meter Replacement." *Proceedings of the 2nd International Conference on Advances in Engineering Technology*, Entebbe, Uganda, February 2011, 649-655, ISBN 978-9970-214-00-7.

Mutikanga, H. E., Sharma, S., and Vairavamoorthy, K. (2009c). "Water Loss Management in Developing Countries: Challenges and Prospects." *Journal AWWA*, 101(12), 57-68.

Mutikanga, H. E., Sharma, S. K., and Vairavamoorthy, K. (2011b). "Assessment of Apparent Losses in Urban Water Systems." *Water and Environment Journal*, 25(3), 327-335.

Mutikanga, H. E., Sharma, S. K., and Vairavamoorthy, K. (2011c). "Investigating water meter performance in developing countries: A case study of Kampala, Uganda." *Water SA*, 37(4), 567-574.

Mutikanga, H. E., Sharma, S. K., and Vairavamoorthy, K. (2011d). "Multi-criteria Decision Analysis: A strategic planning tool for water loss management." *Water Resources Management*, 25(14), 3947-3969.

Mutikanga, H. E., Sharma, S. K., and Vairavamoorthy, K. (2012). "Review of methods and tools for managing losses in water distribution systems." *Journal of Water Resources Planning and Management*, doi:10.1061/(ASCE)WR.1943-5452.0000245.

Mutikanga, H. E., Vairavamoorthy, K., Sharma, S. K., and Akita, C. S. (2011e). "Operational tools for decision support in leakage control." *Water Practice and Technology*, 6(3), doi:10.2166/wpt.2011.057.

Schouten, M., and Halim, R. D. (2010). "Resolving strategy paradoxes of water loss reduction: A synthesis in Jakarta." *Resources Conservation and Recycling*, 54, 1322-1330.

Sharma, S. K., and Vairavamoorthy, K. (2009). "Urban water demand management: prospects and challenges for the developing countries." *Water and Environment Journal*, 23, 210-218.

Thornton, J., Sturm, R., and Kunkel, G. (2008). *Water Loss Control*, McGraw-Hill, New York.

WHO, and UNICEF. (2010). "Progress on Sanitation and Drinking-Water:2010 Update ", World Health Organization and UNICEF, Geneva, Switzerland.

WSP. (2009). "Water Operators Partnerships: African Utility Performance Assessment." Water and Sanitation Program (WSP) - Africa, The World Bank, Nairobi, Kenya.

Wu, Z. Y., Farley, M., Turtle, D., Kapelan, Z., Boxall, J., Mounce, S., Dahasahasra, S., Mulay, M., and Kleiner, Y. (2011). *Water Loss Reduction*, Bentley Institute Press, Exton, Pennsylvania, USA.

Appendix A PM DST Computer Code

Modular Code for Pressure Management Decision Support Tool

```
Sub Run()

'Initialisation of subroutine

Dim N As Integer

'Selection of number of iterations required to stabilise AZP, and calculate CP

For N = 1 To 30

'Calculation of AZP values and critical point pressure in first hour

Range("M15") = Range("N15")

'Calculation of AZP values and critical point pressure in 2nd hour

Range("M16") = Range("N16")

'Calculation of AZP values and critical point pressure in 3rd hour

Range("M17") = Range("N17")

'Calculation of AZP values and critical point pressure in 4th hour

Range("M18") = Range("N18")

'Calculation of AZP values and critical point pressure in 5th hour

Range("M19") = Range("N19")

'Calculation of AZP values and critical point pressure in 6th hour

Range("M20") = Range("N20")

'Calculation of AZP values and critical point pressure in 7th hour

Range("M21") = Range("N21")

'Calculation of AZP values and critical point pressure in 8th hour

Range("M22") = Range("N22")

'Calculation of AZP values and critical point pressure in 9th hour

Range("M23") = Range("N23")

'Calculation of AZP values and critical point pressure in 10th hour

Range("M24") = Range("N24")

'Calculation of AZP values and critical point pressure in 11th hour

Range("M25") = Range("N25")
```

```
'Calculation of AZP values and critical point pressure in 12th hour

Range("M26") = Range("N26")

'Calculation of AZP values and critical point pressure in 13th hour

Range("M27") = Range("N27")

'Calculation of AZP values and critical point pressure in 14th hour

Range("M28") = Range("N28")

'Calculation of AZP values and critical point pressure in 15th hour

Range("M29") = Range("N29")

'Calculation of AZP values and critical point pressure in 16th hour

Range("M30") = Range("N30")

'Calculation of AZP values and critical point pressure in 17th hour

Range("M31") = Range("N31")

'Calculation of AZP values and critical point pressure in 18th hour

Range("M32") = Range("N32")

'Calculation of AZP values and critical point pressure in 19th hour

Range("M33") = Range("N33")

'Calculation of AZP values and critical point pressure in 20th hour

Range("M34") = Range("N34")

'Calculation of AZP values and critical point pressure in 21st hour

Range("M35") = Range("N35")

'Calculation of AZP values and critical point pressure in 22nd hour

Range("M36") = Range("N36")

'Calculation of AZP values and critical point pressure in 23rd hour

Range("M37") = Range("N37")

'Calculation of AZP values and critical point pressure in 24th hour

Range("M38") = Range("N38")

Next N

End Sub
```

Appendix B1 Questionnaire – Survey with DMs

Interview Survey – Preference Thresholds and Weights
NWSC – Kampala Water

Information for the participants:

The information collected during this survey will be treated as confidential.

In responding to the questions, please try to express your views **as stakeholders** on having a fair balance between the following competing objectives;

1. To ensure sound financial viability of the utility
2. To protect the environment by conserving scarce water resources
3. To protect public healthy by supplying water of good quality
4. To enhance system technical efficiency by providing an acceptable level of service
5. To ensure socio-economic responsibility by providing affordable services

Some background information is provided in the attachment. Please kindly read the attachment before answering the questions.

The overall goal for provision of sustainable water services is stated hereunder.
Overall Goal
"To provide safe and reliable water services to our customers at a fair price and in an environmentally friendly manner."

(Q1) Do you agree with the above goal? (Please tick)

Totally agree

Partially agree

Do not agree

Weights Assignment

This will be done in two stages; one for evaluation criteria (EC) and one for higher level objectives.

(a)   Weighting of the Evaluation Criteria (EC)

The relative importance of EC is expressed in terms of weights.

Each of the seven cards has the name of an EC written on it. A small explanatory note is also given at the back of each card.

Step 1: Arrange the cards in a row representing the order of importance starting with the most important EC. For equally important EC, you may group the cards together.

Step 2: To express the gaps in importance, insert any number of blank cards that have been given to you. The greater the difference in importance of the EC, the greater the number of blank cards between them.

Record the pattern on the line given below. The first one is the most important criterion and the last one is the least important.

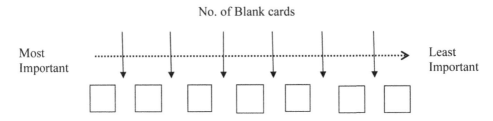

(Q2) How many times is the most important EC compared to the least import one?

(b)   Weighting of the Higher Level Objective

Repeat the procedure outlined in section (a), step 1 and 2 for the five cards representing the general objectives (Financial, Environment, Public Health, Technical and Socio-economic).

Record the pattern on the line given below. The first one is the most important objective and the last one is the least important.

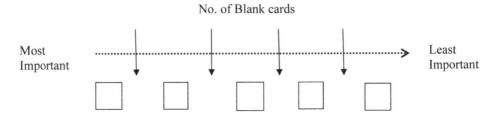

(Q3) How many times is the most important objective compared to the least import one?

(Q4) Indicate a preference function type for each EC in Table B1 and provide the threshold values for q, p and s as deemed appropriate.

Table B1. Preference Functions – DMs Survey

Evaluation Criteria (EC)	Definition	Objective	Acceptable Range (approx.)	Units	Preference Function (select from the six types of generalized criteria)	
					Type	Parameters (q, p & s)
Revenue Generation	Average annual water sales revenue from strategy	Maximise	4-5	Scale (1-5)*		
Investment Costs	Average annual investment cost of strategy	Minimize	1 - 2	Scale (1-5)*		
O & M Costs	Average annual operation and maintenance costs from strategy	Minimize	1 - 2	Scale (1-5)*		
Water Savings	Average annual water savings from strategy	Maximize	4 - 5	Scale (1-5)*		
Water Quality	Improvement in water quality as a result of implemented strategy	Maximize	5	Scale (1-5)*		
Supply Reliability	Ratio of supply hours per day	Maximize	4-5	Scale (1-5)*		
Affordability	Ratio of water bill to household income	Maximize	4-5	Scale (1-5)*		

*(1-very poor, 5-very Good)

Appendix B2 Additional Information–Survey with DMs

Attachment: <u>**Overview of the Research Study and the Interview Survey**</u>

Research Title: Development of a Decision Support Strategic Framework for Water Loss Management in Kampala City Using Multi-criteria Decision Analysis (MCDA).

Introduction

National Water and Sewerage Corporation (NWSC) is still grappling with high levels of non-revenue water (NRW) particularly in Kampala city where NRW is estimated at 40% of water delivered. Water lost means utility revenue is lost, a scarce resource wasted, irregular supply and high O & M costs. Despite various water loss reduction action plans and strategies developed by NWSC-KW, the problem still persists probably due to focusing on a single objective optimisation of reducing water losses. Previous action plans were mainly developed by engineers in the technical department of water supply. Most of these plans are hardly implemented after their formulation. This is partly due to little or no engagement of other stakeholders and decision makers with often conflicting objectives during the planning stage. For example, Finance, Commercial and Water Supply Departments may have cost minimization, revenue maximisation and water loss minimization as their objectives respectively.

To overcome this challenge and arrive at rational operation decisions, this research study is attempting to develop a strategic planning framework (short, medium and long-term) for water loss management in Kampala using MCDA techniques that take into account views of various stakeholders and decision makers (water users, environmentalists and utility managers) to arrive at a compromise solution. Alternative water loss management strategies will be compared and analyzed using a set of Evaluation Criteria (EC). In this study, a list of seven EC categorised under five operational objectives related to financial, environmental, public health, technical and socio-economic aspects for sound and sustainable service delivery are being considered.

Objective 1: Financial-economic: Maximise Utility Financial Viability
- EC1: Revenue Generation (RG)
- EC2: Investment Costs (IC)
- EC3: O & M Costs (OC)

Objective 2: Environmental Protection
- EC4: Water Saved (WS)

Objective 3: Public Health Protection
- EC5: Water Quality (WQ)

Objective 4: Technical: Maximize Customer "Level of Service"
- EC6: Supply Reliability (SR)

Objective 5: Socio-economic
- EC7: Affordability (AF)

A popular outranking method will be used to compare the alternative reduction strategies. The alternatives will be compared pairwise using EC values generated in the above section.

During the interview survey, we will request you to provide two types of information, which are required as inputs for the outranking method. They are:

1. Relative importance of EC (expressed as weights) and
2. Level of preference within each EC (expressed by a "**Preference Function**" which will be explained during the interview).

Information 1 (Inter-criteria): Weights

Your expression of the relative importance of the EC will be aided by a simple procedure. The procedure uses a set of cards; each carrying the name of a criterion. We will request you to lay them on a table and rank them indicating the order of importance that you assign to them by moving the cards around. Further information will be provided to you at the interview session.

Information 2 (Intra-criteria): Preference Function

Within a particular EC the relative preference of one potential value to another can be expressed by a Preference Function (PF). We aim to identify a preference function for each performance measure listed previously.

A typical PF is shown in Figure B1. Further explanation of the PF will be provided at the interview session.

Preference Threshold, $p_j(a,b)$

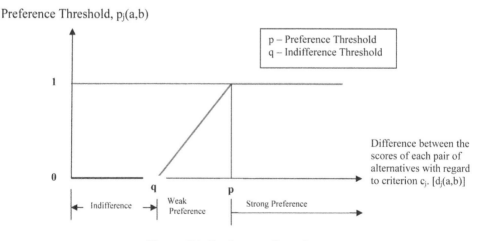

Figure B1: Preference Function

q – Represents the largest difference of EC value which is considered as negligible by the decision maker (DM)

p – Represents the smallest difference of EC value that is considered sufficient to generate a full preference

Example

A simple example of purchasing a car which most stakeholders may have encountered is used here for illustration. Usually the following objectives are considered while purchasing a car: Economy (price); Usage (fuel consumption); performance (power), Space and Comfort. The decision matrix for 3 objectives and 3 alternatives is shown in Table 1. You may realize that it is difficult to decide which alternative is superior to others, without knowing the tolerance levels of the EC.

Table B2 : Decision Matrix Table

Evaluation Criteria	Objective	Alternatives			Preference Function (select from the six types of generalized criteria)	
					Type	Parameters (q, p & s)
		BMW	Corolla	Prado		
Cr1: Price (x 1000 $)	Minimise	30	4	27	III	p = 5,000
Cr 2:Power	Maximize	110	50	95	V	q=20 p = 40
Cr3: Consumption	Minimize	9.0	7.5	8.0	I	

Scenario 1: Assuming we have selected the generalized PF of type III for Price criterion and we need to decide on preference threshold values (i.e. p), we proceed as follows:
If we think that a difference in price of above $5,000 between two selected cars is substantial, then, p = 5,000.

Scenario 2: Assuming we have selected the generalized PF of type V for Power and we need to decide on preference and indifference threshold values (i.e. p and q), we proceed as follows:

If we think that a difference in power of up to 20 between two selected cars is negligible, then, we say, q = 20.

Likewise if a difference above say 40 is decisive (substantial), then, p= 40.

Generalized Preference Function Types

You will be required to select from the six types of generalized preference functions recommended by the authors of the most applied PROMETHEE outranking method in Multi-criteria analysis based on pair-wise comparison. The six PF types are shown in Table B2.

Table B2: Types of generalised criteria (P(d): Preference function)

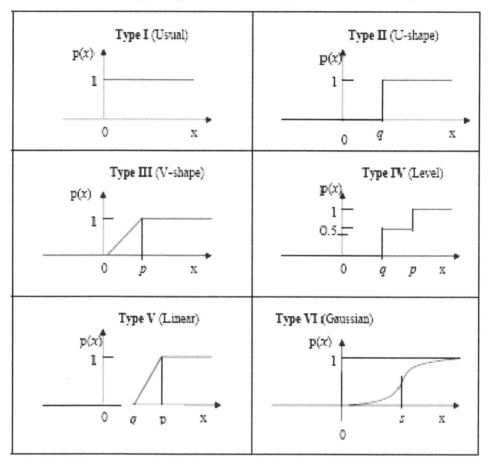

In each case 0,1 or 2 parameters have to be defined, their significance is clear:

        q is a threshold of indifference;

        p is a threshold of strict preference;

        s is an intermediate value between q and p.

Appendix B3: Survey Results of DMs

Table B4: DMs elicitation survey results

Weights Assessment on Objectives:
FE – Financial-economic
EP – Environmental Protection
PP – Public Healthy Protection
LS – Level of Service
SE – Socio-economic

DM Identification No.	Totally Agree	Partially Agree	Do not Agree	RG	IC	OC	WS	WQ	SR	AF	Weights Assessment on ECs (Most Important → Least Important)	Relative Range on EC	Weights Assessment on Objectives (Most Important → Least Important)	Relative Range on Objectives
				FE	EP	PP	LS	SE						
6	√			Type I	Type I	Type I	Type I	Type I	Type I	Type I	SR 2, WQ 2, AF 0, WS 0, RG 0, IC, OC	100	LS 0, SE 0, EP 0, FE, PP	10
1	√			Type I	Type I	Type I	Type I	Type I	Type I	Type I	RG 2, WS 1, SR 1, OC 1, AF 1, IC 1, WQ	7	FE 2, EP 1, LS 1, PP 1, SE	5
2	√			Type I	Type I	Type I	Type I	Type I	Type I	Type I	RG 0, WS 0, IC 1, SR 1, OC 1, WQ 3, AF	3	FE 2, EP 2, LS 3, PP 1, AF	3
3	√			Type I	Type I	Type I	Type I	Type I	Type I	Type I	SR 0, WQ 0, WS 1, IC 1, OC 0, RG 2, AF	3	LS 0, WQ 0, FE 0, EP 0, SE	4
8	√			Type I	Type I	Type I	Type I	Type I	Type I	Type I	RG 1, IC 1, 2, SR 1, WQ, WS, OC, AF	2	FE 1, LS 1, AF 1, EP, PP	2
5		√		Type I	Type I	Type I	Type I	Type I	Type I	Type I	SR 2, AF 1, WQ 4, WS 1, IC 1, OC 1, RG	100	LS 2, PP 1, SE 1, FE 1, EP	5
7	√			Type I	Type I	Type I	Type I	Type I	Type I	Type I	SR 2, WQ 3, WS 2, RG 4, OC 2, AF 2, IC	15	LS 3, FE 2, PP 0, SE 0, SE	5
4	√			Type I	Type I	Type I	Type I	Type I	Type I	Type I	WS 2, WQ 0, LS 2, AF 0, OC 0, 0, RG	8	EP 0, PP 2, T 2, SE 0, FE	5

Appendix B4 Deriving Criteria Weights

The intermediate weights of evaluation criteria (EC) and objectives are calculated using the Revised SIMOS procedure by Figueira and Roy (2002) described in Chapter 7. The intermediate weights of the EC are then multiplied by the corresponding objective weight factor to calculate the final EC weights. The method is illustrated using an example of deriving weight criteria based on DM1's responses.

Step 1 Calculation of EC intermediate Weights

The EC weights are determined using the questionnaire survey responses from DMs and the "Revised SIMOS" computation algorithm.

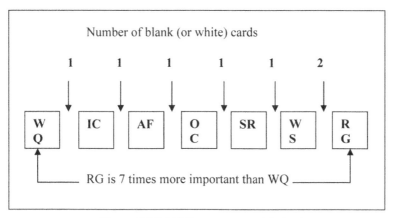

Figure B2 DM1 Responses on EC Cards

The rank (r) of a criterion is defined in the order of increasing importance. "z" is a parameter defined by DMs' responses to the survey questionnaire (Appendix B1). From Figure B2, z = 7.

When there are no blank cards placed between two EC cards, it is interpreted, as one gap between the EC. Similarly, if there are two blank cards between two EC, there are three gaps between them. If X is the total number of gaps between the highest ranked criterion and the lowest ranked criterion (in this case RG and WQ), the parameter u is defined as u = (z-1)/X. The non-normalized weights of EC are calculated using parameter u. Each gap contributes a weight value equal to "u" to the next highest rank.

For DM1, z = 7, X = 13 (Table B5) and u = 0.461.

The non-normalized weight $k(1)$, ..., $k(r)$, ..., $k(\tilde{n})$ associated with each class of equally placed EC, arranged in order of increasing importance is calculated for $r = 1$, ..., $\tilde{n}$ where $\tilde{n}$ = number of ranking levels as:

$k(r) = 1 + u(x_0 + ...x_{r-1})$ with $x_0 = 0$

In case of any equally placed EC on rank r, all the EC are given the same non-normalized weight k(r). The final intermediate weights of EC derived from DM1's responses are shown in Table B5.

Table B5 Revised SIMOS methodology for calculating criteria weights

Rank r	Criteria in the rank r	Number of criteria in rank r, N_r	Number of blank cards following rank r, N_b	Number of gaps between r and (r+1), x_r	Non-normalized intermediate weight, k(r)	Total	Normalized Weight	Intermediate Weight
1	**WQ**	1	1	2	1	1.00	3.724931	4
2	**IC**	1	1	2	1.92308	1.92	7.163326	7
3	**AF**	1	1	2	2.84615	2.85	10.601720	11
4	**OC**	1	1	2	3.76923	3.77	14.040115	14
5	**SR**	1	1	2	4.6923	4.69	17.478509	17
6	**WS**	1	2	3	5.61538	5.62	20.916904	21
7	**RG**	1			6.99999	7.00	26.074495	26
	Sum			13		26.85	100	100

Note: DM1's responses derived from the questionnaire survey are shown in bold

The ranks (or level of importance) assigned by the DMs for each criterion and the resultant intermediate weights of EC within each objective computed using the "Revised SIMOS" procedure is shown is Table B6.

Table B6 Rank and intermediate criteria weights for all DMs

Decision Maker	Rank & Weight	Financial-economic			Environmental Protection	Public Health Protection	Technical	Social-economic
		RG	IC	OC	WS	WQ	SR	AF
DM 1	Rank	7	2	4	6	1	5	3
	Weight	26	7	14	21	4	17	11
DM 2	Rank	7	5	3	6	2	4	1
	Weight	19	17	13	18	11	15	6
DM 3	Rank	2	3	3	5	6	7	1
	Weight	12	13	13	17	18	20	7
DM 4	Rank	7	2	4	6	1	5	3
	Weight	26	8	15	21	3	18	10
DM 5	Rank	1	3	2	4	5	7	6
	Weight	0	8	4	12	21	30	25
DM 6	Rank	3	1	1	4	6	7	5
	Weight	5	0	0	10	28	42	14
DM 7	Rank	4	1	3	5	6	7	2
	Weight	15	2	9	19	23	27	5
DM 8	Rank	7	4	4	7	1	3	1
	Weight	19	16	16	19	9	12	9

Step 2 Calculation of the Objective Weights

The objective weights (normalized) are derived from the DMs' responses on relative importance of the objectives and computed in a similar manner as for EC in step 1 using the "Revised SIMOS" procedure. The computed DMs objectives are shown in Table B7.

Table B7 Rank and objective weights for all DMs

Decision Maker	Rank & Weight	Objectives				
		Financial-economic	Environmental Protection	Public Health Protection	Technical	Social-economic
DM 1	Rank	5	4	2	3	1
	Weight	35	26	13	19	7
DM 2	Rank	5	4	2	3	1
	Weight	31	25	14	20	10
DM 3	Rank	3	2	4	5	1
	Weight	20	14	26	32	8
DM 4	Rank	4	4	1	3	1
	Weight	33	33	7	20	7
DM 5	Rank	1	4	5	7	6
	Weight	13	7	26	35	19
DM 6	Rank	1	2	2	5	4
	Weight	4	15	15	38	27
DM 7	Rank	4	2	3	5	1
	Weight	26	12	15	40	8
DM 8	Rank	5	1	1	4	3
	Weight	29	14	14	24	19

Step 3 Calculation of the Final Evaluation Criteria Weights

The final EC weights are computed by multiplying the normalized EC intermediate weights (Table B6) by the "Objective Weight Factor" defined as the ratio of objective weight to the total aggregated EC weights in the objective. For example, the three EC (RG, IC and OC), all belong to the "Financial-economic" objective. From Table B6, DM1's intermediate weights for RG, IC and OC are 26, 7, 14 respectively. The total aggregated intermediate EC weights within the objective weight of "Financial-economic" is (26+7+14) = 47. The corresponding objective weight for "Financial-economic" is 35. This gives the objective weight factor of 0.745 (or 35/47). DM1's intermediated weight for RG is 26 (Table B6) and this translates into the final weight for RG as 19.36 (or 26 * 0.745). The final rounded off weight is thus 19. Similarly, all other EC weights are computed for all the DMs. Table B8 shows the final EC weight computations for DM1 as an example with final rounded weights in bold for clarity. The final weight values derived in such a manner ensure that the DMs' priority preferences on objectives are explicitly considered in the final decision. These final weights, the preference function were combined with the performance scores in the evaluation matrix table in prioritizing strategy options for water loss reduction.

Table B8 Final (rounded-off) weights of evaluation criteria – DM 1 Example

Objective	Evaluation criteria (EC)	Decision Maker (**DM 1**)			
		Intermediate weight of EC	Objective weight	Final weight	Final rounded weight
Financial-economic	RG	26	35	19.36	**19**
(FE)	IC	7	35	5.21	**5**
	OC	14	35	10.43	**10**
Sub-total		47		35.00	
Environmental Protection (EP)	WS	21	26	26.00	**26**
Sub-total		21		26.00	
Public Health Protection (PH)	WQ	4	13	13.00	**13**
Sub-total		4		13.00	
Technical (TE)	SR	17	19	19.00	**19**
Sub-total		17		19.00	
Social-economic (SE)	AF	11	7	7.00	**7**
Sub-total		11		7.00	
Sum				100	100

Note: All weights in Table 7.5 were derived using this methodology
RG = Revenue Generation; IC = Investment Cost; OC = O & M Costs; WS = Water Saved;
WQ = Water Quality; SR = Supply Reliability; AF = Affordability

Nederlandse Samenvatting (Dutch Summary)

Toegang tot een adequate hoeveelheid drinkwater is een fundamentele menselijke behoefte. Echter, volgens een gezamenlijk monitoring programma van de WHO en UNICEF over het bereiken van de VN-millennium doelstellingen, hebben 884 miljoen mensen in de wereld geen toegang tot een verbeterde voorziening van drinkwater. Dit is bijna uitsluitend in ontwikkelingslanden in Afrika, Azië en Latijns-Amerika. Ironisch genoeg worden grote hoeveelheden veilig drinkwater verspild in de stedelijke waterdistributiesystemen (WDSs) van deze ontwikkelingslanden. Volgens de Wereldbank gaat er bijna 45 miljoen m^3 water dagelijks verloren als lekkage in WDSs – genoeg om ongeveer 200 miljoen mensen van drinkwater te voorzien. Bovendien schat de Wereldbank dat bijna 30 miljoen m^3 water dagelijks geleverd aan klanten niet wordt gefactureerd wegens onnauwkeurigheden in de meting, diefstal, fouten in de facturering en corruptie door werknemers van hulpprogramma's. Deze kosten bedragen ongeveer 6 miljard dollar per jaar voor waterbedrijven in de ontwikkelingslanden.

Waterverlies is niet alleen een economisch verlies en verspilling van een kostbare en schaarse bron, maar levert ook risico's op voor de volksgezondheid. Elk lek is een potentieel inlaatpunt voor vervuiling in geval van een drukverlies van het netwerk. Ook leidt lekkage vaak tot service onderbreking en klachten van klanten, is het kostbaar in termen van energieverliezen en verhoogt het de koolstofvoetafdruk van de dienstverlener. Als gevolg van de groeiende kloof tussen de infrastructuur en investeringen van de watervoorziening, snelle bevolkingsgroei, slechte beheerspraktijken, slecht bestuur en meer extreme gebeurtenissen als gevolg van klimaatverandering, worden deze problemen waarschijnlijk in de toekomst nog erger. Voor de regelgevende instanties en watervoorzieningmaatschappijen heeft deze ongekende druk, in combinatie met de afnemende watervoorraden en de stijgende kosten van het leveren van water, geleid tot het overwegen van serieuze stedelijke watervraag beheersmaatregelen. De hoge waterverliezen in WDSs zijn een uitstekend voorbeeld van een "un-tapped" waterbron die al eerder is gezuiverd en kosteneffectiever zou kunnen zijn. Antwoorden op de vragen waarom, waar en hoeveel water verloren is, zijn noodzakelijk om de winning beter te kunnen regelen door middel van interventiemaatregelen die passen bij de ontwikkelingslanden. De hoofddoelstelling van dit onderzoek is de ontwikkeling van een toolbox die gebruikt kan worden om hulpmiddelen en methoden te beoordelen die nodig zijn om waterbedrijven in ontwikkelingslanden te helpen waterverliezen in hun distributiesystemen te kwantificeren en te minimaliseren.

Watervoorzieningbedrijven in ontwikkelingslanden worstelen om klanten te dienen met een betrouwbare service, vaak vanwege een verouderde waterdistributie infrastructuur, gegevensarme netwerken en beperkte budgetten. Als gevolg kunnen sommige technieken en methoden die worden gebruikt voor waterverliesbeheer (WLM) in ontwikkelde landen niet rechtstreeks toegepast worden in ontwikkelingslanden. Geschikte instrumenten en methoden, van toepassing op WLM in ontwikkelingslanden, zijn nauwelijks beschikbaar. Het einddoel van dit onderzoek is om de efficiëntie van waterdistributie te verbeteren. De belangrijkste aandachtsgebieden van deze studie zijn: analyse van bestaande WLM tools en methoden en hun toepasbaarheid in ontwikkelingslanden, ontwikkeling van het WDS prestatie-evaluatie systeem (PAS), onderzoek naar watermeterprestaties in de case studie WDS, ontwikkeling van methoden voor evaluatie van schijnbare verliezen in stedelijke WDSs, drukbeheerstrategie voor verlaging van lekkage, en een toepassing van het concept van multicriteria besluitvorming (MCDA) voor de beoordeling van en prioriteiten stellen over een waterverlies strategie.

Het onderzoek biedt een overzicht van de state-of-the-art instrumenten en methoden voor WLM om de kennis en onderzoeksbehoeften te identificeren. De belangrijkste opgenomen bevindingen zijn: (i) verschillende hulpmiddelen en methoden kunnen toegepast worden voor WLM en deze variëren tussen eenvoudige leidinggevende hulpmiddelen, zoals prestatie-indicatoren (PIs), en gesofisticeerde optimalisatiemethoden, zoals evolutionaire algoritmen; (ii) waterverliezen variëren tussen 3% van het systeem invoervolume (SIV) in ontwikkelde landen, en 70% in ontwikkelingslanden; (iii) de bestaande instrumenten en methoden kunnen niet rechtstreeks worden toegepast of gelden niet volledig voor alle aspecten van WLM in ontwikkelingslanden; (iv) de meeste bestaande tools en methoden focussen op lekkage en te weinig werk is verricht naar schijnbare verliezen die belangrijk zijn in de WDSs in ontwikkelingslanden; (v) drukbeheer is een krachtige en kosteneffectieve strategie voor verlaging van lekkages; (vi) er is geen duidelijke methode voor het stellen van prioriteiten in waterverlies, en (vii) er is geen duidelijke methode voor de analyse van economische niveaus van waterverliezen. Literatuurstudie wijst op kennisgebrek en behoefte aan het ontwikkelen van meer passende instrumenten en methoden die op een holistische manier de unieke systeemkenmerken van de WDSs in ontwikkelingslanden kunnen aanpakken.

Inzicht in de toestand van een WDS is een belangrijke factor in het minimaliseren van waterverliezen. Hoewel real-time pijpleiding inspectie ideaal is, is het te kostbaar en daardoor buiten bereik van de meeste waterbedrijven in ontwikkelingslanden. Alternatieve indirecte beoordeling van WDSs, zoals de waterbalans en PIs, lijken praktischer te zijn. De International Water Association (IWA) en de American Water Works Association (AWWA) hebben een standaard waterbalans methode ontwikkeld en een alternatief voor PIs. Hoewel de IWA/AWWA waterbalans methodologie en PIs een goede basis bieden, zijn ze onvoldoende en niet direct van toepassing op WDSs in ontwikkelingslanden. Ze vereisen grote hoeveelheden betrouwbare gegevens die kostbaar zijn en nauwelijks gegenereerd kunnen worden door de waterbedrijven die beperk zijn in hun hulpbronnen. In deze studie is een methode ontwikkeld die is gebaseerd op het concept van IWA/AWWA-PI voor selectie en oprichting van nieuwe PIs. Deze methode werd toegepast op selectie van 11 PIs vanuit het IWA/AWWA menu, en verder zijn 14 nieuwe WLM PIs ontwikkeld. De PIs werden getest in een aantal WDSs in Oeganda en geschikt zijn gevonden voor de beoordeling van WDS efficiëntie. Het nut van de resultaten is echter sterk afhankelijk van de nauwkeurigheid van de gegevens. In deze studie is een procedure ontwikkeld voor het schatten van de onderliggende onzekerheid in de waterbalans invoergegevens en hoe deze onzekerheid de NRW-indicator beïnvloedt, alsmede de maatregelen die onzekerheden in de gerapporteerde NRW cijfers minimaliseren. In de afwezigheid van prestatie benchmarks, Data Envelopment Analysis (DEA), is een lineaire programmering techniek toegepast met een Pareto-efficiënte grens als een benchmark, waartegen de prestaties van 25 waterbedrijven in Oeganda werden geëvalueerd en utility rankings werden opgericht. De resultaten geven hoge technische inefficiënties (40-65%) aan in het WDSs met een aanzienlijk potentieel voor waterbesparingen die geraamd worden op 42,600 m3/d. De water utility ranglijst zou kunnen dienen als katalysator voor de vermindering van de hoge inefficiënties in de Oegandese WDSs.

Verder in deze studie is een methode voor de beoordeling van de verschillende componenten van schijnbare verliezen ontwikkeld om de omvang van het probleem te helpen begrijpen en om passende interventiemaatregelen te ontwikkelen om de bijbehorende inkomstenverliezen te minimaliseren. Deze methode werd vervolgens toegepast voor de Kampala WDS (KWDS) en is geschikt voor het schatten van de verschillende componenten van schijnbare water verliezen. De resultaten geven hoge mondiale onnauwkeurigheden in de meting (-22% ± 2%) en illegaal gebruik (-10% ± 2%) uitgedrukt als een percentage van de waterinkomsten. De afleesfouten van de meter (1,4% ± 0,1%), en gegevens en factureringfouten (-3.5 ± 0,5%) bleken laag te zijn. De richtlijnen zijn ook vastgesteld voor de beoordeling van schijnbare verliezen in waterbedrijven met onvoldoende inkomsten en beperkingen op diepgaande beoordeling van gegevens. De invloed van systeemkenmerken, operationele procedures, particuliere verhoogde opslagtanks, sub-metering, lage stroom tarieven, en de invloed van watergebruik profielen op meter nauwkeurigheid werden eveneens onderzocht. De belangrijkste bevindingen wijzen op grote metermislukking (6.6%/jaar), gemiddelde verlaging in registratie van waterinkomsten van 18,0%, toe te schrijven aan sub-metering, en een meer dan US $700.000 van inkomstenverlies per jaar als gevolg van lage stroomtarieven. De gemiddelde meteronderregistratie, dankzij het gecombineerde effect van de vergrijzing van kogelkranen en huishoudelijke watermeters, bleek meer dan 67,2% te zijn. Gebaseerd op deze kennis zijn besluiten genomen over optimale meter grootte, selectie en vernieuwing om de bijbehorende inkomstenverliezen te minimaliseren door toepassing van watervraag profilering en economische optimalisatietechnieken.

Drukbeheer in combinatie met district gemeten gebieden (DMAs) en netwerk hydraulische modellering (NHM) hebben zich bewezen als krachtige engineering tools voor het verminderen van lekkage in ontwikkelde landen. Ondanks hun succes hebben deze hulpmiddelen geen brede toepassing in ontwikkelingslanden. Dit is vooral te wijten aan: (i) gebrekkige informatie met betrekking tot kosten-batenanalyse ter ondersteuning van besluitvorming in uitvoering van PM beleid, en (ii) gebrek en/of ontoereikende zonering van het netwerk. In deze studie werd een decision support tool (DST) ontwikkeld voor het voorspellen van potentiële netto voordelen van de uitvoering van een PM regeling om investeringsbeslissingen te rechtvaardigen. Om de gebruikers vertrouwen te geven in de resultaten van de planning tool, werd NHM toegepast op de doeltreffendheid van de DST validatie. Beide methoden werden vervolgens toegepast op de potentiële netto voordelen van drukbeheer om een DMA in de KWDS te voorspellen. De voorspellingen op basis van de DST en NHM geven aan dat een vermindering van de gemiddelde zonale druk van 7 m resulteert in een waterbesparing van 254 m^3/dag en 302 m^3/dag respectievelijk, zonder gevolgen voor het serviceniveau van de klant. De resultaten wijzen erop dat de besparingen van het voorspelde water met behulp van beide technieken vrij goed te vergelijken is. Dit is gelijk aan een jaarlijks netto financieel voordeel van meer dan €56.000. Hoewel conservatief in haar voorspellingen, is de DST een waardevol instrument voor ingenieurs en besluitvormers (DMs) voor drukbeheer strategieën in ontwikkelingslanden met onvoldoende middelen voor het rekenkundig veeleisende NHM.

Duurzame controle van het waterverlies is een multi-dimensionaal probleem dat de toepassing van strategische planningstechnieken op basis van meerdere criteria vereist. Hoewel er verschillende waterverliesstrategieën bestaan, is het beslissen welke optie moet worden gekozen, rekening houdend met vaak tegenstrijdige meerdere doelstellingen en verschillende belangen van stakeholders, een uitdagende taak voor water-utility managers. In deze studie werd een geïntegreerde multi-criteria decision support methode voor strategische planning van WLM ontwikkeld. De PROMETHEE methode van de MCDA familie werd toegepast in het kader van het beoordelen van waterverliesvermindering tussen de strategische opties voor de KWDS. De methode werd geselecteerd vanwege de helderheid in het besluitvormingsproces en de mogelijkheid om onnauwkeurige gegevens te behandelen. De methode is gericht op het bereiken van grootst mogelijke financieel-economische, milieu, volksgezondheid, technische en sociaal-economische voordelen op basis van zeven criteria (inkomsten generatie, investeringskosten, O&M kosten, waterbesparing, waterkwaliteit, levering betrouwbaarheid en betaalbaarheid) en voorkeuren van de DMs. Een strategisch plan is ontwikkeld dat combineert (i) de vervanging van selectieve pijp en aansluitleidingen, (ii) drukbeheer, en (iii) verbeterde snelheid en kwaliteit van reparaties als prioriteit. Deze studie heeft aangetoond dat de besluitvormingstheorie en operationele onderzoekstechnieken gekoppeld kunnen worden toegepast in de praktijk, om problemen met de duurzame planning van complexe WLM op te lossen.

Kortom, deze studie heeft een decision support toolbox (tools en methoden) ontwikkeld voor WLM in ontwikkelingslanden. De werkset omvat de volgende kerncomponenten:

1. PAS voor waterdistributieverantwoordingsplicht (waterbalans en PI berekening tool; Richtsnoeren voor de raming van de onzekerheid in het waterbalans model; en DEA benchmarking methodologie voor WDS efficiëntie meting en verbetering).

2. Geïntegreerde Water Meter Management kader (IWMM) als steun voor waterbedrijven om inkomstenverliezen als gevolg van onjuistheden te minimaliseren. Het bestaat uit watervraag profiling optimale selectie, grootte en vervanging van watermeters, en richtlijnen voor het schatten van waterverlies als gevolg van minder-registratie en falen van watermeter.

3. Methode voor de beoordeling van schijnbare verliezen in stedelijke WDSs; richtlijnen voor de beoordeling van schijnbare verliezen in gegevensarme WDSs; richtlijnen voor het kwantificeren en herstel van AL vanwege onjuistheden op ultra-lage stroomsnelheid meting.

4. Drukbeheer kosten-batenanalyse tool (PM-COBT) voor drukbeheer strategieën planning door het uitvoeren van de analyse van de lekkagevermindering in verband met verschillende PRV instellingen.

5. Multicriteria besluithulp methodologie voor strategische WLM planning.

Hoewel de werkset is getest en gevalideerd met behulp van WDSs in Oeganda, zijn de tools en methodes van de werkset generiek en gemakkelijk aan te passen aan andere WDSs in ontwikkelingslanden. De bedoeling is dat deze thesis een "advocacy document" wordt dat goed beheer van water (specifiek de efficiëntie van water distributie systemen) en een duurzame levering van diensten van de watervoorziening in ontwikkelingslanden steunt. De studie zal ook van belang zijn voor praktijkmensen, onderzoekers, regelgevers en financiële instellingen die werken aan het verminderen van waterverliezen in WDSs, vooral in ontwikkelingslanden.

Trefwoorden: Beslissingsondersteunende tools en methoden; Ontwikkelingslanden; Urban waterdistributiesystemen; Water- en inkomstenverliezen verantwoordingsplicht; Waterbesparing.

About the Author

Mutikanga Harrison holds a Bachelor's degree in Civil Engineering with honor obtained from Makerere University in 1994. He obtained a Master's degree in Sanitary Engineering from the International Institute for Infrastructure, Hydraulic and Environmental Engineering (now UNESCO-IHE), Delft, The Netherlands in 1999.

He is a registered Professional Engineer with the Uganda's Engineers Registration Board (ERB- Uganda). He is a member of the Uganda Institution of Professional Engineers (UIPE), a member of Environment and Water Resources Institute (EWRI of ASCE), a member of the International Water Association (IWA) Strategic Council (2011/2012) representing Developing Countries, a member of the Science Advisory Committee for Uganda Christian University (Mukono–Uganda) and a member of the Rotary International. He was the founding chair of the Private Water Operators Association in Uganda (2003-2005). Before starting his PhD study, he was the General Manager for Kampala Water (2006-2007) at the National Water and Sewerage Corporation, Uganda.

Mutikanga's research focuses on water distribution systems, and his current research interests include performance evaluation and benchmarking, multi-criteria decision analysis, and water loss management.

Mutikanga Harrison currently holds the position of Water Loss Control Manager, Kampala Water at the National Water and Sewerage Corporation, Uganda. He has over 18 years experience in urban water utility services operations and management.

Publications List

Peer Reviewed International Journals

Mutikanga, H.E, Sharma, S.K., and Vairavamoorthy, K. (2012). "Review of methods and tools for managing losses in water distribution systems". *Journal of Water Resources Planning and Management;* doi:10.1061/(ASCE)WR.1943-5452.0000245.

Mutikanga, H.E, Sharma, S.K., and Vairavamoorthy, K. (2011). "Multi-criteria Decision Analysis: A strategic planning tool for water loss management." *Water Resources Management*, 25(14), 3947-3969.

Mutikanga, H.E., Sharma, S.K., and Vairavamoorthy, K. (2011). "Assessment of Apparent Losses in Urban Water Systems." *Water and Environment Journal*, 25(3), 327-335.

Mutikanga, H.E, Sharma, S.K., and Vairavamoorthy, K. (2011). "Investigating Water Meter Performance in Developing Countries: A Case Study of Kampala, Uganda." *Water SA*, 37(4), 567-574.

Mutikanga, H.E., Vairavamoorthy, K., Sharma, S.K., and Akita, C.S (2011). "Operational Tools for Decision Support in Leakage Control". *Water Practice and Technology*, 6(3), doi:10.2166/wpt.2011.057.

Mutikanga, H.E, Sharma, S.K, Vairavamoorthy, K., and Cabrera Jr., E. (2010). "Using Performance Indicators as a Water Loss Management Tool in Developing Countries." *Journal of Water Supply: Research and Technology-AQUA*, 59 (8), 471-481.

Mutikanga, H.E, Sharma, S.K, and Vairavamoorthy, K (2009). "Water Loss Management in Developing Countries: Challenges and Prospects." *Journal of American Water Works Association (AWWA)*, 101 (12), 57-68.

Publications in Conference Proceedings

Mutikanga, H.E, and Sharma, S.K. (2012). "Strategic Planning for Water Loss Reduction with Imprecise Data." *Proceedings of the 7th IWA Water Loss Reduction Specialist Conference*, Manila, Philippines, February 26-29, CD-ROM.

Mutikanga, H.E, Vairavamoorthy, K., Kizito, F., and Sharma, S.K. (2011). "Decision Support Tool for Optimal Water Meter Replacement." *Proceedings of the 2nd International Conference on Advances in Engineering and Technology (AET 2011)*, Entebbe, Uganda, January 30-February 1, pp 649-655, ISBN 978-9970-214-00-7.

Mutikanga, H.E, Nantongo O., Wozei,E., Sharma,S.K., and Vairavamoorthy,K., (2011). "Investigating the Impact of Utility Sub-metering on Revenue Water." *Proceedings of the 2nd International Conference on Advances in Engineering and Technology (AET 2011)*, Entebbe, Uganda, January 30-February 1, pp 633-639, ISBN 978-9970-214-00-7.

Mutikanga, H.E, Sharma, S.K., and Vairavamoorthy, K. (2010). "A Comprehensive Approach for Estimating Non-revenue Water in Urban Water Supply Systems." *Proceedings of the IWA World Water Congress and Exhibition*, Montreal, Canada, September 19-24, CD-ROM.

Mutikanga, H.E, Sharma, S.K., and Vairavamoorthy, K. (2010). "Customer Demand Profiling for Apparent Water Loss Reduction." *Proceedings of the 6th IWA Water Loss Reduction Specialist Conference*, Sao Paulo, Brazil, June 6-9, CD-ROM.

Mutikanga, H.E, Akita C.S., Sharma, S.K., and Vairavamoorthy K. (2010). "Pressure Management as a Tool for Water Leakage Reduction." *Proceedings of the 3rd International Perspective on Current & Future State of Water Resources & the Environment (EWRI of ASCE) conference,* IIT Madras, Chennai, India, January 5-7, CD-ROM.

Mutikanga, H.E, Nantongo O., Wozei, E., Sharma, S.K., and Vairavamoorthy, K. (2009). "Assessing Meter Accuracy for Reduction of Non-revenue Water." *Proceedings of the IWA's Efficient Water Use Conference Proceedings*, Sydney, Australia, October 26-28, *CD-ROM.*

Mutikanga, H.E, Sharma, S.K., and Vairavamoorthy, K. (2009). "Performance Indicators as a Tool for Water Loss Management in Developing Countries." *Proceedings of the 5th IWA Water Loss Reduction Specialist Conference*, Cape Town, South Africa, April 26-30, pp 22 – 28, ISBN 978-1-920017-38-5.

Mutikanga, H.E, Sharma, S.K., and Vairavamoorthy, K. (2009). "Apparent Water Losses Assessment: The case of Kampala City, Uganda." *Proceedings of the 5th IWA Water Loss Reduction Specialist Conference*, Cape Town, South Africa, April 26-30, pp 36 – 42, ISBN 978-1-920017-38-5.

Invited Oral Presentations

Water Loss Control through Metering. *Itron Users Conference –Africa 2012*, Cape Town (Spier), South Africa, March 5-8, 2012.

Apparent (Commercial) losses: The Ugandan Case. 4th *Global Leakage Summit*, Grand Connaught Rooms, London, UK, January 27-28, 2010.

Water Metering: A Uganda Case Study. *Customer Metering Summit*, Grand Connaught Rooms, London, UK, September 14-15, 2009.